A Short Course in
Modern Organic Chemistry

A Series of Books in Organic Chemistry
Andrew Streitwieser, Jr., Editor

A Short Course in
Modern Organic Chemistry

John E. Leffler

Florida State University

The Macmillan Company · New York
Collier-Macmillan Publishers · London

COPYRIGHT © 1973, JOHN E. LEFFLER

PRINTED IN THE UNITED STATES OF AMERICA

All rights reserved. No part of this book may be reproduced or transmitted in any form or by any means, electronic or mechanical, including photocopying, recording, or any information storage and retrieval system, without permission in writing from the Publisher.

THE MACMILLAN COMPANY
866 Third Avenue, New York, New York 10022

COLLIER-MACMILLAN CANADA, LTD., Toronto, Ontario

Library of Congress catalog card number: 72-80180

Printing: 1 2 3 4 5 6 7 8   Year: 3 4 5 6 7 8 9

Contents

	Introduction	xiii
1	Ionic and Covalent Compounds	1
2	Acids and Bases	17
3	Isomers and the Structure of Molecules	30
4	Alkanes and Cycloalkanes	48
5	Small Rings, Alkenes, and Alkynes	67
6	Aromatic Hydrocarbons	95
7	Alcohols, Ethers, and Phenols	116
8	Aldehydes, Ketones, and Quinones	138
9	Optical Isomerism	162
10	Carboxylic Acids, Esters, and Related Compounds	189
11	Nitrogen Compounds	219
12	Introduction to Large Molecules	250
13	Carbohydrates	277
14	Polyamides, Polypeptides, and Proteins	301
15	Nucleic Acids	335
	Index	361

Detailed Contents

	Introduction	xiii
1.	**Ionic and Covalent Compounds**	1
	Organic chemistry	1
	Ionic compounds	2
	Covalent compounds	3
	Orbitals and the covalent bond	4
	The octet rule	4
	The covalent bond	7
	Valence of charged atoms	9
	Functional groups	13
	Behavior of functional groups	14
	SUMMARY	15
	EXERCISES	16
2.	**Acids and Bases**	17
	Acid-base reactions	18
	Dissociation constants	19
	Electronegativity	20
	Acidity and the separation of mixtures	24
	Acid catalysis	26
	Lewis acids	26
	SUMMARY	27
	EXERCISES	27
3.	**Isomers and the Structure of Molecules**	30
	Structural formulas	30
	Conformations	32
	Isomers	33
	Dimethyl ether and ethyl alcohol	36
	Instrumental methods	36

	Ultraviolet and visible light	39
	Infrared light	40
	Radio waves and nuclear magnetic resonance	42
	SUMMARY	45
	EXERCISES	46

4. Alkanes and Cycloalkanes 48

Methane	48
The bond orbitals of methane	49
Other alkanes	50
Branched alkanes	53
Source of alkanes	55
Cycloheyane	58
Other cycloalkanes	59
Reactions of alkanes	62
Chlorination	62
Combustion	63
SUMMARY	64
EXERCISES	65

5. Small Rings, Alkenes, and Alkynes 67

The Small Rings	67
Alkenes	70
Trigonal carbon atoms and the π bond	70
Consequences of the π bond	72
The butenes and cis-trans isomerism	73
Double bond position isomerism	73
Cis-trans isomerism	73
The role of the π electrons in reactions	75
Reactions of alkenes	76
Addition of halogens	76
Hydrogenation	77
Addition of hydrogen halides	78
Addition of water	79
Preparation of alkenes	80
Dehydration	80
Dehydrohalogenation	81
Double bonds in larger molecules	82
Ips confusus, the bark beetle	82
The pigments of the eye	83
Alkynes	85
The triple bond	85
Reactions of alkynes	87
Water	87
Hydrogen halides	88
The terminal $\equiv$C—H	88
Preparation of alkynes	89
From alkenes	89
From acetylides	89
Preparation of acetylene	90
SUMMARY	91
EXERCISES	92

6. Aromatic Hydrocarbons 95

The bonds in benzene	95
Reactions of benzene	99

Combustion and oxidation	99
Addition reactions	99
Substitution reactions	100
Chlorine and bromine substitution	101
Other aromatic substitution reactions	102
Alkylbenzenes	103
Polynuclear hydrocarbons	110
Naphthalene	110
Anthracene and higher members	111
SUMMARY	114
EXERCISES	114

7. Alcohols, Ethers, and Phenols — 116

Alcohols	116
Nomenclature	117
Physical properties	118
Reactions of alcohols	120
Alcohols as bases or acids	120
Replacement of the hydroxyl group by chlorine	120
Replacement of the hydroxyl group by sulfate or nitrate	121
Preparation of alcohols	123
Hydrolysis of alkyl halides	123
Hydration of alkenes	124
Hydroboration	124
Enols	125
Ethers	126
Preparation of ethers	128
Reactions of ethers	129
Phenols	129
Reaction of phenols	130
Preparation of phenols	133
SUMMARY	136
EXERCISES	136

8. Aldehydes, Ketones, and Quinones — 138

The carbon-oxygen double bond (C=O)	138
Nomenclature of aldehydes and ketones	140
Preparation of aldehydes and ketones	143
Addition reactions of the carbonyl group	143
Addition of hydrogen	143
Addition of water or alcohols	144
Addition of Grignard reagents	145
Addition of amines	146
Addition of HCN	148
Enolization	149
Oxidation of aldehydes and ketones	152
Quinones	153
Other quinonoid compounds	157
SUMMARY	158
EXERCISES	159

9. Optical Isomerism — 162

Introduction	162
Enantiomers	162
The Fischer projection	163

Detailed Contents

Plane polarized light	165
The polarimeter	168
Optical activity	168
How to predict optical activity	169
Synthesis of asymmetric molecules	171
The R and S system	173
The priority rules	175
Diastereomers	176
Meso compounds—a special case	178
Resolution of racemic mixtures	178
Compounds with many asymmetric carbon atoms	180
Asymmetric carbon atoms in living systems	182
SUMMARY	184
EXERCISES	185

10. Carboxylic Acids, Esters, and Related Compounds — 189

Carboxylic Acids	189
Nomenclature of carboxylic acids	189
Naming substituted branched-chain acids	190
Dissociation and physical properties of carboxylic acids	191
Acidity	191
Solubility and boiling point	193
Preparation of carboxylic acids	197
Oxidation	197
Hydrolysis	198
The carbonation of Grignard reagents	198
Formation of acid chlorides and anhydrides	199
Acid chlorides	199
Acid anhydrides	200
Esters	204
Nomenclature of esters	204
Preparation of esters	205
Reactions of esters	205
Hydrolysis	205
Ammonolysis of esters	208
Reduction of esters	208
Esters and Grignard reagents	208
Occurrence and properties of esters	209
Amides	211
Relationships between reactions	213
SUMMARY	214
EXERCISES	215

11. Nitrogen Compounds — 219

Nitrogen fixation	219
Ammonia and amines	220
Classification of amines	222
Preparation	223
Addition to aldehydes and ketones	224
Amines and nitrous acid	225
Diazonium salts and azo compounds	226
Nitrogen heterocycles	227
Five-membered rings	228
Six-membered rings	230
The purine ring system	233
Psychomimetic drugs	234
Epinephrine and related compounds	234

Serotonin and related compounds	237
The tropane alkaloids	238
The morphine alkaloids	240
Barbiturates	242
An art of synthesis	243
SUMMARY	244
EXERCISES	247

12. Introduction to Large Molecules — 250

Some new terms	250
Polymers and monomers	250
Molecular weight	252
End groups and crosslinks	252
Asymmetry in polymers	255
Vinyl polymers	257
Polyethylene	257
Polypropylene	257
Polyvinyl chloride	257
Polyvinyl acetate	258
Polyacrylates	258
Polytetrafluoroethylene	260
Diene polymers and rubber	262
The structure of rubber	262
Synthetic elastomers	264
Oxygen-containing chains	265
Polyacetals and polyethers	265
Polyesters	267
Epoxy resins	271
Silicones	272
Paints and coatings	274
SUMMARY	274
EXERCISES	275

13. Carbohydrates — 277

Monosaccharides	277
Aldohexoses	277
Aldopentoses and deoxyaldopentoses	284
Ketohexoses	285
Glycosides	286
Disaccharides	288
Reducing and nonreducing disaccharides	288
Cellulose	291
Starch	293
Glycogen	294
Chitin	296
Inulin	296
Agar	296
Other polysaccharides	297
Energy and carbohydrates	297
SUMMARY	298
EXERCISES	298

14. Polyamides, Polypeptides, and Proteins — 301

Nylon	302
Silk fibroin	303
Amino acids	304

Detailed Contents

xi

Structure	304
Optical activity	308
Essential amino acids	309
The primary structure of polypeptides	309
Analysis of the hydrolysate	309
End-group analysis	311
The amino acid sequence	313
Secondary structure of polypeptides	317
Tertiary structures	319
Denaturation	319
Conjugated proteins	319
Enzymes	319
How catalysts work	320
Enzyme-substrate complexes	321
Enzyme inhibitors and antimetabolites	321
Enzyme specificity	322
The nature of the active site	324
Polypeptide synthesis	325
The protein-making machine	328
SUMMARY	331
EXERCISES	332
15. Nucleic Acids	**335**
Introduction	335
Nucleic acids defined	336
Nucleotides and nucleosides	337
Nomenclature of the nucleosides and nucleotides	339
Adenosine triphosphate (ATP)	340
Deoxyribonucleic acid (DNA)	343
Nucleotide composition	343
The double helix	344
Replication	346
Nucleic acids and protein synthesis	349
The code dictionary and mutations	352
Mutations	354
Viruses as nucleoproteins	356
Tobacco mosaic virus (TMV)	356
SUMMARY	357
EXERCISES	358
Index	**361**

A Short Course in
Modern Organic Chemistry

Introduction

Organic chemistry is the chemistry of the element carbon and its compounds. The other elements most frequently found in carbon compounds are hydrogen, oxygen, and nitrogen. Many organic compounds are found in plants and animals; in fact, a single cell may contain several thousand of them.

Organic compounds found in living organisms perform an extremely wide variety of functions. They are responsible for the fragrances of flowers and skunks, the colors of flowers and butterfly wings, and the toxicity of arrow poisons. They serve as chemical signals to bring various animals together for mating, as visual pigments in our eyes, as vitamins, drugs, and hormones. The cellulose that acts as the major structural material of plants and the polypeptides that serve the same and other functions in animals are organic chemicals. Most important of all, they include the nucleic acids in which our genetic inheritance is recorded.

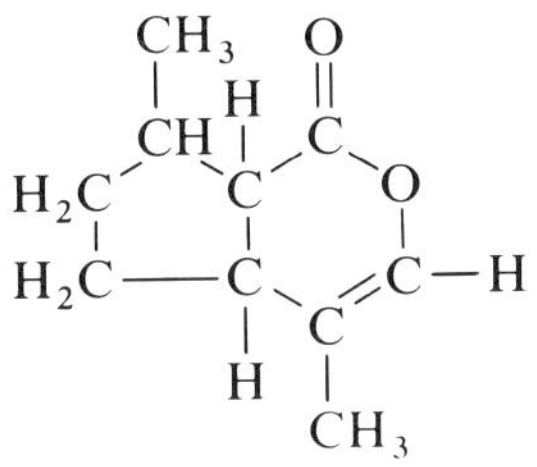

nepetalactone (catnip)

Nepetalactone is an ingredient of catnip, a plant that gives joy to all cats from house cats to tigers.

The reason for the versatility of organic compounds is the tendency of carbon to form covalent bonds, not only with atoms of other elements but also with other carbon atoms. Thus we have carbon compounds ranging from small molecules like methane (CH_4) or the five isomeric hexanes (C_6H_{14}) to molecules like nucleoproteins that may contain a million atoms. The diamond in a diamond ring is a molecule that is actually big enough to see. It consists of a network of covalently linked carbon atoms, typically about 10^{21} of them.

Even more important than the great range in size possible in organic molecules is the phenomenon of isomerism; that is, the same atoms can be put together in different ways to make entirely different compounds with entirely different physical or biological properties. For example, C_2H_6O is the empirical formula of both ethyl alcohol and the gas dimethyl ether.

$$\begin{array}{cc} \text{H} \quad \text{H} & \text{H} \quad \text{H} \\ | \quad | & | \quad | \\ \text{H}-\text{C}-\text{C}-\text{OH} & \text{H}-\text{C}-\text{O}-\text{C}-\text{H} \\ | \quad | & | \quad | \\ \text{H} \quad \text{H} & \text{H} \quad \text{H} \\ \text{ethyl alcohol} & \text{dimethyl ether} \\ \text{liquid; bp 78.5°C} & \text{gas; bp } -23.65°\text{C} \end{array}$$

Another more subtle kind of isomerism is particularly important in compounds found in living organisms. This is mirror-image isomerism. Just as a nut with a left-hand thread is of no use with a right-hand bolt, living organisms require molecules with the proper "handedness." For example the amino acids that make up the proteins of our bodies all have to be of the L or "left-handed" series.

$$\begin{array}{cc} \text{COOH} & \text{COOH} \\ | & | \\ \text{H}_2\text{N}-\text{C}-\text{H} & \text{H}-\text{C}-\text{NH}_2 \\ | & | \\ \text{CH}_3 & \text{CH}_3 \\ \text{L-alanine} & \text{D-alanine} \end{array}$$

A "left-handed" amino acid molecule and its mirror image.

To give other examples: Because of the "handedness" of the molecules that form the odor-detecting physiological structures in the nose, left- and right-handed molecules often have quite different odors. It is also found that some right- or left-handed molecules are potent drugs, whereas their mirror-image molecules have little or no effect.

The planet Earth is a closed ecological system based on L-amino acids. These molecules are traded back and forth and reused as one plant or animal lives on or devours another. Presumably a system based on the mirror image or D-amino acids would work just as well, but a visitor from an all-L-amino acid planet like ours would starve to death in the midst of apparent plenty on a D-amino acid planet. This is what *should* have happened to Alice when she went through the looking-glass. Because Alice had to hold the Jabberwockey verse up to a mirror before she could read it, we can take it that Alice and her L-amino acids were unchanged by their passage through the mirror. A mirror-image Alice should have been able to read mirror-image writing without a mirror. Lewis Carroll can be excused, however, since the theory of structural organic chemistry was in its infancy in 1872 when he wrote *Through the Looking-Glass*.

Up until about 1826 organic chemistry was still exclusively the chemistry of compounds from living organisms and, in that sense, truly "organic." In contrast to inorganic compounds, organic compounds were widely believed to contain a mysterious metaphysical ingredient that could be supplied only by the living organism. This belief began to be abandoned (although reluctantly) when Friedrich Wöhler made urea by heating ammonium cyanate.

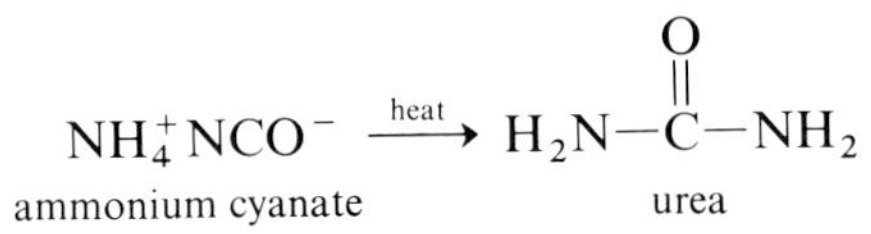

$$NH_4^+ NCO^- \xrightarrow{\text{heat}} H_2N-\underset{\text{urea}}{\overset{\overset{\displaystyle O}{\|}}{C}}-NH_2$$

ammonium cyanate

Ammonium cyanate was a typical inorganic compound, and urea, found in urine, was a typical "organic" compound. The conversion of an inorganic compound into an organic compound suggested that organic molecules are governed by essentially the same laws as inorganic molecules. Today's controversy is at a higher level: where does the domain of complicated molecules such as nucleic acids or nucleoproteins end and that of simple living organisms begin? Are living organisms governed by laws different from those of mere molecules?

Many compounds that we make today are organic only by the modern definition and are not found in nature at all. They include pain-relieving drugs such as aspirin

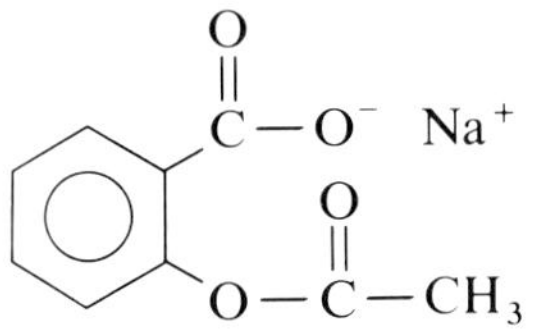

labor-saving explosives such as TNT

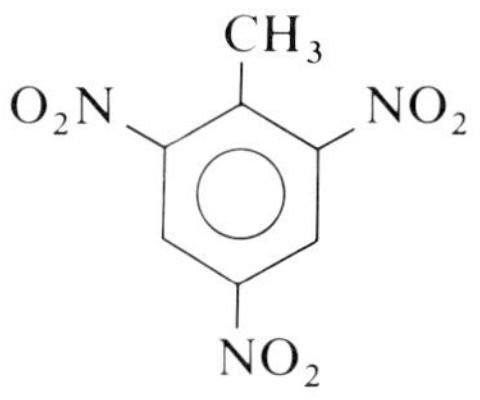

and various substances such as nylon that are used for purposes once served by natural fibers.

To understand the widely different things that make up organic chemistry it is necessary to analyze the subject systematically. As a starting point, one needs to know what holds the atoms of a molecule together. The first chapter therefore is concerned with ionic valence and covalent bonds.

Alice and L-alanine

Next, simple molecules and parts of molecules are examined. It turns out, for example, that the hydroxyl group (HO—) has about the same chemical properties in a complicated nucleic acid molecule as it has in a simple molecule like methyl alcohol (CH_3—OH). This is good news because it means that chemists do not need to learn about all of the millions of organic molecules separately;

A short segment of a nylon molecule. The rest of the molecule repeats the same parts over again many times.

A Short Course in Modern Organic Chemistry

instead it is possible to concentrate on a relatively small number of standard parts. These standard parts, or functional groups, usually show very much the same chemical behavior, even when they are located in a wide variety of different molecules.

Alice and D-alanine

The situation of the chemist is like that of a builder who needs to know about various kinds of bricks and lumber and something about the ways in which they can be combined. The builder can then put up almost any kind of building, even though he may never have seen that particular set of plans before. After finishing this course, the reader should have a good understanding of a very large number of organic molecules, some simple, some quite complicated. A few of these molecules will have been described in this book; however, the reader will also be able to cope with others that were not included (in some cases because they have not yet been discovered) because most of the functional groups and fundamental principles governing the chemistry of these undiscovered molecules will be familiar. And, the molecules yet to be discovered will probably turn out to be more exciting and more important than the ones we already know about.

J. E. L.

Introduction

Chapter 1
Ionic and Covalent Compounds

Organic Chemistry

Organic chemistry is largely the chemistry of **covalently bonded molecules** and of the element carbon. Carbon readily forms covalent bonds not only to atoms of other elements but also to other carbon atoms. As a result, the number of known carbon compounds vastly exceeds those of the other elements. For example, only two compounds containing just hydrogen and oxygen (H_2O and H_2O_2) are known, but we know at least 3000 hydrocarbons.

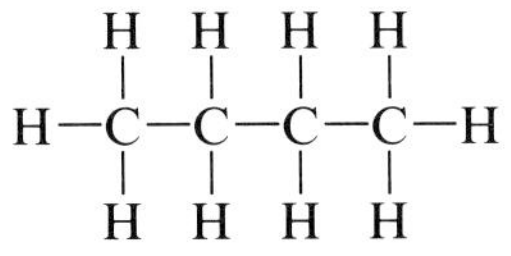

butane

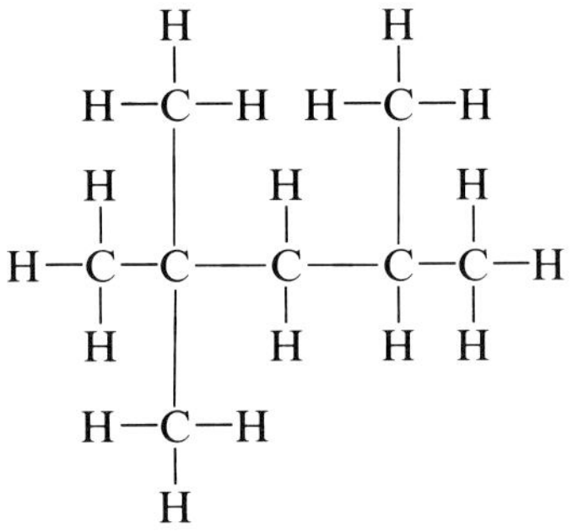

isooctane

Two hydrocarbons

Butane is a constituent of natural gas from oil fields and is used as fuel for heating and cooking. Isooctane is used as a standard fuel for measuring the knock or "octane" ratings of gasolines.

Figure 1-1. A gasoline refinery. [Courtesy of the Standard Oil Co. of New Jersey]

Ionic Compounds

Before concentrating on covalent compounds, it is worthwhile to compare these with ionic compounds and note the differences.

A typical ionic compound such as sodium chloride (Na^+Cl^-) exists as a crystal containing separate ions of sodium (Na^+) and chlorine (Cl^-) and *not* molecules of sodium chloride (NaCl). Each Na^+ has several Cl^- neighbors and no particular Na^+ is married to a particular Cl^- (Figure 1-2). The formula Na^+Cl^- or NaCl merely expresses the fact that in the crystal the number of positive charges must equal the number of negative charges; hence, the number of Na^+ must equal the number of Cl^-. In a crystal of calcium fluoride (CaF_2), each calcium ion (Ca^{2+}) has a charge of $+2$ and each fluoride ion (F^-) has a charge of -1. Electrical

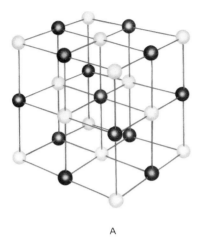

A   B

FIGURE 1-2. A crystal of sodium chloride. **(A)** The black spheres are Na^+ and the light colored ones are Cl^-. **(B)** Another representation, showing the relative sizes of the ions.

neutrality therefore requires twice as many F^- as Ca^{2+} ions and the formula is CaF_2.

The ionic crystal owes its strength, hardness, and high melting point to the force of attraction between the oppositely charged ions. Many ionic compounds dissolve in water; the reason is that the force of electrical attraction between objects with opposite charges is decreased when they are immersed in water.[1] Once dissolved, the ions become separated and move about in the solution almost independently. That is to say, there are Na^+ particles moving through a solution of NaCl and there are also Cl^- particles, but there are no particles or molecules "NaCl."

Covalent Compounds

Crystals of most covalent compounds are soft, have low melting points, and do not tend to be particularly soluble in water. The reason is that they do not contain ions but instead consist of neutral molecules. The atoms in a given molecule are firmly connected together, but there is only a very slight tendency for the molecules to stick to each other.

An example of a covalent compound is **methyl alcohol**, CH_3OH. Methyl alcohol (or methanol) has a low melting point and a low boiling point. The low temperatures at which this compound melts

$$\begin{array}{c} H H \\ | / \\ H-C-O \\ | \\ H \end{array}$$

methyl alcohol
bp 64.65°C; mp −97°C

---

[1] Empty space has a **dielectric constant** of 1; water has a dielectric constant of 78. The force of attraction between negative and positive ions in water is decreased proportionately.

and boils reflect the ease with which the *molecules* can be separated from each other, in contrast to the difficulty of separating oppositely charged *ions*.

Methanol dissolved in water consists of molecules CH_3OH, not ions. The *atoms* of a given CH_3OH molecule belong together and remain together when the methanol dissolves.

The number of particles (whether ions or molecules) dissolved in water can be counted by the number of degrees by which the freezing point of water is lowered. A formula weight (6.02×10^{23} molecules) of CH_3OH will depress the freezing point of water only half as much as a formula weight of NaCl. The reason is that NaCl supplies twice as many particles as CH_3OH (6.02×10^{23} Na^+ and 6.02×10^{23} Cl^-).

Orbitals and the Covalent Bond

The best way to understand both ionic valence and covalence is in terms of **orbitals**.[2] An **atomic orbital** is essentially a region in space centered about the nucleus of an atom. Each atomic orbital can contain from zero to two electrons, but never more than two. Orbitals have shapes that can be calculated by means of quantum mechanics, and the shapes of some of the orbitals of the hydrogen atom are shown in Figures 1–3 and 1–4.

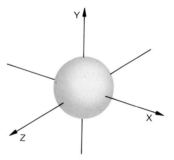

FIGURE 1-3. The 1s orbital of a hydrogen atom. The 1s orbital is closest to the nucleus and has the lowest energy.

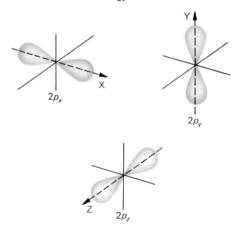

FIGURE 1-4. The $2p_x$, $2p_y$ and $2p_z$ orbitals of a hydrogen atom.

[2] At one time it was thought that electrons orbited around the nucleus like the moon going around the earth. Hence the name orbital.

THE OCTET RULE

Orbitals also differ in energy; the lowest energy orbitals are those that put the electron closest to the nucleus. The orbital of lowest energy is the 1s orbital (Figure 1-3), and there is a considerable gap in energy between it and the next group of orbitals (Figure 1-4). Thus the 1s orbital, which can hold two electrons, corresponds to the first complete "shell" of electrons. Atoms (and also ions) whose outer shell of electrons is complete are particularly stable and unreactive. One example is the helium atom, He, with a charge of +2 on the nucleus and two electrons filling the 1s shell. Another example is the lithium ion, Li^+, with a charge of +3 on the nucleus, which also has two electrons filling the 1s shell (Figure 1-5).

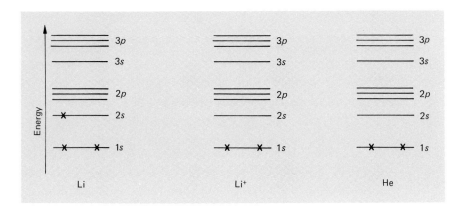

FIGURE 1-5. Orbital energy levels and electrons of Li, Li^+, and He.

The next group of orbitals, called the second shell, consists of the 2s orbital, and the $2p_x$, $2p_y$, and $2p_z$ orbitals. The $2p_x$, $2p_y$, and $2p_z$ orbitals are actually equal in energy; they are represented by three separate lines to indicate that there are three of them. Eight electrons are needed to fill all four orbitals of the second shell ($2s$, $2p_x$, $2p_y$, $2p_z$), and again an atom (or ion) with a filled outer shell is unusually stable and unreactive. Examples are the neon atom, Ne (nuclear charge +10), and the fluoride ion, F^- (nuclear charge +9), shown in Figure 1-6.

Lithium is reactive and readily loses just one electron to form Li^+ because this is just what is required to give a configuration having a complete 1s shell and no partly completed shells (equation 1-1).

$$Li \rightarrow e^- + Li^+ \qquad (1\text{-}1)$$
3 electrons
+3 charge on
nucleus

The helium atom is unreactive because it already has a filled outer shell and needs neither to gain nor to lose any electrons. Fluorine is reactive and readily *gains* just one electron because it has seven

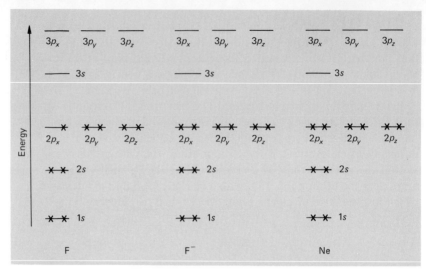

FIGURE 1-6. Fluorine needs one more electron to fill its octet, whereas F^- and Ne have filled outer shells of eight electrons each.

electrons in its outer shell (the second shell) and needs only one more to make a complete octet (equation 1-2).

$$F + e^- \rightarrow F^- \tag{1-2}$$

9 electrons   10 electrons
+9 charge on   +9 charge on
the nucleus   the nucleus

If F^- accepted a second electron to give F^{2-}, the second electron would have to be put in the much more energetic $3s$ orbital; therefore, this kind of reaction does not occur. Similar explanations using the octet rule can be given for the fact that magnesium, Mg (+12 on the nucleus) forms an ion Mg^{2+}, whereas oxygen, O, (+8 on the nucleus) forms an ion O^{2-}, and so forth.

Ionic compounds are formed by *transfer* of electrons from atoms with a surplus to atoms with a deficit, both ions winding up with filled shells (equation 1-3).

$$Li + F \xrightarrow{\text{one electron transferred}} Li^+ + F^- \tag{1-3}$$

The rules for predicting the main ionic valence of an atom in the top rows of the periodic table may be summarized as follows:

1. Look up the atomic number, Z, equal to the charge on the nucleus.
2. Put that number of electrons in the energy levels (as in Figure 1-5), two in each level, starting at the bottom.
3. Either add a *small* number of electrons (one to three) to give a filled shell, or
4. Remove a *small* number of electrons (one to three) to give a cation (positive ion) with a filled shell.

Note that the electrons in the outer shell are sometimes represented by dots. Using this convention, the reaction of lithium with fluorine is given by equation 1-4.

$$\cdot Li + \cdot \ddot{\underset{\cdot\cdot}{F}}\colon \rightarrow \colon Li^+ + \colon \ddot{\underset{\cdot\cdot}{F}}\colon^- \tag{1-4}$$

THE COVALENT BOND

Covalent compounds and molecules like H_2 are held together by pairs of electrons shared between adjacent atomic nuclei. The dashes in the structures for butane and methyl alcohol, for example, each represent a pair of shared electrons, a single covalent bond.

The covalent bond in the hydrogen molecule is the simplest example because each hydrogen atom has only one electron to begin with.

$$H\cdot + \cdot H \rightarrow H-H \qquad (1\text{-}5)$$

Before reaction, each hydrogen atom consists of its nucleus and one electron in a $1s$ orbital. If the two hydrogen atoms get close enough together, the two orbitals should overlap as shown in Figure 1-7. Instead of the orbitals just overlapping, however,

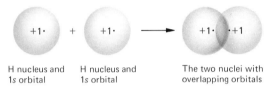

H nucleus and   H nucleus and   The two nuclei with
$1s$ orbital $1s$ orbital   overlapping orbitals

FIGURE 1-7. Overlap of $1s$ hydrogen orbitals.

something new happens: the two $1s$ hydrogen *atomic* orbitals are replaced by a hydrogen *molecular* orbital (Figure 1-8). The **molecular orbital** is like an atomic orbital except that it is centered around two nuclei instead of just one. The negatively charged electrons are now influenced by the attractive force of *both* of the positively charged hydrogen nuclei, whereas before each electron was influenced by just one nucleus. It is this force of attraction of *both* electrons for *both* nuclei that holds the molecule together.

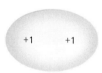

FIGURE 1-8. The hydrogen molecular orbital or bonding orbital.

A molecular orbital, like an atomic orbital, can hold just two electrons. Since the original hydrogen atoms supplied only two electrons, there are just enough to fill the new molecular orbital. The filled hydrogen molecular orbital is like the filled outer shell of a stable and unreactive helium atom. It is much more stable than two separated hydrogen atoms, each with just one electron and neither with a filled shell.

Because orbitals like the one shown in Figure 1-8 correspond to covalent bonds holding the atoms together, they are often called **bonding orbitals**. Molecular orbitals, like atomic orbitals, have shapes: the shape of the $H-H$ bonding orbital is shown in Figure 1-8. Ordinarily chemists are not particularly interested in the shape of the bonding orbital and simply represent it by a dash or bond as in the **structural formulas** given for butane, methanol, and hydrogen. However, it is important to remember that each dash or bond represents two shared electrons.

$$H-H \quad \text{or} \quad H:H$$
hydrogen   hydrogen

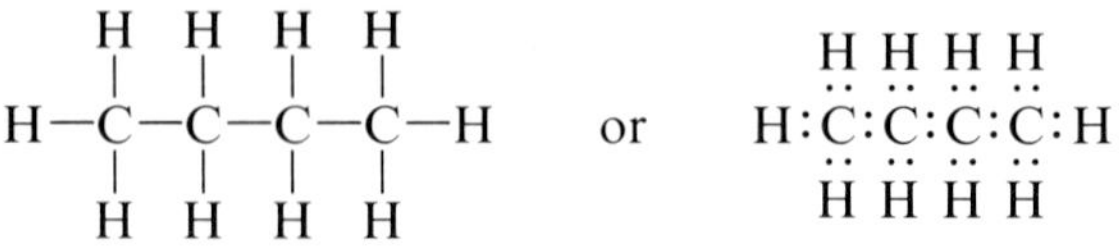

$$\begin{array}{cccc} H & H & H & H \\ | & | & | & | \\ H-C-C-C-C-H \\ | & | & | & | \\ H & H & H & H \end{array} \quad \text{or} \quad \begin{array}{c} H\;H\;H\;H \\ H:C:C:C:C:H \\ H\;H\;H\;H \end{array}$$
butane   butane

If two pairs of electrons are shared between two nuclei, the result is a double bond, represented by two dashes.

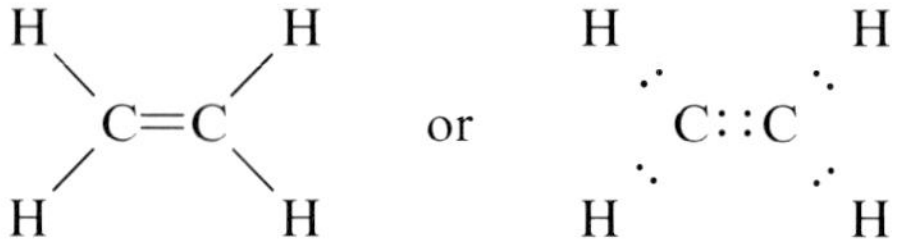

$$\begin{array}{c} H \quad\quad\quad H \\ \;\;\backslash\quad\quad/ \\ \;\;\;C=C \\ \;\;/\quad\quad\backslash \\ H \quad\quad\quad H \end{array} \quad \text{or} \quad \begin{array}{c} H \quad\quad\quad H \\ \quad C::C \\ H \quad\quad\quad H \end{array}$$
ethylene   ethylene

Note that carbon in the structures just shown always has four bonds, a **covalence** of 4. The outer shell of a carbon atom ($+6$ charge on the nucleus) has four electrons. These electrons plus four more supplied by four other atoms or groups are just enough to make four electron-pair bonds.

Oxygen in stable uncharged molecules always has a covalence of 2, just as the oxide ion O^{2-} has an **ionic valence** of -2. Each of these bonds uses one of the electrons of the outer shell of the oxygen atom and one electron from the atom bonded to it. This leaves four **unshared electrons** in the outer shell of oxygen electrons. The structural formula for H_2O may be written as

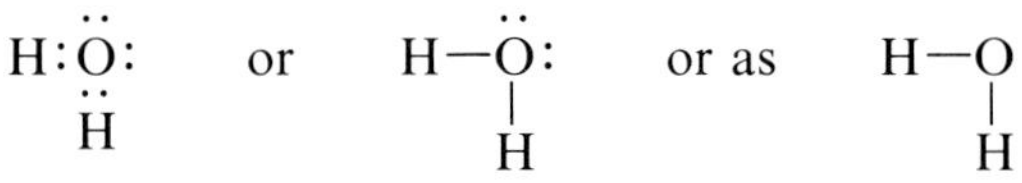

$$\begin{array}{c} H:\ddot{O}: \\ H \end{array} \quad \text{or} \quad \begin{array}{c} H-\ddot{O}: \\ | \\ H \end{array} \quad \text{or as} \quad \begin{array}{c} H-O \\ | \\ H \end{array}$$

Note that, in writing structural formulas, unshared pairs of electrons are often omitted.

Similarly, Cl^- always has an ionic valence of -1, and chlorine in covalent compounds has a covalence of 1.

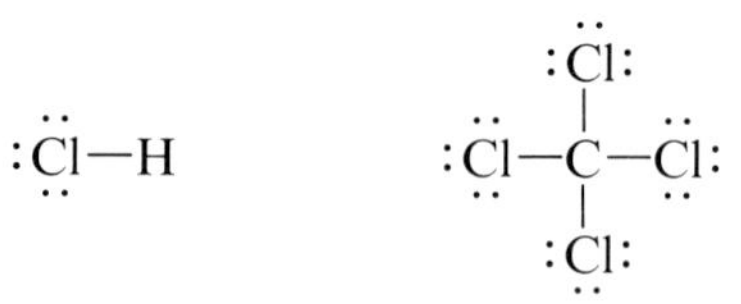

$$:\ddot{Cl}-H \quad\quad :\ddot{Cl}-\underset{\underset{:\ddot{Cl}:}{|}}{\overset{\overset{:\ddot{Cl}:}{|}}{C}}-\ddot{Cl}:$$

Because covalent bonds can be linked together to form chains, including branched chains and rings, the number of possible covalent molecules is almost infinite. In some ways organic chemistry is like art, not only allowing but even demanding the creative use of the imagination. If an invented structural formula obeys the rules of chemical valence, it is usually possible to make the corresponding compound in the laboratory, although it may not be very stable. Of the structures shown below, bicyclooctane is a stable, known compound; carbon suboxide is stable but highly reactive; and 1,1-

dihydroxyethane can be made but tends to fall apart immediately, giving a molecule of acetaldehyde and a molecule of water. Muscalure is a substance used by female houseflies (*Musca domestica*) to attract the male.

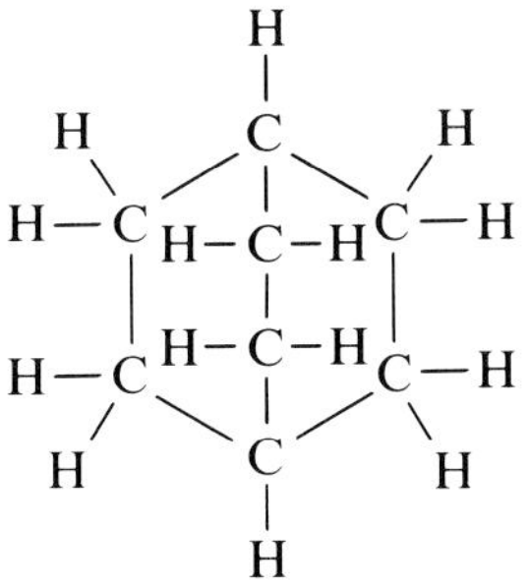

bicyclooctane
mp 169°C

O=C=C=C=O
carbon suboxide
bp 6.8°C

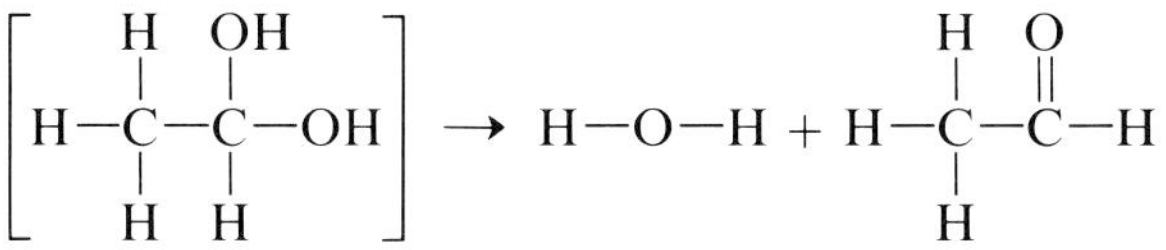

$$\left[\begin{array}{c} H\ OH \\ H-C-C-OH \\ H\ H \end{array} \right] \rightarrow H-O-H + H-C-C-H$$

1,1-dihydroxyethane

acetaldehyde
bp 20.8°C

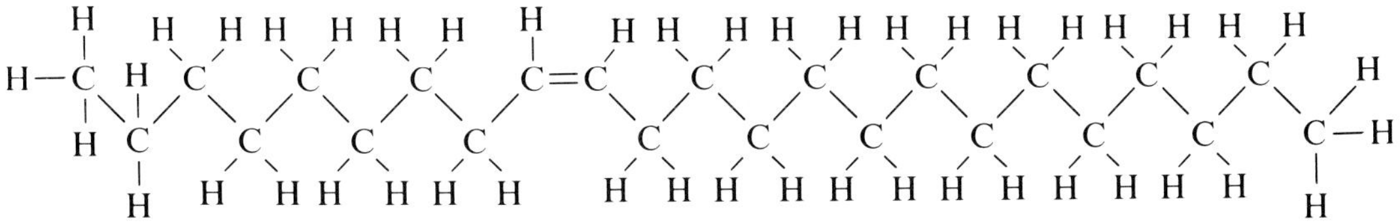

muscalure

The structures shown below are examples of compounds that have not yet been made (or found existing in nature), but which should be capable of existence. Many more could easily be invented.

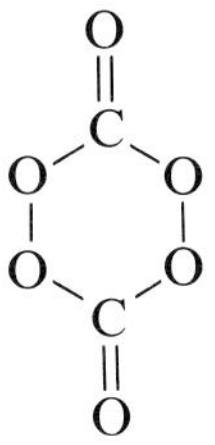

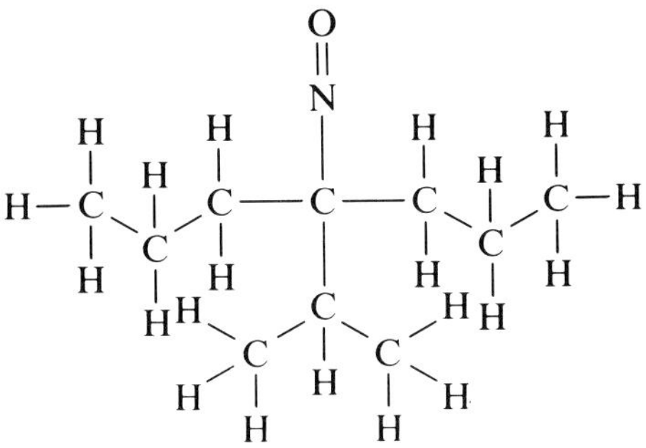

Valence of Charged Atoms

Most of the compounds of interest to organic chemists contain only elements from a rather restricted list: hydrogen, carbon, nitrogen, oxygen, and the halogens. The number of covalent bonds to an atom of a given element in a compound depends on whether or not the atom is also carrying a charge. For example, each

Ionic and Covalent Compounds

hydrogen atom in the structural formula already given for butane has one and only one covalent bond, while H⁺ has no covalent bonds. Similarly, the halogens fluorine, chlorine, bromine, and iodine have one covalent bond in molecules like carbon tetrachloride and no covalent bonds in halide ions such as Cl⁻.

$$\begin{array}{c} Cl \\ | \\ Cl-C-Cl \\ | \\ Cl \end{array}$$

carbon tetrachloride

Nitrogen has a covalence of 3 in uncharged molecules, and its outer shell of eight electrons consists of three covalent bonds (of two shared electrons each) and one pair of unshared electrons.

H:N̈:H   or   N̈
 H   H H H

ammonia   ammonia

However, it is possible to use the unshared pair of electrons to form a fourth covalent bond. An example is the ammonium ion, NH_4^+, which can be made by adding a fourth proton (H⁺) to NH_3.

$$H-\overset{H}{\underset{H}{N}}: + H^+ \rightarrow H-\overset{H}{\underset{H}{N^\pm}}-H \qquad (1\text{-}6)$$

The charges on the atoms of a molecule (or ion) such as NH_4^+ can easily be predicted by noting the atomic number of the atom in question and counting the electrons around it. Thus each hydrogen in NH_4^+ has an atomic number or nuclear charge of $+1$, but each hydrogen has a one-half share of the two electrons of the covalent bond. This is just enough to equal the nuclear charge; hence, there is no net charge on the hydrogen atoms of NH_4^+. The positive charge has to be somewhere, of course, and it can be shown to be on the nitrogen atom by the same process of reasoning.

The atomic number of nitrogen is 7; hence, it has a charge of $+7$ on the nucleus. Two units of this positive charge are balanced by the two electrons of the $1s$ orbital. This leaves five units to be balanced by the remaining electrons, either unshared or in covalent bonds. Because the electron pairs of the bonds are shared with another atom, only one of the two electrons is counted for each bond. Hence the nitrogen atom in NH_4^+ has a charge of $+7-2-4 = +1$. Similar calculations are shown for other molecules in Table 1-1.

Unlike most of the other examples in Table 1-1, **carbonium ions**, which have a positive charge on carbon and only three covalent bonds instead of the usual four, are highly reactive and there are no stable compounds such as $CH_3^+Cl^-$, for example. The reason for the instability of such compounds is the fact that the outer electron shell of CH_3^+ has only six electrons (two in each of three bonds), leaving a fourth low-energy orbital totally unoccupied. Attempts at making compounds like $CH_3^+Cl^-$ give completely covalent

TABLE 1-1. *Charge on an Atom and the Number of Covalent Bonds*

Charge	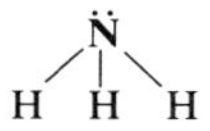 $H-\overset{..}{N}(-H)(-H)$	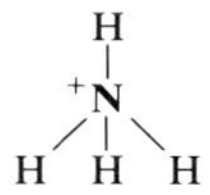 $H-\overset{+}{N}(H)(H)(H)$	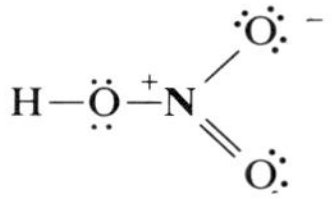 $H-\overset{+}{\overset{..}{O}}-\overset{+}{N}(=\overset{..}{O})(\overset{..}{\overset{..}{O}}{}^-)$	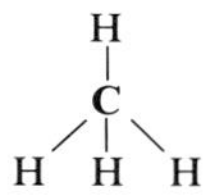 $H-C(H)(H)(H)H$	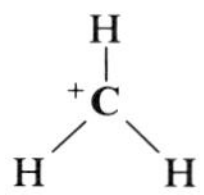 $H-\overset{+}{C}(H)(H)H$	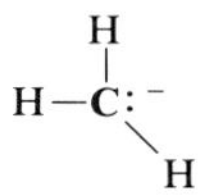 $H-\overset{..}{C}{}^-(H)(H)$	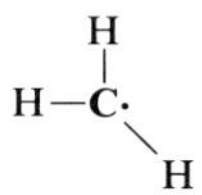 $H-\overset{.}{C}(H)(H)$	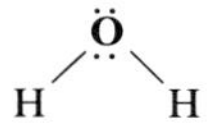 $H-\overset{..}{O}-H$	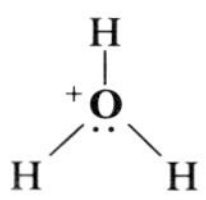 $H-\overset{+}{O}(H)(H)$	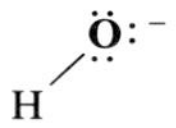 $H-\overset{..}{\overset{..}{O}}{}^-$
Of the nucleus (in bold face) less 2 for the 1s electrons	+5	+5	+5	+4	+4	+4	+4	+6	+6	+6
Of the unshared electrons	−2	0	0	0	0	−2	−1	−4	−2	−6
Of ½ of the shared (bonding) electrons	−3	−4	−4	−4	−3	−3	−3	−2	−3	−1
Net charge on atom	0	+1	+1	0	+1	−1	0	0	+1	−1

compounds like methyl chloride instead. In methyl chloride,

$$\begin{array}{c} H \\ | \\ H-C-Cl \\ | \\ H \end{array}$$

methyl chloride

the carbon atom has a share in four pairs of electrons, and all of the low-energy orbitals are occupied. Carbonium ions R_3C^+ will be encountered in later chapters, but such ions are rarely found in stable, isolatable compounds. They are postulated as unstable intermediates to explain certain reactions.

For example, equation 1-7 shows a reaction that is unusually fast because the reactants form the unusually stable carbonium ion shown in brackets in equation 1-8 as an intermediate. As will be noted in later chapters, carbonium ions that have other carbon atoms directly attached to the charged carbon atom are much more stable and more easily formed than CH_3^+.

$$Cl^- + \begin{array}{c} H \\ | \\ H-C \\ | \\ H \end{array} \begin{array}{c} H \\ | \\ H-C-H \\ | \\ C \\ | \\ H-C-H \\ | \\ H \end{array} Br \rightarrow \begin{array}{c} H \\ | \\ H-C \\ | \\ H \end{array} \begin{array}{c} H \quad H \\ \diagdown | \diagup \\ C \\ | \\ C \\ \diagup | \diagdown \\ H \; H \; H \end{array} Cl + Br^- \quad (1\text{-}7)$$

Explanation:

$$\begin{array}{c} H \\ | \\ H-C \\ | \\ H \end{array} \begin{array}{c} H \quad H \\ \diagdown | \diagup \\ C \\ | \\ C \\ \diagup | \diagdown \\ H \; H \; H \end{array} Br \rightarrow \left[\begin{array}{c} H \\ | \\ H-C \\ | \\ H \end{array} \begin{array}{c} H \\ | \\ H-C-H \\ | \\ C^+ \\ | \\ C \\ \diagup | \diagdown \\ H \; H \; H \end{array} \right] + Br^- \quad (1\text{-}8)$$

$$Cl^- + \left[\begin{array}{c} H \\ | \\ H-C \\ | \\ H \end{array} \begin{array}{c} H \\ | \\ H-C-H \\ | \\ C^+ \\ | \\ C \\ \diagup | \diagdown \\ H \; H \; H \end{array} \right] \rightarrow \begin{array}{c} H \\ | \\ H-C \\ | \\ H \end{array} \begin{array}{c} H \\ | \\ H-C-H \\ | \\ C \\ | \\ C \\ \diagup | \diagdown \\ H \; H \; H \end{array} Cl \quad (1\text{-}9)$$

Structures with three bonds to carbon and a pair of electrons occupying the otherwise vacant orbital are called **carbon anions**, or **carbanions**. Structures with three bonds to carbon and only one unshared electron ($CH_3 \cdot$ for example) are called **free radicals**. Like carbonium ions, carbanions and free radicals are unstable and highly reactive but their existence will be used in later chapters to explain why certain reactions take place.

Functional Groups

In **hydrocarbons** such as butane and isooctane, a skeleton of carbon atoms is covered with hydrogen atoms, just enough to use up the remaining valences of each carbon and ensure that each carbon atom has a covalence of 4.

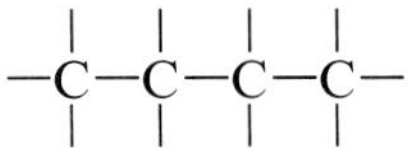

carbon skeleton

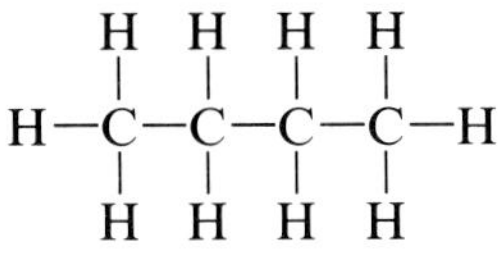

butane

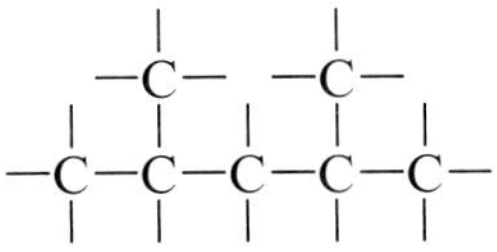

carbon skeleton

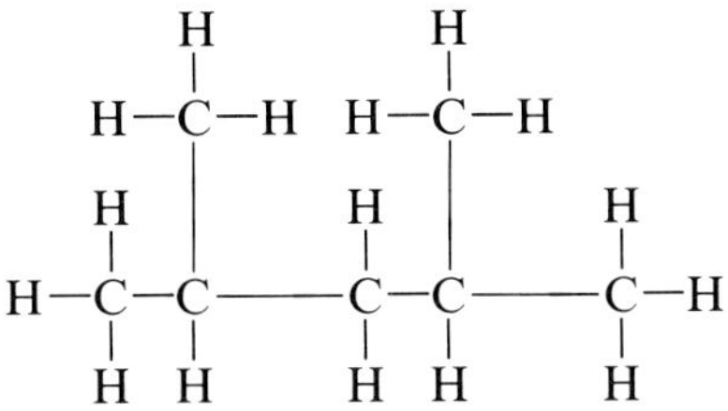

2,4-dimethylpentane

Most biologically important molecules have one or more groups of atoms other than hydrogen attached to the main carbon skeleton. Such structures are called functional groups because they are chemically the most important parts of the molecule.

The names and structures of most of the important functional groups are shown in Table 1-2.

TABLE 1-2. *Important Functional Groups*

Functional Group	Example (shown in boxes)	Functional Group	Example (shown in boxes)
Double bond, found in alkenes	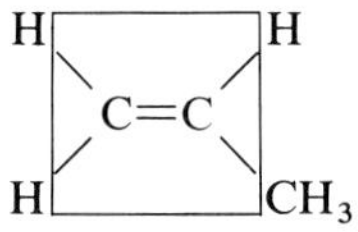 (1-propene)	Oxygen, found in ethers	$CH_3CH_2-O-CH_2CH_3$ (diethyl ether)
Triple bond, found in alkynes	$CH_3-CH_2-C\equiv C-CH_3$ (2-pentyne)	Aldehyde group, found in aldehydes	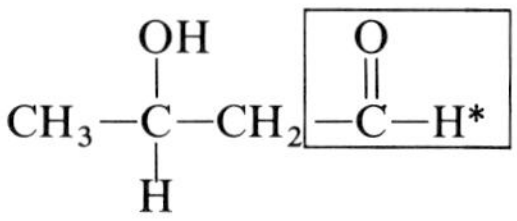 (aldol)
Hydroxyl, found in alcohols and phenols	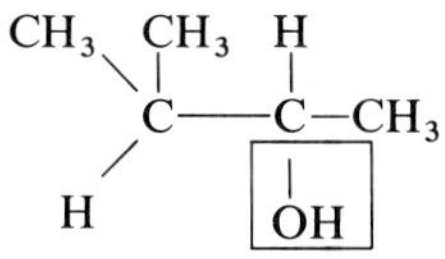 (3-methyl-2-butanol)	Keto group, found in ketones	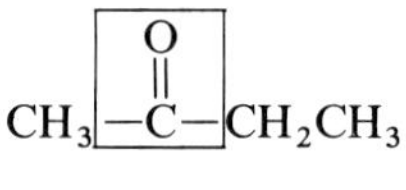 (2-butanone)

13

TABLE 1-2. *Important Functional Groups* (*contd.*)

Functional Group	Example (shown in boxes)	Functional Group	Example (shown in boxes)
Hydroxyl, continued	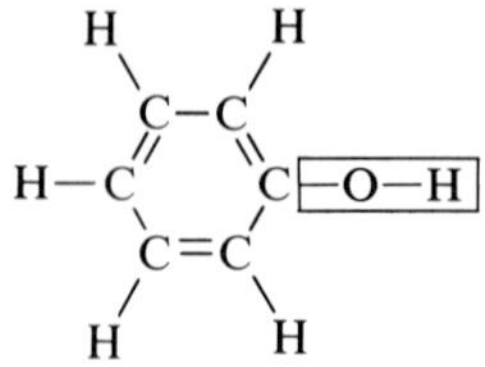 (phenol) — benzene ring with $-$O$-$H boxed	Carboxyl group, found in carboxylic acids	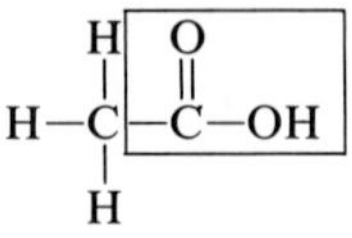 (acetic acid) $H-C(H)(H)-$[$C(=O)-OH$]
Amino group, found in amines	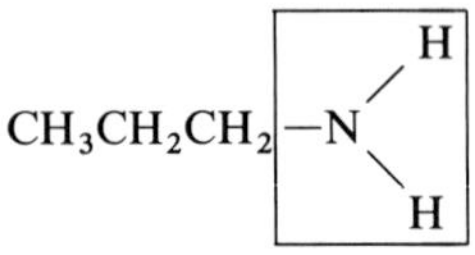 $CH_3CH_2CH_2-$[NH_2] (1-aminopropane)	Amide group, found in amides	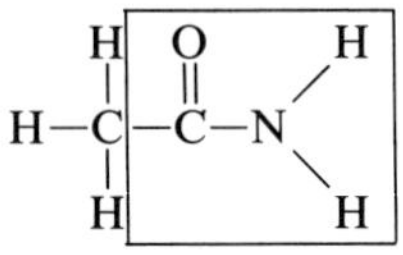 (acetamide) $H-C(H)(H)-C(=O)-$[NH_2]
Ester group, found in esters	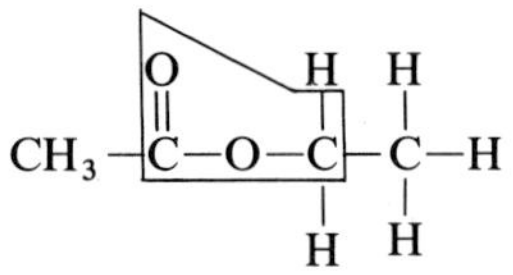 CH_3-[$C(=O)-O-C$]$(H)(H)-C-H$ (ethyl acetate)	Sulfonic acid group	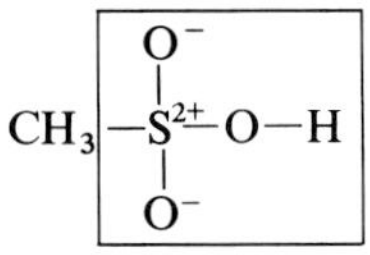 CH_3-[$S^{2+}(O^-)(O^-)-O-H$] (methanesulfonic acid)
Nitro group, found in nitro compounds	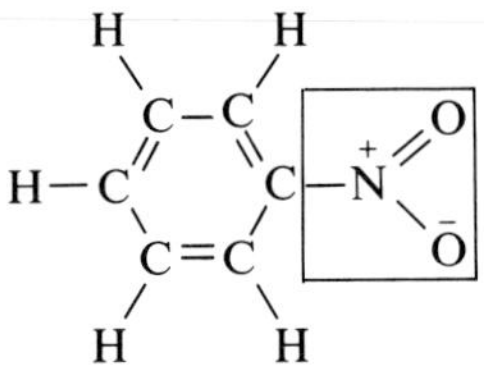 (nitrobenzene) — benzene ring with [$-N^+(=O)(O^-)$]	Sulfate ester group	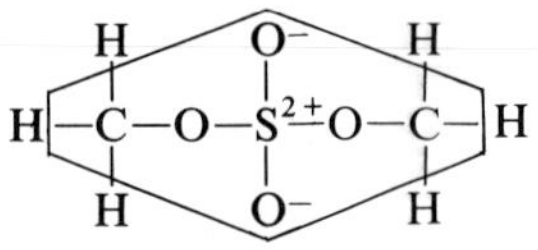 (dimethyl sulfate) $H-C(H)(H)-$[$O-S^{2+}(O^-)(O^-)-O$]$-C(H)(H)-H$
Nitrate ester group	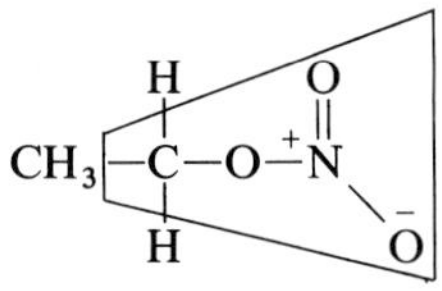 $CH_3-C(H)(H)-$[$O-N^+(=O)(O^-)$] (ethyl nitrate)	Phosphate ester group	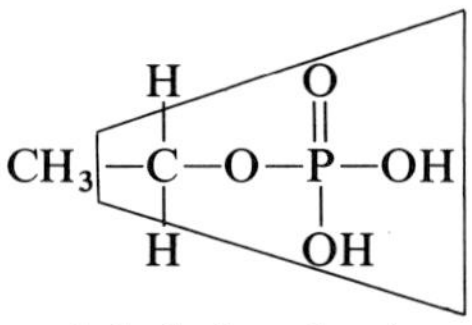 $CH_3-C(H)(H)-$[$O-P(=O)(OH)(OH)$] (ethyl phosphate)

* Note that several functional groups can occur in the same molecule. In this example there is a hydroxyl (alcohol) group as well as an aldehyde group.

Behavior of Functional Groups

In most cases the behavior of a given functional group in different molecules is so very nearly the same that it is useful and appropriate to think of the reaction as a reaction of the functional group rather than of the molecule. An example is the reaction of the hydroxyl functional group with sodium to give sodium compounds and hydrogen. When methyl alcohol reacts with sodium to give **sodium methoxide** and hydrogen (equation 1-10), the important

A Short Course in Modern Organic Chemistry

changes take place at the hydroxyl group. The methyl group (CH_3) in sodium methoxide is not very different from the methyl group in the original methyl alcohol molecule.

$$2\,H-\underset{\underset{H}{|}}{\overset{\overset{H}{|}}{C}}-O-H + 2\,Na \rightarrow 2\,H-\underset{\underset{H}{|}}{\overset{\overset{H}{|}}{C}}-O^-Na^+ + H_2 \quad (1\text{-}10)$$

If the reaction of methyl alcohol with sodium is really just a reaction of the hydroxyl group, it is not surprising that water and ethyl alcohol react in the same way (equations 1-11 and 1-12).

$$2\,H-O-H + 2\,Na \rightarrow 2\,HO^-Na^+ + H_2 \quad (1\text{-}11)$$

$$2\,H-\underset{\underset{H}{|}}{\overset{\overset{H}{|}}{C}}-\underset{\underset{H}{|}}{\overset{\overset{H}{|}}{C}}-OH + 2\,Na \rightarrow 2\,H-\underset{\underset{H}{|}}{\overset{\overset{H}{|}}{C}}-\underset{\underset{H}{|}}{\overset{\overset{H}{|}}{C}}-O^-Na^+ + H_2 \quad (1\text{-}12)$$

ethyl alcohol sodium ethoxide

A large part of organic chemistry is just the reactions of functional groups. This is something for which all chemists are thankful, because it eliminates the need for memorizing the reactions of each new molecule separately. All that is necessary in most cases is to look for the functional group in the molecule and recall its typical reactions.

SUMMARY

1. Ionic compounds are essentially just mixtures of ions.
2. Melting points and solubility: Ionic compounds tend to have higher melting points, higher boiling points, and to be more soluble in water than covalent compounds.
3. Covalent compounds, molecules: Strong forces hold the atoms together in a molecule, only weak forces hold the molecules together in a crystal.
4. Orbitals: There are several kinds differing in shape and in energy. One orbital can hold two electrons.
5. Covalent bonds
    (a) A covalent bond is formed by two electrons in a bonding orbital.
    (b) A bonding orbital is formed by the overlap of two atomic orbitals.
    (c) The covalently bonded nuclei are held together by the attraction of the shared, negatively charged electrons for both of the positively charged nuclei.
6. Valence
    (a) Hydrogen and the halogens have a covalence of 1 in purely covalent compounds, oxygen a covalence of 2, nitrogen a covalence of 3, and carbon a covalence of 4.
    (b) Nitrogen can have four covalent bonds in compounds in which the nitrogen atoms bears a positive charge. Oxygen can have three covalent bonds if it is positively charged or just one if it is negatively charged.
    (c) Carbon can have a covalence of 3 in unstable intermediates such as R_3C^+, $R_3C\cdot$, or $R_3C:^-$. These have only a fleeting existence.
7. Functional groups. Most reactions take place at certain parts of a molecule, called functional groups.

EXERCISES

1. Describe the physical properties (hardness, melting point, boiling point, solubility in water) of (a) a typical ionic compound and (b) a typical covalent compound.

2. Make an intelligent guess about the compounds in petroleum and gasoline. Are they likely to be ionic or covalent? Justify your answer.

3. If a certain number of formula weights of ethyl alcohol (C_2H_6O) or the same number of formula weights of ethylene glycol ($C_2H_6O_2$) are dissolved in the water of an automobile radiator, the freezing point of the water is depressed from its normal 0 to $-5°C$. If the same number of formula weights of NaCl are dissolved in the water, the freezing point is depressed twice as much, to $-10°C$. Why?

4. If one formula weight each of NaCl and LiBr are dissolved in water, the solution is absolutely the same as the one obtained by dissolving NaBr and LiCl in water. Why?

5. If CH_4 (methane, bp $-161.5°C$) and CCl_4 (carbon tetrachloride, bp $76.8°C$) are mixed, the resulting mixture is quite different from the mixture obtained from $CHCl_3$ (chloroform, bp $61°C$) and CH_3Cl (methyl chloride, bp $-24°C$). Explain why. *Hint:* See the previous question and notice the low boiling points of the compounds in this question.

6. Make a sketch of (a) a $1s$ orbital; (b) a $2p_x$ orbital; (c) a $2p_z$ orbital.

7. Make a sketch showing a $1s$ orbital, a $2p_x$ orbital, and a $2p_y$ orbital *on the same atom*.

8. How many electrons can be accommodated by one atomic orbital? How many can be accommodated by one molecular orbital?

9. How many electrons are represented by each bond (line) in the structural formula for methane?

$$\begin{array}{c} H \\ | \\ H-C-H \\ | \\ H \end{array}$$
methane

10. Carbon has a charge of $+6$ on the nucleus, hydrogen a charge of $+1$. How many electrons are there altogether in the methane molecule? Where are the electrons that are not in the bonds?

11. Invent four covalent molecules and draw their structural formulas.

12. Explain why one of the oxygen atoms of $HONO_2$ in Table 1-1 has a negative charge.

13. Why does ammonia ($:NH_3$) readily add a proton, whereas methane (CH_4) does not?

14. Would methide ion (CH_3^-) readily add a proton?

Chapter 2
Acids and Bases

In this chapter we will be concerned with the fundamental concept of acid and base reactions. The reactions will be illustrated mainly with molecules containing one or another of three kinds of functional groups. These will be the hydroxyl group, HO— (usually written without the bond, simply as OH or HO) of **alcohols** such as methyl alcohol and glycerol; the amino group —NH$_2$ (NH$_2$) of **amines** such as methylamine, and the carboxyl group —COOH (COOH) of **carboxylic acids** such as acetic and formic acids. The same functional groups will be discussed separately and at greater length in later chapters, at which time processes other than acid-base reactions will be taken up.

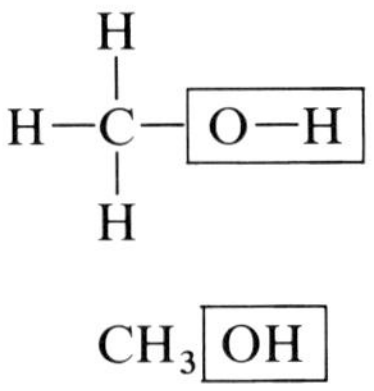

$$\begin{array}{c} H \\ | \\ H-C-\boxed{O-H} \\ | \\ H \end{array} \qquad \begin{array}{c} H\ H\ H \\ |\ \ |\ \ | \\ H-C-C-C-H \\ |\ \ |\ \ | \\ \boxed{\begin{array}{c} O\ O\ O \\ |\ \ |\ \ | \\ H\ H\ H \end{array}} \end{array} \qquad \begin{array}{c} H\qquad H \\ |\quad\ \ \diagup \\ H-C-\boxed{N} \\ |\quad\ \ \diagdown \\ H\qquad H \end{array}$$

CH$_3$ $\boxed{\text{OH}}$ $\boxed{\begin{array}{c} O\ O\ O \\ H\ H\ H \end{array}}$ CH$_3$ $\boxed{\text{NH}_2}$

methyl alcohol $\boxed{\text{HO}}$ CH$_2$CH $\boxed{\text{OH}}$ CH$_2$ $\boxed{\text{OH}}$ methylamine

glycerol

$$\begin{array}{c} H\quad O \\ |\quad\ \ \| \\ H-C-\boxed{C-O-H} \\ | \\ H \end{array} \qquad\qquad \begin{array}{c} O \\ \| \\ H-\boxed{C-O-H} \end{array}$$

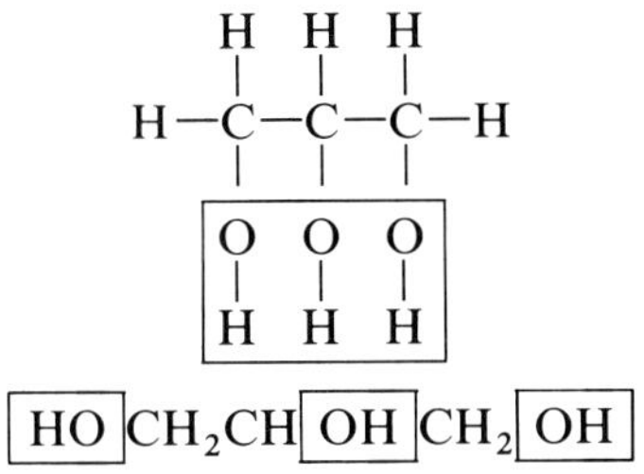

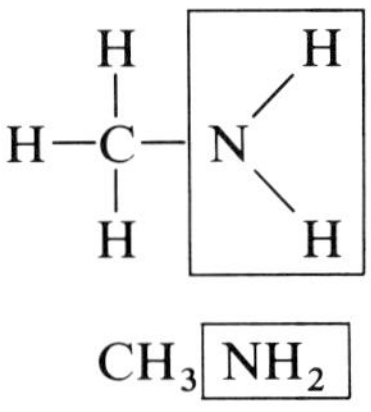

CH$_3$ $\boxed{\text{COOH}}$ H $\boxed{\text{COOH}}$

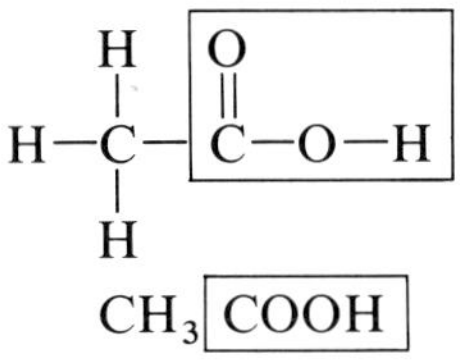

acetic acid 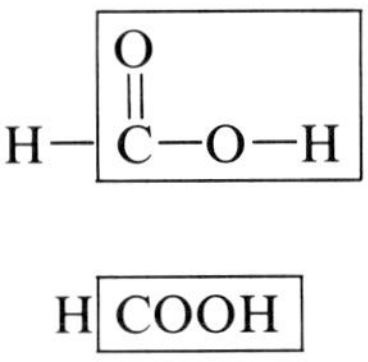   formic acid

Acid-Base Reactions

A Brønsted **acid** (usually just called an acid without any qualification) is a molecule that can donate a proton or H^+ to a base. A **base** is any molecule that can accept a proton and hold it by a covalent bond. To act as a base, a molecule need only have an accessible unshared pair of electrons.

In the **autoprotolysis** reaction of water, one molecule acts as a base by virtue of a pair of unshared electrons in one of the $2p$ orbitals of the oxygen atom. A second molecule acts as an acid, or proton donor.

$$H-\ddot{O}\cdot + H-\ddot{O}-H \rightleftharpoons \overset{H}{\underset{H}{\overset{+}{O}}}\overset{H}{\diagup} + :\ddot{O}H \qquad (2\text{-}1)$$

The reaction shown in equation 2-1 is responsible for an equilibrium H_3O^+ (or hydronium ion) concentration of only about 10^{-7} molar (M) in pure water, because water is neither a very strong acid nor a very strong base.

The reaction can also be written as in equation 2-2. The first of the curved arrows indicates that an unshared electron pair of the first water molecule becomes a bonding electron pair. The second curved arrow indicates that the electron pair that previously formed the bond to the proton is released to become an unshared electron pair of OH^-.

$$H-\ddot{O}\overset{\curvearrowright}{} \quad H\overset{\curvearrowleft}{-}\ddot{O} \rightarrow H-\ddot{O}\overset{^+H}{\diagup} \quad :\ddot{O} \qquad (2\text{-}2)$$

Water is an example of an **amphoteric** substance, that is, a substance that can act either as an acid or as a base. An amphoteric substance has both an unshared electron pair and a detachable (acidic) proton.

Water as a base

$$H-\underset{H}{\overset{..}{O}}: + HO-\overset{O}{\overset{\|}{C}}-CH_3 \rightleftharpoons H-\underset{H}{\overset{+}{O}}\overset{H}{\diagup} + :\ddot{O}-\overset{O}{\overset{\|}{C}}-CH_3 \qquad (2\text{-}3)$$

Water as an acid

$$H-\ddot{O}-H + :N\overset{H}{\underset{H}{\diagup}}-H \rightleftharpoons H-\ddot{O}:^- + H-\overset{H}{\underset{H}{\overset{+}{N}}}-H \qquad (2\text{-}4)$$

A pair of molecules, one an acid and the other a base, that differ only by the presence of a proton are known as a conjugate acid-base pair. Acetic acid, for example, has acetate ion for its conjugate base.

To give another example, the conjugate acid of the water molecule is the hydronium ion, H_3O^+. A few other such pairs are shown in Table 2-1.

TABLE 2-1. *Some Conjugate Acid-Base Pairs*

Acid	Base	Acid	Base
$H\overset{+}{-}\underset{H}{O}-H$	$H-\underset{H}{\ddot{O}}:$	$CH_3-\underset{H}{\overset{H}{N}}:$	$CH_3-\underset{}{\overset{H}{\ddot{N}}}:^-$
$H-\underset{H}{\ddot{O}}:$	$H-\ddot{\ddot{O}}:^-$	$CH_3\overset{+}{-}\underset{H}{\overset{H}{O}}:$	$CH_3-\ddot{\ddot{O}}:$
$CH_3-\overset{O}{\overset{\|}{C}}-O-H$	$CH_3-\overset{O}{\overset{\|}{C}}-\ddot{\ddot{O}}:^-$	$H-\ddot{\ddot{O}}:^-$	$:\ddot{\ddot{O}}:^{2-}$
$CH_3\overset{+}{-}\underset{H}{\overset{H}{N}}-H$	$CH_3-\underset{H}{\overset{H}{N}}:$	$CH_3-\ddot{O}\diagdown^H$	$CH_3-\ddot{\ddot{O}}:^-$

Dissociation Constants

The dissociation constant in water as solvent is a frequently used measure of the relative strengths of acids and bases. For acetic acid, the dissociation constant K_A is given by equation 2-5.

$$H-\underset{H}{\overset{H}{C}}-\overset{O}{\overset{\|}{C}}-OH + H_2O \rightleftharpoons H-\underset{H}{\overset{H}{C}}-\overset{O}{\overset{\|}{C}}-O^- + H_3O^+$$

$$K_A = \frac{[CH_3COO^-][H_3O^+]}{[CH_3COOH]} \approx 10^{-5} \qquad (2\text{-}5)$$

The concentration of the water is a constant and is not included in the expression for K_A. The dissociation constant for acetic acid, a moderately weak acid typical of a whole series of carboxylic acids, RCOOH,[1] is about 10^{-5}. The value for ammonium ion, NH_4^+, is about 10^{-9}, making this "acid" about a ten-thousandth as strong as acetic acid. The K_As of methylammonium ion ($CH_3NH_3^+$) and other protonated amines (RNH_3^+) are in the neighborhood of 10^{-11}.

$$NH_4^+ + H_2O \rightleftharpoons NH_3 + H_3O^+$$

$$K_A = \frac{[NH_3][H_3O^+]}{[NH_4^+]} \approx 10^{-9} \qquad (2\text{-}6)$$

[1] R stands for any **alkyl** group such as methyl (CH_3-) or ethyl (CH_3CH_2-).

A rather strong acid such as phosphoric acid (first hydrogen) has a K_A of about 10^{-2}, making it about 1000 times as strong as acetic acid.

$$\text{HO}-\underset{\underset{\text{OH}}{|}}{\overset{\overset{\text{O}}{\|}}{\text{P}}}-\text{OH} + \text{H}_2\text{O} \rightleftharpoons \text{HO}-\underset{\underset{\text{OH}}{|}}{\overset{\overset{\text{O}}{\|}}{\text{P}}}-\text{O}^- + \text{H}_3\text{O}^+$$

phosphoric acid

$$K_A = \frac{[\text{H}_2\text{PO}_4^-][\text{H}_3\text{O}^+]}{[\text{H}_2\text{O}]} = 10^{-2} \qquad (2\text{-}7)$$

Alcohols (ROH) such as methyl alcohol (CH_3OH) and ethyl alcohol (CH_3CH_2OH) are very weak acids, $K_A \approx 10^{-18}$, which makes them weaker acids than water.

One measure of the strength of a base is the K_A for dissociation of its conjugate acid. A strong conjugate acid means a weak base and vice versa. A second measure of base strength is the base dissociation constant, K_B.

$$\text{NH}_3 + \text{H}_2\text{O} \rightleftharpoons \text{NH}_4^+ + \text{HO}^-$$

$$K_B = \frac{[\text{NH}_4^+][\text{HO}^-]}{[\text{NH}_3]} \approx 10^{-5} \qquad (2\text{-}8)$$

The two measures of base strength are inversely proportional to each other.

$$K_B = \frac{10^{-14}}{K_A}$$

For example, if K_A for phosphoric acid is 10^{-2}, then K_B for dihydrogen phosphate ion is 10^{-12}. If K_A for acetic acid is 10^{-5}, then K_B for acetate ion is 10^{-9}.

$$\text{CH}_3-\overset{\overset{\text{O}}{\|}}{\text{C}}-\text{O}^- + \text{H}_2\text{O} \rightleftharpoons \text{CH}_3-\overset{\overset{\text{O}}{\|}}{\text{C}}-\text{OH} + \text{HO}^- \qquad (2\text{-}9)$$

acetate ion $\qquad K_B \approx 10^{-9} \qquad$ acetic acid

Acetate ion is thus a stronger base than dihydrogen phosphate ion but a much weaker base than the ammonia molecule.

Electronegativity

Much of the differences in acid or base strengths between one molecule and another are due to differences in the electronegativity of the atom to which the proton is attached in the acid (or conjugate acid of the base). The electronegativity of an atom is the tendency for that atom to attract electrons. Differences in electronegativity are due to differences in the energies of the atomic orbitals. In general, the most electronegative elements are to the right and up in the periodic table. Going across the top row, for example, the sequence of *increasing* electronegativities is

$$\text{Li} < \text{C} < \text{N} < \text{O} < \text{F}$$

Going down the halogen column, the electronegativities of the halogens *decrease* in the sequence

$$F > Cl > Br > I$$

Extreme differences in electronegativity cause a reacting pair of elements to form ionic rather than covalent compounds (for example, Li^+Cl^-). Lesser differences still permit covalent bond formation, but the electrons are shared unequally. The more electronegative of the two atoms has the greater share of the electrons and tends to have a slight negative charge, whereas the less electronegative atom has a slight positive charge. The sum of these two slight charges is of course zero.

One way of suggesting this **polarization**, or inequality of sharing of the electrons, is to write the dots representing the electron pair closer to one of the atoms than to the other, as in Figure 2-1. Another way is to place the symbol δ^+ near the more positive end of the bond in question and δ^- near the more negative end.

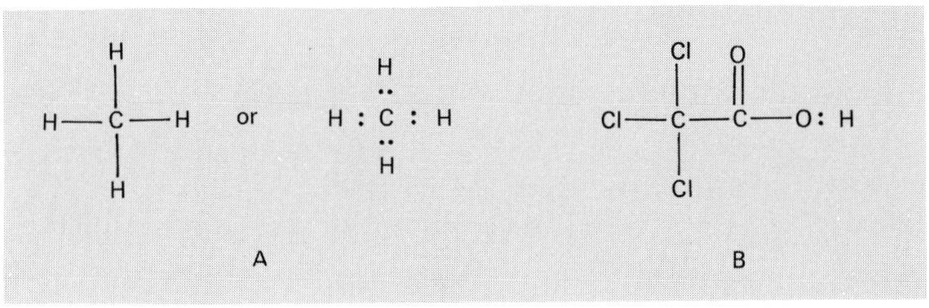

FIGURE 2-1. **(A)** Methane. The electron pairs are shared equally. **(B)** Trichloroacetic acid. The electrons of the O—H bond are shared unequally.

In the methane molecule (Figure 2-1) the electrons of the C—H bond are shared almost equally because there is very little difference in electronegativity (tendency to attract electrons) between the carbon and the hydrogen of a methane molecule. In trichloroacetic acid, the electrons of the O—H bond are shared unequally because of the greater electronegativity of the oxygen atom. When trichloroacetic acid is dissolved in water, the covalent trichloroacetic acid molecule dissociates into hydrogen ions (actually H_3O^+ ions) and trichloroacetate ions.

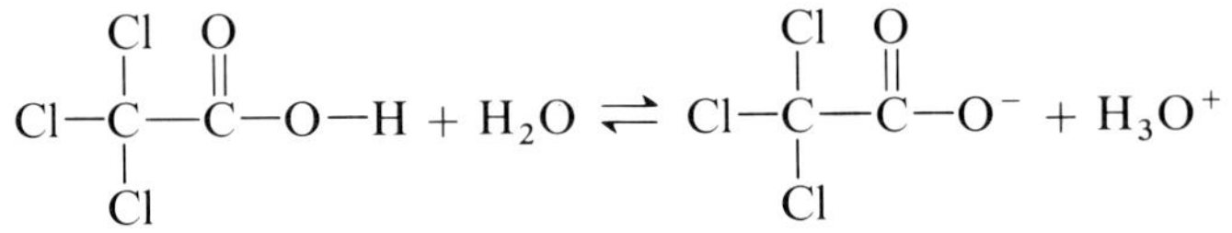

$$Cl_3C\text{—}C(=O)\text{—}O\text{—}H + H_2O \rightleftharpoons Cl_3C\text{—}C(=O)\text{—}O^- + H_3O^+ \quad (2\text{-}10)$$

The dissociation of trichloroacetic acid into ions is just an exaggeration of the tendency already present in the polarized O-H bond of the original molecule. When methane is dissolved in water (it is almost insoluble), none of the molecules gives up a hydrogen ion. Methane, therefore, is not considered to be an acid at all. Trichloroacetic acid is a **strong acid** in water because virtually every molecule is found to have dissociated into the solvated hydrogen (H_3O^+) and trichloroacetate ions. Trichloroacetic acid is stronger than acetic acid because the electronegative chlorine substituents also help to pull electrons away from the proton and make the oxygen more electronegative than usual.

In a series of hydrogen compounds, acidity always increases as the electronegativity of the atom to which the proton is attached increases. At the carbon end of the series of increasing electro-

$$\underset{\underset{H}{|}}{\overset{\overset{H}{|}}{R-C-H}} < \underset{}{\overset{\overset{H}{|}}{R-N-H}} < R-O-H < \overset{\overset{O}{\|}}{R-C-OH} < Cl-H$$

Sequence of increasing electronegativity and acidity

negativity there is essentially no tendency at all to lose a proton. Amines, RNH_2, and alcohols (ROH) are also very weak acids. In fact, they demonstrate their acidity only by reacting with metal-carbon compounds such as methylmagnesium bromide. The carbon-to-magnesium bond in this compound is highly polarized and reacts as though it were ionic in the sense CH_3^- ^+MgBr and a source of carbanions, CH_3^-.

$$CH_3CH_3 + \overset{\delta^-}{C}H_3-\overset{\delta^+}{M}gBr \rightarrow \text{no reaction} \tag{2-11}$$

$$\underset{\underset{H}{|}}{\overset{\overset{H}{|}}{H-C-N}}\overset{H}{\underset{H}{\diagdown}} + \overset{\delta^-}{C}H_3-\overset{\delta^+}{M}gBr \rightarrow \underset{\underset{H}{|}}{\overset{\overset{H\ H}{|\ |}}{H-C-N^-}} \ ^+MgBr + CH_4 \tag{2-12}$$

$$\underset{\underset{H}{|}}{\overset{\overset{H}{|}}{H-C-\ddot{O}-H}} + \overset{\delta^-}{C}H_3-\overset{\delta^+}{M}gBr \rightarrow \underset{\underset{H}{|}}{\overset{\overset{H}{|}}{H-C-\ddot{O}:^-}} \ ^+MgBr + CH_4 \tag{2-13}$$

The carboxylic acids are distinctly more acidic and will react not only with methylmagnesium bromide but also with weak bases such as sodium bicarbonate ($NaHCO_3$).

$$R-\overset{\overset{O}{\|}}{C}-\ddot{O}-H + \ ^-:\ddot{O}-\overset{\overset{O}{\|}}{C}-O-H \rightarrow R-\overset{\overset{O}{\|}}{C}-\ddot{O}:^- + H-O-\overset{\overset{O}{\|}}{C}-O-H$$
$$\text{bicarbonate ion} \tag{2-14}$$

$$HO-\overset{\overset{O}{\|}}{C}-OH \rightarrow H_2O + CO_2$$

The halogen acids are so strongly acidic that they react completely even with water (solvent) to give hydronium ions and halide ions.

$$HCl + H_2O \rightarrow H_3O^+ + Cl^- \tag{2-15}$$

In a series of bases with the electron pair situated on atoms of different electronegativity, the base strength increases as the electronegativity decreases. That is to say, a strongly electron-attracting atom is reluctant to share its electrons in a bond to a proton. The sequence of increasing base strength is thus the reverse of the sequence of increasing acidity.

$$:\ddot{\underset{..}{Cl}}:^- \;<\; R-\overset{\overset{O}{\|}}{C}-\ddot{\underset{..}{O}}:^- \;<\; R-\ddot{\underset{..}{O}}:^- \;<\; R-\overset{H}{\underset{..}{N}}:^- \;<\; H-\overset{H}{\underset{H}{C}}:^-$$

Sequence of increasing basicity and decreasing electronegativity

In this series, Cl^- will add a proton only in environments like concentrated sulfuric acid (H_2SO_4), acetate ion (CH_3COO^-) is weakly basic and is partly protonated in water, whereas the rest of the series is strongly basic and the reactions with water go to completion.

$$R-\ddot{\underset{..}{O}}:^- + H-O-H \rightarrow R-\ddot{\underset{..}{O}}-H + H-\ddot{\underset{..}{O}}:^- \qquad (2\text{-}16)$$
alkoxide ion　　　　　　　　　alcohol

$$R-\overset{H}{\underset{..}{N}}:^- + H-O-H \rightarrow R-\overset{H}{\underset{..}{N}}-H + H\ddot{\underset{..}{O}}:^- \qquad (2\text{-}17)$$
substituted　　　　　　　　amine
nitride ion

Compounds such as RCOOH and ROH are much less basic than their conjugate bases $RCOO^-$ and RO^- even though they still have unshared electron pairs. Thus carboxylic acids and alcohols are extensively protonated only in strongly acidic media such as H_2SO_4.

$$R-\overset{\overset{\cdot\cdot}{O}}{\underset{\|}{C}}-O-H + H_2SO_4 \rightarrow R-\overset{\overset{\cdot\cdot}{O}\!\!-\!\!H}{\underset{\|}{C}}-\overset{+}{\underset{..}{O}}-H \;\;\text{and}\;\; R-\overset{\overset{\cdot\cdot}{O}}{\underset{\|}{C}}-\overset{+}{\underset{..}{O}}\!\!-\!\!H + HSO_4^-$$
(2-18)

$$R-\ddot{\underset{..}{O}}-H + H_2SO_4 \rightarrow R-\overset{H}{\underset{..}{\overset{+}{O}}}-H + HSO_4^- \qquad (2\text{-}19)$$

Amines, RNH_2, are moderately strong bases and will partly remove protons from water as in equation 2-20. In dilute HCl the

$$H-\overset{H}{\underset{H}{\overset{|}{C}}}-\overset{..}{\underset{H}{N}}-H + H_2O \rightleftharpoons H-\overset{H}{\underset{H}{\overset{|}{C}}}-\overset{H}{\underset{H}{\overset{|}{\overset{+}{N}}}}-H + H\ddot{\underset{..}{O}}:^- \qquad (2\text{-}20)$$
methylamine

amines are completely converted to their hydrochloride salts as in equation 2-21.

$$H-\overset{H}{\underset{H}{\overset{|}{C}}}-\overset{..}{\underset{H}{N}}-H + H^+ + Cl^- \rightarrow H-\overset{H}{\underset{H}{\overset{|}{C}}}-\overset{H}{\underset{H}{\overset{|}{\overset{+}{N}}}}-H \;\; :\ddot{\underset{..}{Cl}}:^- \qquad (2\text{-}21)$$

Another example of the reaction of an amine with an acid, in which the —NH$_2$ functional group is part of a much larger and more complicated molecule, is shown in equation 2-22.

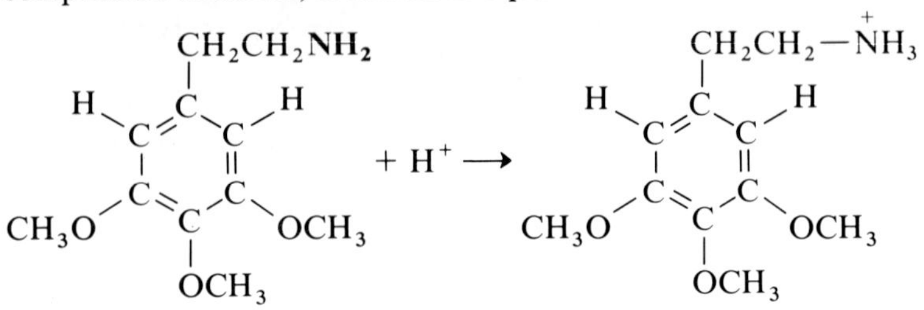

mescaline
mp 35–36°

Mescaline is one of a large number of naturally occurring basic substances called **alkaloids** (Chapter 11). Its basic properties are shown by the fact that it readily forms a salt with HCl, and this property can be used to extract it from the buttons of the mescal or peyote cactus.

Intoxication by mescaline produces a kaleidoscopic display of brightly colored visual halucinations. The dried mescal buttons were used in Aztec religious ceremonies in pre-Columbian times.

The lethal dose of mescaline (for mice) is 500 mg/kg

Acidity and the Separation of Mixtures

The generalization that carboxylic acids are more acidic than alcohols provides an easy way to separate mixtures. Suppose, for example, that a mixture of cyclohexanol and cyclohexanecarboxylic acid is to be separated. The mixture will be an oily liquid insoluble or almost insoluble in water. However, cyclohexanecarboxylic acid,

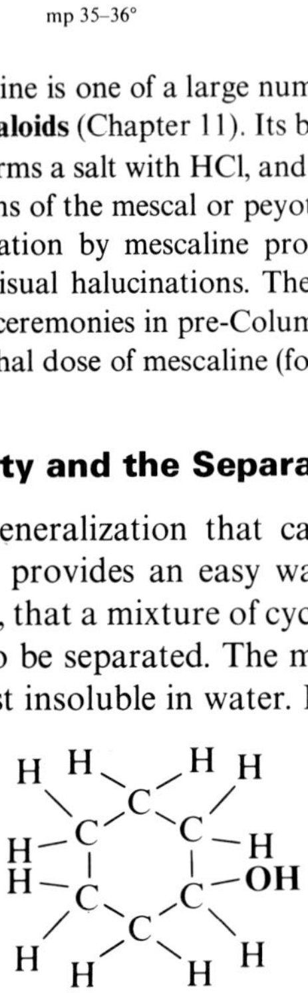

cyclohexanol
liquid, bp 161.5°C

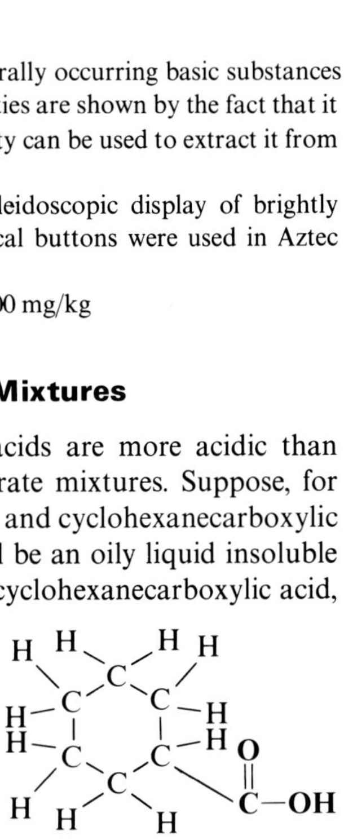

cyclohexanecarboxylic acid
colorless solid, mp 31°C

like most carboxylic acids, dissolves easily in a solution of NaHCO$_3$ in water (equation 2-23). The reason for the "solubility" of cyclohexanecarboxylic acid in NaHCO$_3$ solution is that the acid reacts to form a salt. Like most salts of carboxylic acids, sodium cyclohexanecarboxylate is soluble in water.

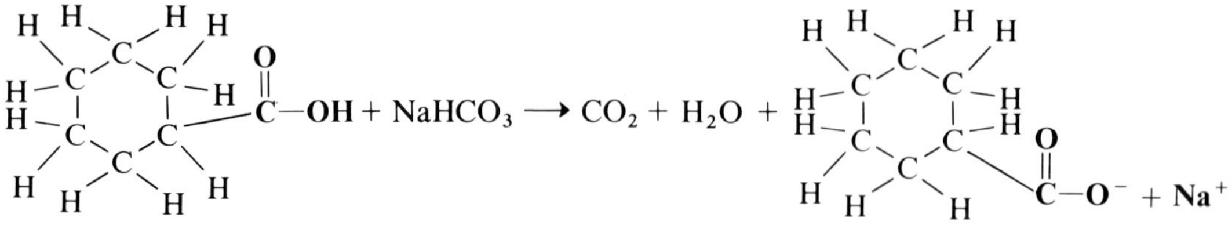

the acid; not soluble in H$_2$O

sodium cyclohexanecarboxylate; soluble in H$_2$O

(2-23)

The cyclohexanol does not dissolve in $NaHCO_3$ solution. Like other alcohols, it is not acidic enough to react with the base $NaHCO_3$ and is not converted to a soluble salt.

If, then, a mixture of cyclohexanecarboxylic acid and cyclohexanol is shaken with aqueous $NaHCO_3$, the acid reacts to form a soluble salt in the water layer and the alcohol remains undissolved. The shaking would probably be done in a separatory funnel (Figure 2-2) so that the lower layer can be drawn off into another

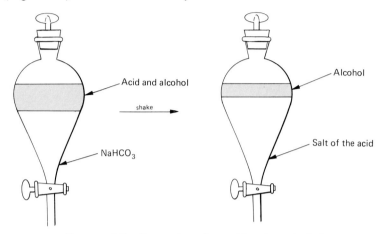

FIGURE 2-2. Separation of an acid and an alcohol.

container. After draining off the layer containing the sodium cyclohexanecarboxylate into a beaker, the pure cyclohexanecarboxylic acid can be made to reappear as a precipitate by adding HCl (equation 2-24). In general, the salt of a carboxylic acid will react with the very much stronger acid, HCl, to give the free carboxylic acid and the salt (NaCl) of the stronger acid.

$$\text{C}_6\text{H}_{11}\text{-C(=O)-O}^-\text{Na}^+ + \text{HCl} \rightarrow \text{C}_6\text{H}_{11}\text{-C(=O)-OH} + \text{NaCl} \tag{2-24}$$

Just as water-insoluble carboxylic acids can be separated from water-insoluble mixtures by temporarily converting the acids to their water-soluble sodium salts, water-insoluble amines, such as mescaline, can be separated from such mixtures by temporarily converting them to their water-soluble salts with a strong acid such as hydrochloric acid. In general, if R is permitted to stand for the rest of the molecule, the process can be summarized as in equations 2-25 and 2-26.

$$\text{R-NH}_2 + \text{H-Cl} \rightarrow \text{R-}\overset{+}{\text{N}}\text{H}_3\ \text{Cl}^- \tag{2-25}$$

insoluble amine   alkylammonium chloride, water soluble, washes out

Acids and Bases

The new saltlike functional group is usually enough to make the substance water soluble. An excess of hydroxide ion from NaOH can be used to liberate the amine from its water-soluble salt after it has been separated from nonbasic components of the mixture.

$$R-\overset{H}{\underset{H}{\overset{|}{N}}}-H\ Cl^- + HO^- \longrightarrow R-\overset{H}{\underset{..}{\overset{|}{N}}}-H + H_2O + Cl^- \quad (2\text{-}26)$$
excess NaOH

Acid Catalysis

In later chapters of this book we shall encounter reactions that are made to go faster by adding strong acids as catalysts. Some of the reactions of carboxylic acids, for example, go faster in the presence of a trace of a strong mineral acid such as H_2SO_4. The reason for this is the conversion of part of the carboxylic acid into a more reactive species such as

$$R-\overset{\overset{\displaystyle \overset{+}{:}\overset{H}{O}}{\|}}{C}-OH$$

Lewis Acids

In the reactions discussed so far, the acids have been substances that donate a proton which bonds to an unshared pair of electrons on the base. These substances are called Brønsted acids. A Lewis acid is a compound that can *itself* bond to a base, using the unshared pair of electrons of the base. Examples of Lewis acids are ferric chloride, $FeCl_3$, and aluminum chloride, $AlCl_3$.

$$\begin{array}{cc} \text{Cl} & \text{Cl} \\ | & | \\ \text{Cl}-\text{Fe} & \text{Cl}-\text{Al} \\ | & | \\ \text{Cl} & \text{Cl} \\ \text{ferric chloride} & \text{aluminum chloride} \end{array}$$

Ferric chloride and aluminum chloride are electron-deficient in that the metal atom has only six electrons (the three covalent bonds) in its outer shell. They therefore tend to be **electrophilic** (Greek: electron-loving) in their reactions. For example, they will react with the weak base, or **nucleophilic**, reagent, Cl^-.

$$\underset{\underset{\text{Cl}}{|}}{\overset{\overset{\text{Cl}}{|}}{\text{Cl}-\text{Fe}}} + Cl^- \longrightarrow \underset{\underset{\text{Cl}}{|}}{\overset{\overset{\text{Cl}}{|}}{\text{Cl}-\overset{-}{\text{Fe}}-\text{Cl}}} \quad (2\text{-}27)$$

$$\underset{\underset{\text{Cl}}{|}}{\overset{\overset{\text{Cl}}{|}}{\text{Cl}-\text{Al}}} + Cl^- \longrightarrow \underset{\underset{\text{Cl}}{|}}{\overset{\overset{\text{Cl}}{|}}{\text{Cl}-\overset{-}{\text{Al}}-\text{Cl}}} \quad (2\text{-}28)$$

Lewis acids are often used as catalysts. For example, some reactions of alkyl halides, RCl, require the conversion of the alkyl halide into an alkyl cation or carbonium ion, R^+.

$$R-\ddot{\underset{..}{C}l}: + \underset{\underset{Cl}{|}}{\overset{\overset{Cl}{|}}{Al}}-Cl \rightarrow R^+ + Cl-\underset{\underset{Cl}{|}}{\overset{\overset{Cl}{|}}{Al}}-Cl \qquad (2\text{-}29)$$

The carbonium ion, R^+, is more reactive than is the alkyl halide RCl; hence, the presence of the Lewis acid, $AlCl_3$, catalyzes the reaction.

An example (Chapter 6) is the Friedel–Crafts reaction of methyl chloride with benzene.

$$CH_3Cl + AlCl_3 \rightarrow CH_3^+ \; AlCl_4^-$$

CH_3^+ + benzene $\rightarrow$ (methylbenzene) + H^+

SUMMARY

1. Acids and bases
    (a) Brønsted acids are compounds with a proton attached to an atom of an electronegative element. The proton is detached from the acid and bonded to a base, using an unshared electron pair of the base.
    (b) All bases have unshared electron pairs that can be used to form new bonds to protons. A compound can be both an acid and a base if it can both give up a proton and accept one.
    (c) Acidity and bond polarity increase in the sequence —C—H ≪ —N—H ≪ —OH ≪ ClH.
    (d) Basicity, other things being equal, increases in the sequence Cl ≪ O ≪ N.
2. Separation of mixtures
    (a) Carboxylic acids can be separated from mixtures with insoluble nonacidic compounds by converting the acid to a water-soluble sodium salt by reaction with $NaHCO_3$.
    (b) Amines (bases) can be separated from mixtures with nonbasic compounds by converting the amine to a water-soluble salt such as a hydrochloride.
3. Lewis acids: The reaction of a Lewis acid with a base is very much like the reaction of a proton with a base. In both cases the acid has a vacant atomic orbital and the base supplies the pair of electrons for the bond.
    (a) Lewis acids are used as catalysts.

EXERCISES

1. What functional groups are found in (a) alcohols, (b) amines, (c) carboxylic acids?
2. Write formulas for (a) two alcohols, (b) two amines, (c) three carboxylic acids.

3. Complete the following reactions. Write "no reaction" if none occurs.

(a) 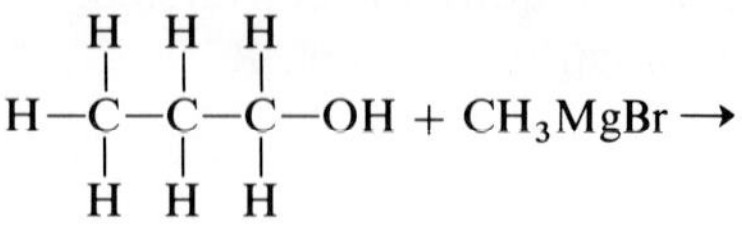 $\text{H}-\underset{\underset{\text{H}}{|}}{\overset{\overset{\text{H}}{|}}{\text{C}}}-\underset{\underset{\text{H}}{|}}{\overset{\overset{\text{H}}{|}}{\text{C}}}-\underset{\underset{\text{H}}{|}}{\overset{\overset{\text{H}}{|}}{\text{C}}}-\text{OH} + \text{CH}_3\text{MgBr} \rightarrow$

(b) $\text{H}-\underset{\underset{\text{H}}{|}}{\overset{\overset{\text{H}}{|}}{\text{C}}}-\underset{\underset{\text{H}}{|}}{\overset{\overset{\text{H}}{|}}{\text{C}}}-\underset{\underset{\text{H}}{|}}{\overset{\overset{\text{H}}{|}}{\text{C}}}-\text{H} + \text{CH}_3\text{MgBr} \rightarrow$

(c) $\text{CH}_3-\underset{\underset{\text{CH}_3}{|}}{\overset{\overset{\text{CH}_3}{|}}{\text{C}}}-\overset{\overset{\text{O}}{\|}}{\text{C}}-\text{O}-\text{H} + \text{NaHCO}_3 \rightarrow$

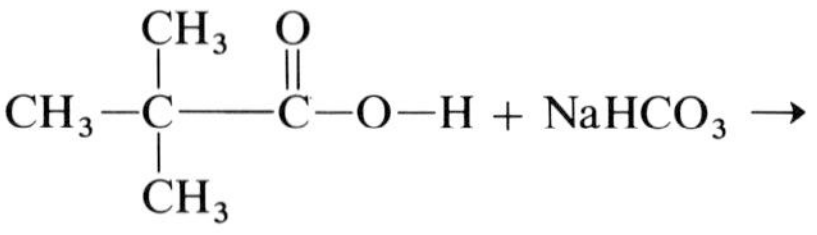

(d)  $\text{CH}_3-\overset{\overset{\text{O}}{\|}}{\text{C}}-\text{OH} + \text{NaHCO}_3 \rightarrow$

(e) $\text{CH}_3\text{CH}_2\text{CH}_2\text{CH}_2-\text{OH} + \text{NaHCO}_3 \rightarrow$

(f) 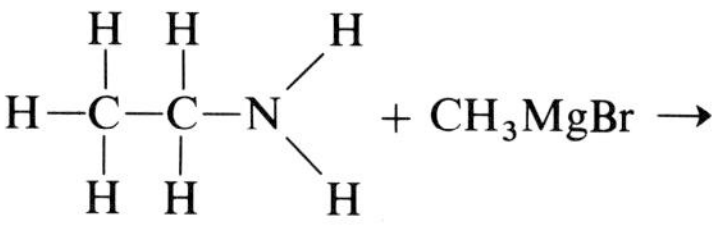 $\text{H}-\underset{\underset{\text{H}}{|}}{\overset{\overset{\text{H}}{|}}{\text{C}}}-\underset{\underset{\text{H}}{|}}{\overset{\overset{\text{H}}{|}}{\text{C}}}-\text{N}\underset{\text{H}}{\overset{\text{H}}{\diagup}} + \text{CH}_3\text{MgBr} \rightarrow$

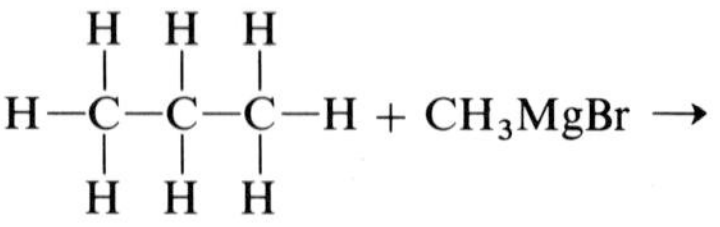

(g) $\text{H}-\underset{\underset{\text{H}}{|}}{\overset{\overset{\text{H}}{|}}{\text{C}}}-\underset{\underset{\text{H}}{|}}{\overset{\overset{\text{H}}{|}}{\text{C}}}-\overset{..}{\text{N}}\underset{\text{H}}{\overset{\text{H}}{\diagup}} + \text{HCl} \rightarrow$

(h) [substituted benzene with $\text{CH}_2\text{CH}_2-\overset{..}{\text{N}}\text{H}_2$, $\text{H}_3\text{C}-\text{O}$, OCH_3, and $\text{O}-\text{CH}_3$ substituents] + HCl →

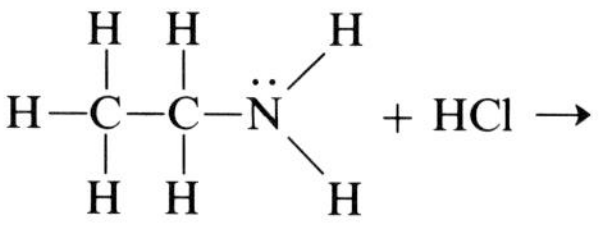

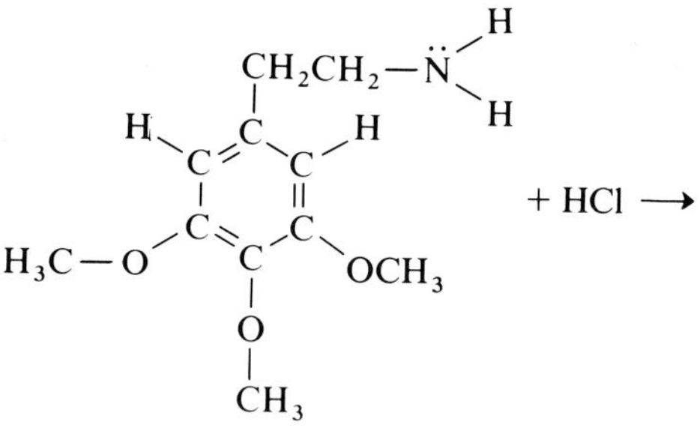

(i) 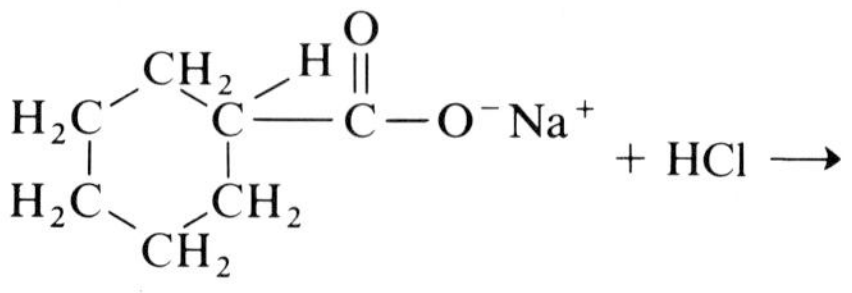 [cyclohexane with $-\overset{\overset{\text{O}}{\|}}{\text{C}}-\text{O}^-\text{Na}^+$ and H] + HCl →

(j) 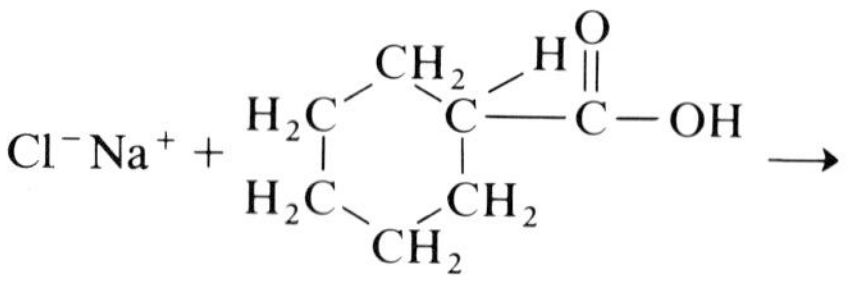 Cl^-Na^+ + [cyclohexane with $-\overset{\overset{\text{O}}{\|}}{\text{C}}-\text{OH}$ and H] →

(k) 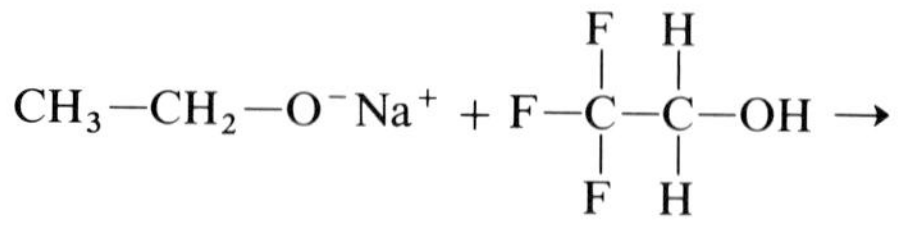 $\text{CH}_3-\text{CH}_2-\text{O}^-\text{Na}^+ + \text{F}-\underset{\underset{\text{F}}{|}}{\overset{\overset{\text{F}}{|}}{\text{C}}}-\underset{\underset{\text{H}}{|}}{\overset{\overset{\text{H}}{|}}{\text{C}}}-\text{OH} \rightarrow$

Hint: Fluorine is more electronegative than hydrogen.

(l) 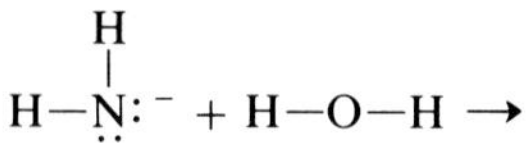 $H-\underset{\underset{H}{|}}{\overset{..}{N}}:^- + H-O-H \rightarrow$

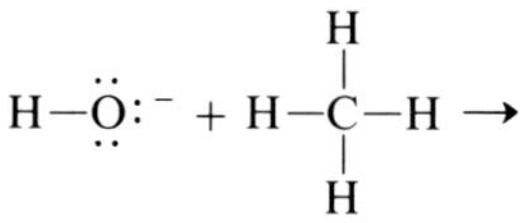

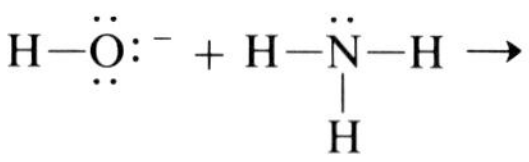

(m) $H-\overset{..}{\underset{..}{O}}:^- + H-\underset{\underset{H}{|}}{\overset{\overset{H}{|}}{C}}-H \rightarrow$

(n) $H-\overset{..}{\underset{..}{O}}:^- + H-\underset{\underset{H}{|}}{\overset{..}{N}}-H \rightarrow$

(o) $H-\underset{\underset{H}{|}}{\overset{..}{N}}:^- + H-\underset{\underset{H}{|}}{\overset{\overset{H}{|}}{C}}-O-H \rightarrow$

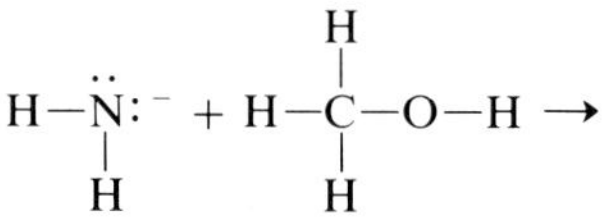

4. Explain how to separate a mixture of a water-insoluble alcohol and a water-insoluble carboxylic acid.

5. Define "conjugate acid" and "conjugate base" and give five examples.

6. Explain how to separate a mixture of

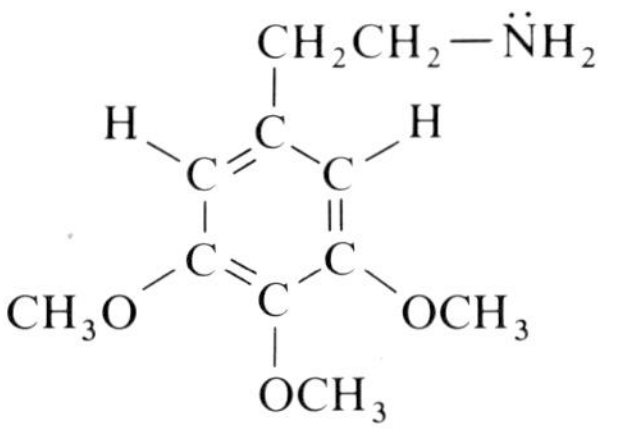

[structure with $CH_2CH_2-\overset{..}{N}H_2$ substituent on a ring with CH_3O, OCH_3, OCH_3, H, H substituents] and [structure with CH_2CH_2-OH substituent on a ring with CH_3O, OCH_3, $O-CH_3$, H, H substituents]

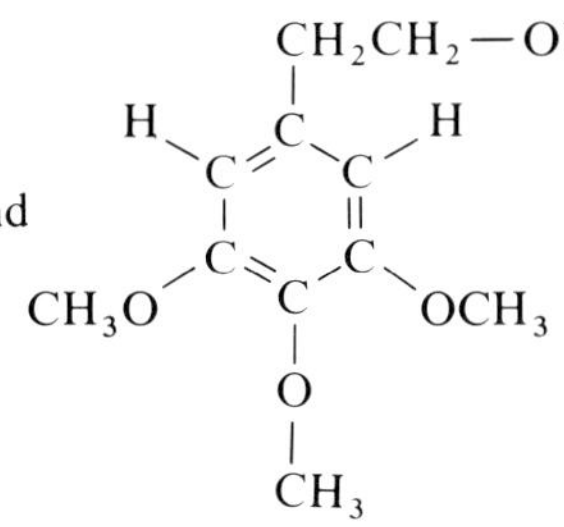

7. Write the equation for the reaction of the Lewis acid BF_3 with the base F^-.

Acids and Bases

Chapter 3
Isomers and the Structure of Molecules

In the preceding chapter several molecules were introduced that had different kinds of functional groups, such as hydroxyl (HO—), carboxyl (HOOC—), and amino (H$_2$N—). In later chapters still other functional groups will be encountered, but before going any further it is desirable to discuss the two main ways in which the structure of a molecule is determined.

The first, and historically the more important, of the methods used for determining the structure of a molecule is to study the reactions of the molecule. A second method uses modern instruments to measure the absorption of light by the molecule. The particular wavelength or color of the light that is absorbed tells not only what functional groups are present but, in many cases, something about the structure of the carbon skeleton as well.

Structural Formulas

A structural formula like those given in Chapter 2 for methane, butane, and ethyl alcohol is meant to tell **which atoms are directly connected** to which other atoms. Thus the structure

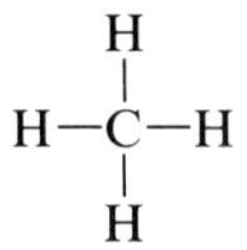

$$\begin{array}{c} H \\ | \\ H-C-H \\ | \\ H \end{array}$$

for methane says that a methane molecule has four hydrogen atoms, **each directly connected to one carbon atom** rather than to each other.

After noting that a structural formula is meant to tell which atoms are directly connected to each other in the molecule, it is also important to understand that an ordinary structural formula is not meant to tell anything about the *shape* of the molecule. Methane, for example, is not flat and square even though it is convenient to draw the structural formula that way. Figure 3-1 shows what is believed to be the three-dimensional shape of a methane molecule.

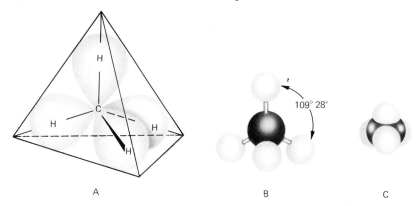

FIGURE 3-1. Methane. **(A)** and **(B)** The positions of the nuclei. **(C)** The shapes of the electron clouds surrounding the nuclei.

The angles between the C—H bonds of methane are all equal to 109° 28′, the same as the angles between the center and the corners of a regular tetrahedron. This is illustrated in Figure 3-1A in which a methane molecule is shown inside a tetrahedron. An important consequence of the tetrahedral arrangement of the bonds to carbon is that **all four hydrogen atoms of methane are equivalent, just as are all four corners of a tetrahedron**. Thus, if any one of the four hydrogen atoms is singled out for replacement by another kind of atom, chlorine, for example, only one product is obtained and not four different products (Figure 3-2). Each of the apparently different

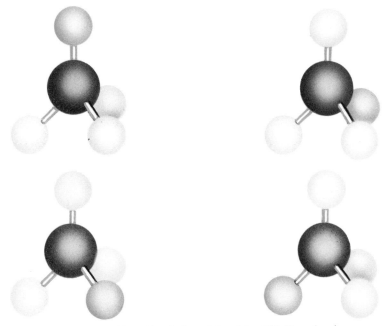

FIGURE 3-2. Four identical models of the CH$_3$Cl molecule.

Isomers and the Structure of Molecules

products can be seen to be the same as the first just by rotating it.

The pentane molecule affords another good example of the difference between the structural formula of a molecule and its actual shape. All of the formulas shown below are equally good structural formulas for pentane, but none of them really represents the shape. They merely tell which atoms are directly connected to one another by covalent bonds.

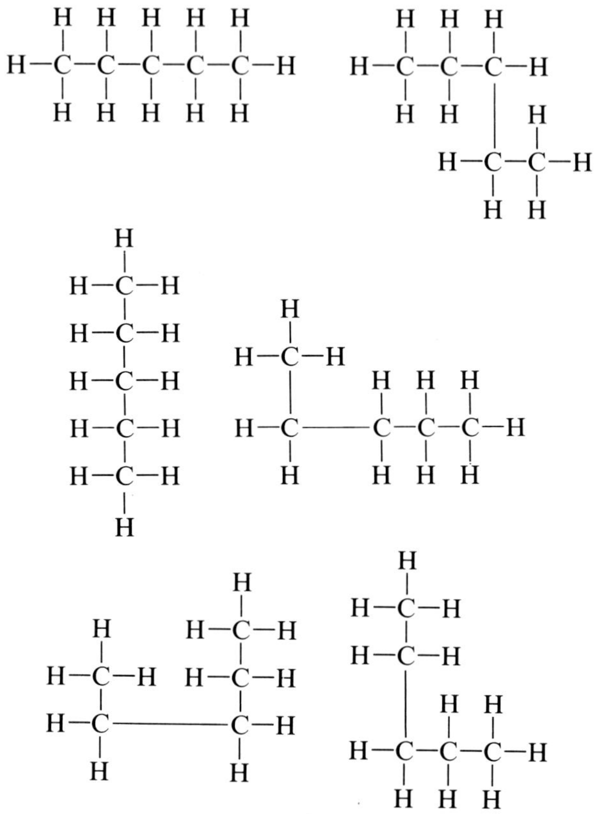

pentane

Many more equivalent structural formulas for pentane could be written. If it had been practical to do so, this book would have included a structural formula printed on a rubber sheet which the reader could pull and stretch so as to make the structural formula *look* different without really being different.

CONFORMATIONS

Although simple hydrocarbons such as pentane all have tetrahedral or 109° 28′ angles between the bonds to a given carbon atom, they can have many different shapes or conformations. In an actual sample of pentane all of the possible shapes will be represented. The various pentane molecules that differ from one another only in shape or conformation are called **conformers**. The conformers change rapidly and continually into other conformers of pentane, a process represented in the models shown in Figure 3-3 by twisting the pegs in their holes. The various conformers of pentane are equivalent to one another in the same sense that a person standing up is still the same person when he sits down momentarily.

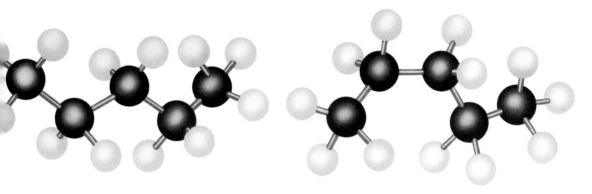

FIGURE 3-3. Models of pentane conformers.

Some conformations require less energy, are more stable, and at any given instant are represented by more molecules than are other conformations. The staggered and eclipsed conformations of ethane (Figure 3-4) exemplify this. In the eclipsed conformation, pairs of hydrogens on different carbon atoms are too close together and crowded. In the staggered conformation, a hydrogen on one carbon atom is opposed to a gap between hydrogens on the other carbon atom and the crowding is much less severe. Hence, although rotation about the C—C bond of ethane occurs almost constantly, on the average the molecule is most often to be found in the less energetic staggered conformation rather than in the more energetic eclipsed conformation. If the reader will permit another human analogy, not even a contortionist spends very much of his time tied up like a pretzel. Because the change from one conformation to another requires so little energy and is so rapid, mixtures of conformers cannot be separated. A substance such as pentane is therefore considered to be a single compound rather than a mixture.

It is particularly important to distinguish the mere changes in shape that convert one conformer into another from more fundamental changes that require bonds to be broken and remade. Molecules that differ from one another in this more fundamental way are the topic of the next section.

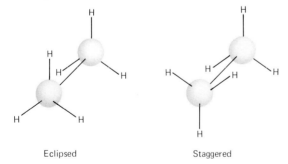

Eclipsed   Staggered

FIGURE 3-4. Two conformations of ethane.

ISOMERS

Isomers are molecules having the same numbers of each kind of atom, but assembled in different ways. They cannot be converted into one another merely by twisting about a bond.

The C_5H_{12} Isomers. When the petroleum from an oil field is distilled, it is found to contain many different compounds of different boiling points. Among them are three compounds, *all three consisting of molecules with five carbon atoms and twelve hydrogens*; that is, all three compounds have the same molecular formula, C_5H_{12}, but nevertheless have different boiling points and other physical properties; *therefore they cannot possibly be the same compound*. The structural formulas and names of the three isomeric C_5H_{12} hydrocarbons are shown in the equations that follow and illustrate the reactions of the isomers with chlorine.

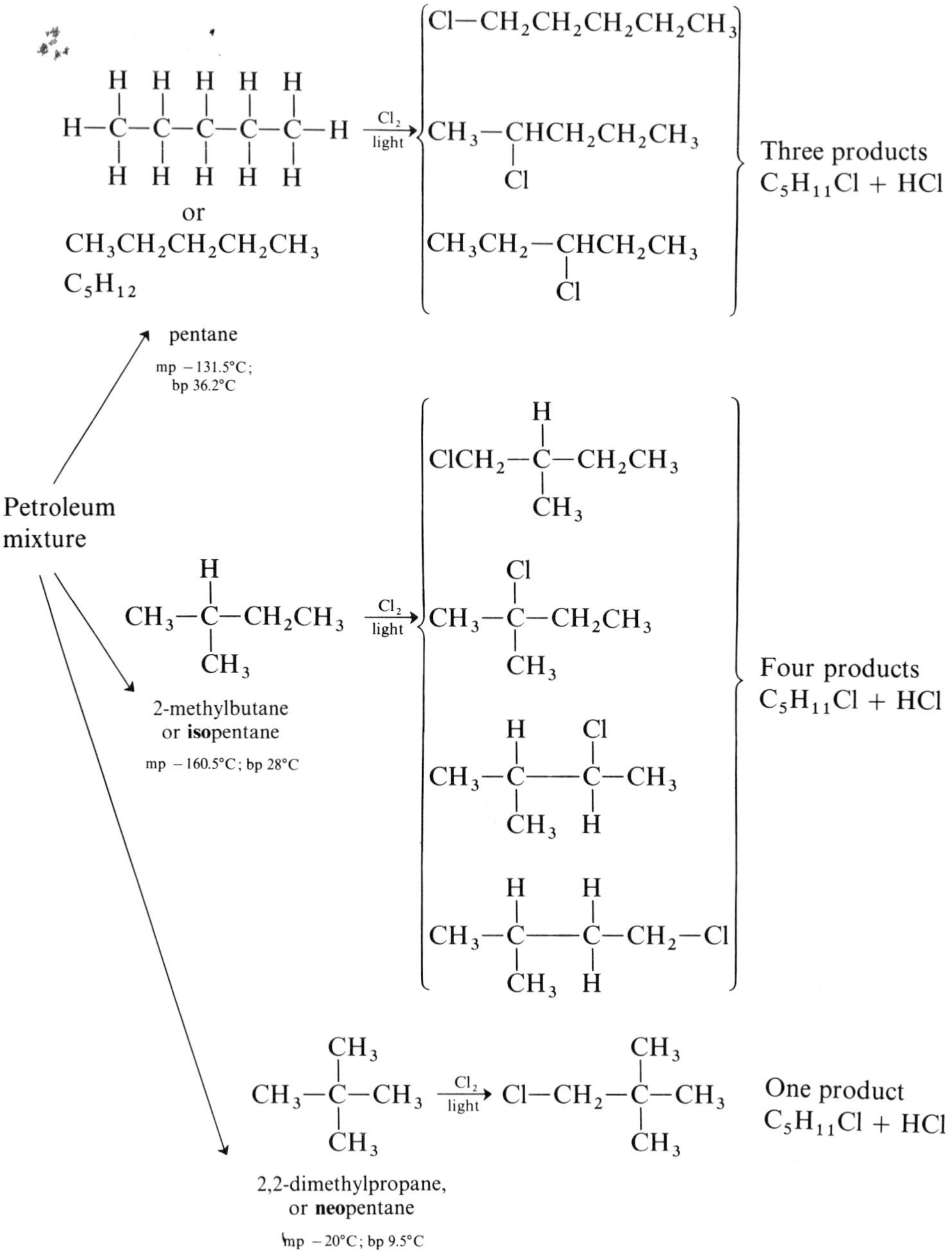

Petroleum mixture →

pentane
mp −131.5°C; bp 36.2°C

$$H-\underset{H}{\overset{H}{C}}-\underset{H}{\overset{H}{C}}-\underset{H}{\overset{H}{C}}-\underset{H}{\overset{H}{C}}-\underset{H}{\overset{H}{C}}-H \xrightarrow[\text{light}]{Cl_2}$$

or $CH_3CH_2CH_2CH_2CH_3$
C_5H_{12}

Products:
- $Cl-CH_2CH_2CH_2CH_2CH_3$
- $CH_3-\underset{Cl}{CH}CH_2CH_2CH_3$
- $CH_3CH_2-\underset{Cl}{CH}CH_2CH_3$

Three products $C_5H_{11}Cl + HCl$

2-methylbutane or **iso**pentane
mp −160.5°C; bp 28°C

$$CH_3-\underset{CH_3}{\overset{H}{C}}-CH_2CH_3 \xrightarrow[\text{light}]{Cl_2}$$

Products:
- $ClCH_2-\underset{CH_3}{\overset{H}{C}}-CH_2CH_3$
- $CH_3-\underset{CH_3}{\overset{Cl}{C}}-CH_2CH_3$
- $CH_3-\underset{CH_3}{\overset{H}{C}}-\underset{H}{\overset{Cl}{C}}-CH_3$
- $CH_3-\underset{CH_3}{\overset{H}{C}}-\underset{H}{\overset{H}{C}}-CH_2-Cl$

Four products $C_5H_{11}Cl + HCl$

2,2-dimethylpropane, or **neo**pentane
mp −20°C; bp 9.5°C

$$CH_3-\underset{\underset{CH_3}{|}}{\overset{\overset{CH_3}{|}}{C}}-CH_3 \xrightarrow[\text{light}]{Cl_2} Cl-CH_2-\underset{\underset{CH_3}{|}}{\overset{\overset{CH_3}{|}}{C}}-CH_3$$

One product $C_5H_{11}Cl + HCl$

One chemical way of deciding that the compound boiling at 36.2°C is in fact the one that has the structure labeled pentane, that the compound boiling at 28°C is in fact isopentane, and that the compound boiling at 9.5°C is in fact neopentane makes use of the reaction of each of these isomers with chlorine. The chlorination reaction, which will be discussed at greater length in Chapter 4, is a reaction that replaces the hydrogen of a C—H bond with a chlorine atom. It is quite unselective, and a given molecule of neopentane, for example, might react at *any* of its twelve hydrogen atoms.

$$\begin{array}{c} H \\ | \\ H-C-H \\ H \quad | \quad H \\ | \quad | \quad | \\ H-C-C-C-H \\ | \quad | \quad | \\ H \quad | \quad H \\ H-C-H \\ | \\ H \end{array}$$

neopentane
All twelve hydrogens are equivalent

Free rotation about the C—C bond makes all of the hydrogen atoms of each **methyl group** (CH$_3$—, usually written without the bond as CH$_3$) equivalent; that is, any hydrogen can occupy any of the possible rotational positions. Not only that, but all four methyl groups are equivalent just as are all four hydrogens of a methane molecule. (Neopentane can also be called tetramethylmethane.) As a consequence, all $4 \times 3 = 12$ hydrogen atoms of neopentane are equivalent. Hence the monochlorination of neopentane gives only *one* product, $C_5H_{11}Cl$, just as the monochlorination of methane gives only one product, CH_3Cl.

The case of the other isomers is different. Thus pentane has *three* different nonequivalent positions for replacing a hydrogen atom with a chlorine atom. These are the end methyl groups (CH$_3$), the next-to-the-end **methylene** groups (—CH$_2$—, usually written without the bonds as CH$_2$), and the middle methylene group. Within each of these categories the hydrogens are equivalent to one another; therefore pentane gives three and only three $C_5H_{11}Cl$ products. Similarly, isopentane has four nonequivalent positions for chlorination and gives a mixture of four $C_5H_{11}Cl$ products.

To summarize: the three isomeric C_5H_{12} hydrocarbons can be identified with the proper structural formulas merely by chlorinating each isomer *and counting the number of different $C_5H_{11}Cl$ compounds produced*. With the advent of modern instruments, this method of counting isomeric reaction products has largely fallen into disuse. It can be quite difficult technically because the boiling points of some of the $C_5H_{11}Cl$ products, from isopentane for example, are so close together that it is not easy to separate the mixture by distillation. On the other hand, it is good practice for developing the knack of recognizing equivalent positions in a molecule, and has been included mainly for that reason.

There is another chemical method that is quite easy: synthesis from smaller molecules by means of well-known reactions. Later chapters will include ways of making isopentane and neopentane by sequences of reactions known to give **branched** carbon skeletons, and normal pentane by a sequence of reactions known to give an **unbranched** carbon skeleton. It turns out that neopentane, the most highly branched product, is indeed the compound boiling at 9.5°C, whereas pentane, the unbranched product, is the compound boiling at 36.2°C.

DIMETHYL ETHER AND ETHYL ALCOHOL

Dimethyl ether is a highly flammable substance, boiling at −23.65°C; it is a gas at room temperature. Ethyl alcohol boils at 78.5°C; it is a liquid at room temperature. Both compounds have the molecular formula C_2H_6O and therefore are isomers. Their structural formulas are easily determined from their chemical reactions alone.

First, because oxygen has to have a valence of two, there are only two possible covalently bonded structures. One of these has an oxygen atom bonded to two carbons. The other has an oxygen atom bonded to a carbon and a hydrogen; that is, it has a hydroxyl functional group, like water.

$$\begin{array}{cc} H & H \\ | & | \\ H-C-O-C-H \\ | & | \\ H & H \end{array} \qquad \begin{array}{cc} H & H \\ | & | \\ H-C-C-O-H \\ | & | \\ H & H \end{array}$$

The compound known as ethyl alcohol, bp 78.5°C, is a liquid at room temperature (like water) and it reacts with sodium (like water). Therefore it must have the structure with the hydroxyl functional group.

$$2\,H-\underset{\underset{H}{|}}{\overset{\overset{H}{|}}{C}}-\underset{\underset{H}{|}}{\overset{\overset{H}{|}}{C}}-O-H + 2\,Na \rightarrow H_2 + 2\,H-\underset{\underset{H}{|}}{\overset{\overset{H}{|}}{C}}-\underset{\underset{H}{|}}{\overset{\overset{H}{|}}{C}}-O^-Na^+ \quad (3\text{-}1)$$
ethyl alcohol

$$2\,H-O-H + 2\,Na \rightarrow H_2 + 2\,H-O^-Na^+ \quad (3\text{-}2)$$
water

The compound called dimethyl ether, a gas at room temperature, does not react with sodium and therefore does not have any hydroxyl group. It must be

$$\begin{array}{cc} H & H \\ | & | \\ H-C-O-C-H \\ | & | \\ H & H \end{array}$$
dimethyl ether

a conclusion confirmed by still other reactions and physical properties.

Instrumental Methods

The structures of the pentanes, ethyl alcohol, and dimethyl ether were determined originally from a study of their chemical reactions and syntheses alone. Today this could be done more easily using light-absorption methods.

Before our rather brief discussion of the uses of light absorption in identifying structures of molecules, it is desirable to recall something about the nature of light, or rather about **electromagnetic radiation** in general. Visible light (Figure 3-7) is only a small part or "band" in a wide range of wavelengths (Figure 3-5) extending from long radio waves, through the somewhat shorter waves used for transmitting television signals, then to microwaves (radar), then to infrared (so-called heat waves), then to visible light (red, yellow, green, blue, violet), then to ultraviolet light, and finally to x-rays and gamma (γ) rays (Figure 3-6). All of these are essentially the same thing: waves of oscillating electrical fields. They differ only in their wavelengths and in their energy.

As an electromagnetic wave passes a charged particle such as a nucleus or an electron, the charged particle tries to move with the changing electrical field in much the same way that a boat tries to bob up and down in response to the passage of a water wave.

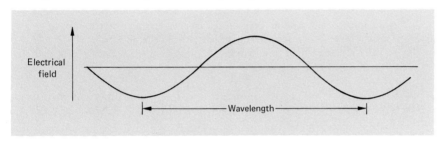

FIGURE 3-5. The wavelength is the distance between troughs or between crests.

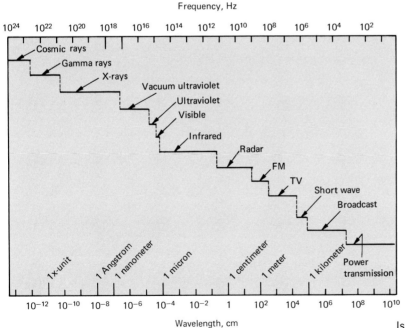

FIGURE 3-6. The electromagnetic spectrum.

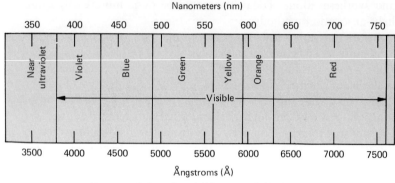

FIGURE 3-7. The visible part of the spectrum.

When a radio-frequency wave passes a radio antenna, the electrons in the antenna move back and forth in response to the undulations of the electrical field. "Electrons moving back and forth" is a definition of an alternating current; hence, the radio wave generates an alternating current in the antenna that is detected by the radio receiver. A transmitter works the other way around: it generates an alternating current in the antenna which in turn causes the antenna to emit a radio wave. Note that a wave can be described equally well either in terms of its wavelength or in terms of its **frequency** (waves per second). They are related by the expression: wavelength (λ) = velocity of light (c) divided by frequency (v), or $\lambda = c/v$.

In order to keep the numbers small, different units are often used for the wavelengths in different parts of the spectrum. The relationships between the various units are shown in Table 3-1. A unit not shown in the table but often used to identify the radiation is the **wavenumber** or reciprocal of the wavelength (cm^{-1}). It is proportional to the frequency.

TABLE 3-1. *Conversion Table*

Multiply → number of ↓ To obtain number of	Ångstroms, Å	Nanometers, nm	Microns, μ	Millimeters, mm	Centimeters, cm	Meters, m
Ångstroms, Å	1	10	10^4	10^7	10^8	10^{10}
Nanometers, nm	10^{-1}	1	10^3	10^6	10^7	10^9
Microns, μ	10^{-4}	10^{-3}	1	10^3	10^4	10^6
Millimeters, mm	10^{-7}	10^{-6}	10^{-3}	1	10	10^3
Centimeters, cm	10^{-8}	10^{-7}	10^{-4}	0.1	1	10^2
Meters, m	10^{-10}	10^{-9}	10^{-6}	10^{-3}	10^{-2}	1

Thus far the wavelike nature of light has been emphasized. Now, however, it must be noted that light also acts as though it consists of a stream of individual particles called photons.

Each photon carries a certain amount of energy. When a photon is absorbed by a molecule, the photon disappears and the molecule gains an amount of energy equal to that previously associated with the photon. The amount of energy carried by a photon is proportional to its frequency or to the reciprocal of its wavelength, as shown in equation 3-3.

$$\text{Energy (of a photon)} = \underset{\substack{\text{Planck's}\\\text{constant}}}{h} \times \underset{\substack{\text{(frequency of}\\\text{the photon)}}}{\nu} \qquad (3\text{-}3)$$

ULTRAVIOLET AND VISIBLE LIGHT

The fate of the energy of the photon once it has been absorbed by a molecule depends on the broad range of wavelength to which the photon belongs. Photons of visible or ultraviolet light move electrons out of orbitals of lower energy and put them in orbitals of higher energy, as shown in Figure 3-8. Since the exact amount of energy

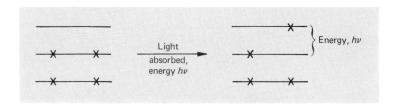

FIGURE 3-8. When a photon of light in the visible or ultraviolet range is absorbed by a molecule, the energy, $h\nu$, is used to elevate an electron to an orbital of higher energy.

required to promote an electron to a higher energy level depends on the structure of the molecule, the frequency or wavelength of the absorbed light is a clue to the structure.

The **absorption spectrum** of a substance can be presented as a graph of the amount of light absorbed at each wavelength against the wavelength. A typical spectrometer (Figure 3-9) includes a strip-chart recorder which produces such a graph automatically. A maximum, or peak, in a particular range of wavelengths in the absorption spectrum is an indication of the presence of a particular kind of molecular structure. Acetic acid, for example, has an absorption maximum at 204 nm (nanometers) in the ultraviolet due to its carboxyl group (COOH).

In later chapters compounds will be encountered that absorb light in the visible region of the spectrum, the region from 400 (violet) to about 750 nm (red). As in the case of the absorption of ultraviolet light, the absorbed energy is used to promote an electron to an orbital of higher energy. A compound will have a color visible to human eyes only if the required energy happens to correspond with that of a photon whose wavelength is in the 400–750 nm region.

Our ability to see by this particular range of frequencies is due partly to the fact that these are the frequencies that the pigments in our eyes absorb and partly to the fact that these are the frequencies to which the lenses of our eyes are transparent. Persons whose lenses have been removed surgically can see part way into the ultraviolet.

White light with one of the colors of the spectrum removed from it is seen by the eye as the "complementary" color. Thus, if a compound absorbs blue

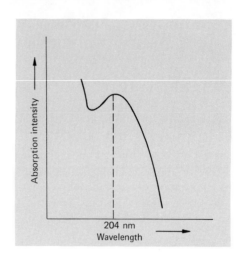

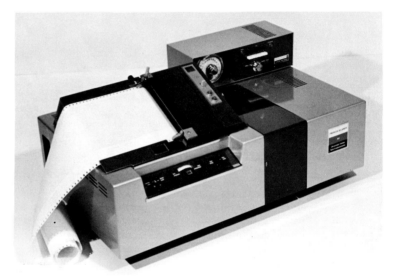

FIGURE 3-9. **(A)** Ultraviolet absorption spectrum of a solution of acetic acid. **(B)** A Perkin-Elmer Ultraviolet Spectrophotometer. [Courtesy Perkin-Elmer Corporation]

light, it looks yellow; if a compound absorbs yellow light, it looks blue. For physiological reasons the eye gives the same response to yellow light as it does to white-light-minus-blue and the same response to blue light as to white-light-minus-yellow.

INFRARED LIGHT

The light of frequency adjacent to the red end of the visible spectrum (invisible to human eyes) is called infrared light. Because infrared photons have lower energies than photons of ultraviolet or visible light, their absorption is no longer able to promote electrons to higher energy molecular orbitals. However, infrared photons do have enough energy to make the molecule vibrate.

It is not oversimplifying very much to say that, when a molecule vibrates as a result of absorbing a photon of infrared light, it is actually a particular bond or functional group that vibrates as suggested by the spring in place of the C—H bond in Figure 3-10.

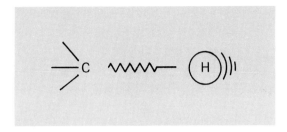

FIGURE 3-10. The absorption band at 3500 nm is due to a **stretching** vibration of the C—H bonds in the molecule in which the distance between the carbon and hydrogen nuclei becomes alternately shorter and longer.

For example, molecules with C—H bonds have **infrared absorption bands** near 3500 nm. This absorption band occurs in the infrared spectrum of isopentane (Figure 3-11).

A careful inspection of the spectrum in Figure 3-11 will reveal that the absorption band at 3500 nm is really at least two bands very close together and partly overlapping. The reason for this is that C—H bonds differ slightly in their frequencies of vibration, and in the wavelength of the infrared light that they absorb. The C—H bonds of the methyl groups (CH_3) of isopentane absorb light of slightly different wavelength than do the C—H bonds of the methylene group (CH_2) or the C—H bond of the **methine** group ($-\overset{|}{\underset{|}{C}}-H$, usually written CH). The absorption bands at 6900 and 7300 nm are due to **bending** vibrations of the C—H bonds in CH_3 and CH_2 groups rather than to stretching vibrations.

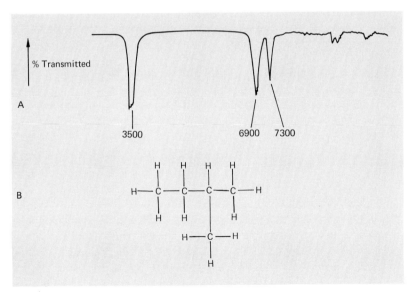

FIGURE 3-11. **(A)** Infrared spectrum of isopentane. The wavelengths are in nanometers. **(B)** Structural formula of isopentane.

Other types of bond also have typical absorption bands in the infrared. The band due to the stretching vibration of the O—H bond of the hydroxyl group, for example, is found at about 2770 nm. This band is present in the infrared spectrum of ethyl alcohol but not in that of its isomer, dimethyl ether.

FIGURE 3-12. An infrared spectrophotometer. [Courtesy of the Perkin-Elmer Corporation]

A C=O double bond betrays its presence in a molecule by causing an absorption band to appear in the 5000–6000 nm range. The exact position depends on other details of the molecule and can give an expert a great deal of information about the structure. In the case of the C=O double bond of the carboxyl group (COOH) in compounds like acetic acid, the absorption is at about 5850 nm.

RADIO WAVES AND NUCLEAR MAGNETIC RESONANCE

The absorption of radio waves by organic molecules is studied by means of instruments like the one shown in Figure 3-13. The instrument, which includes a large magnet, is called a nuclear magnetic resonance spectrometer. Protons, and also nuclei of several other kinds, have a property called "spin" that makes them act like magnets. In the case of a proton, the nuclear magnet can be aligned in either of two different directions: with the field of the large external magnet or against it. These two orientations of the proton magnet differ in energy, and the absorption of a radio-frequency photon supplies just enough energy to flip the spin of the proton from the less energetic orientation to the more energetic orientation.

The spectrometer uses a fixed radio frequency—usually one in the 60–200 megahertz (MHz) range. By varying the external magnetic field, a field strength can be found that will cause a proton in any part of the molecule to absorb the fixed radio frequency.

The field strength at which a given proton will begin to absorb the radio-frequency photons depends on the location of the proton in the molecule. This is illustrated by the nuclear magnetic resonance spectrum of ethyl alcohol (Figure 3-14).

FIGURE 3-13. A nuclear magnetic resonance spectrometer. [Courtesy of Varian Associates, Palo Alto, Calif.]

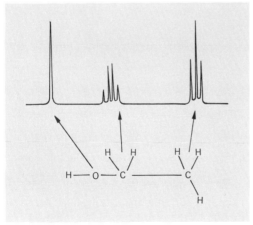

FIGURE 3-14. Nuclear magnetic resonance spectrum of ethyl alcohol. Magnetic field strength increases from left to right.

The line at the left, or low magnetic field side of the spectrum, is an absorption due to the protons of the OH groups in the alcohol. The reason that these protons absorb at a lower magnetic field than the protons attached to carbon is the greater electronegativity of

Isomers and the
Structure of Molecules

oxygen. The oxygen atom tends to pull electrons away from the proton, and these electrons would otherwise have served as a kind of shield against the magnetic field. The next group of lines is due to the CH_2 ($-CH_2-$) protons. These protons require a stronger magnetic field because they are further from the oxygen atom, and their electrons shield them more effectively from the external magnet. Finally, we come to a group of lines due to the CH_3 protons. These protons are still further from the oxygen and still more shielded by their electrons; therefore they require a still higher field to make them absorb the radio-frequency energy.

By measuring the intensities of the three groups of lines it is possible to tell that the relative numbers of OH, CH_2, and CH_3 protons are in the ratio 1:2:3.

We can also obtain evidence for the structure from the multiplicities of the absorptions. The absorption due to the CH_3 group is split into three lines. The reason for this is the magnetic influence of the two protons of the adjacent CH_2 group (Figure 3-15). The

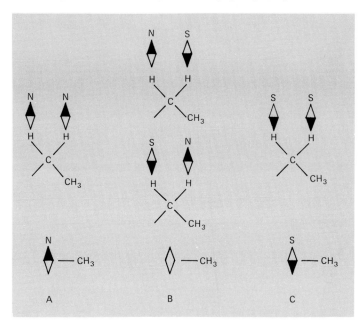

FIGURE 3-15. Three possible combinations of the magnetic effects of the two protons of the CH_2 group. In **(A)** the two CH_2 spins combine to produce an effect in one direction, in **(C)** they produce an effect in the opposite direction, and in **(B)** they cancel each other and produce no effect.

magnetic moments of the two protons of the CH_2 can be either both down, one down and one up, or both up. Correspondingly, the magnetic fields due to these protons either subtract from, have no effect on, or add to the field of the external magnet. Thus the CH_3 protons in some of the alcohol molecules exist in a decreased field, the CH_3 protons in other molecules exist in an unchanged field, and the CH_3 protons in still other alcohol molecules exist in an increased magnetic field. The mixture of alcohol molecules in all three conditions therefore produces not one line but a triplet of lines due to the CH_3 group.

Similarly, the protons of the CH_2 groups are affected by the adjacent CH_3 protons. They may be influenced by three proton magnets down, two down and one up, one down and two up, or three up. This causes the CH_2 signal to be split into four lines.

To summarize the implications of the ethyl alcohol spectrum, this molecule has

1. Three different locations for protons.
2. One on oxygen, one on carbon adjacent to oxygen, and one on a carbon atom further away from the oxygen.
3. The numbers of protons in these three different parts of the molecule are in the ratio $1:2:3$.
4. The molecule has to include a structure in which a CH_2 group is *directly attached* to a CH_3 group.

The magnetic resonance spectrum of dimethyl ether (Figure 3-16), the isomer of ethyl alcohol, is absurdly simple in comparison: it consists of just one line. The two CH_3 groups of CH_3-O-CH_3 are equivalent, so there is only one kind of proton.

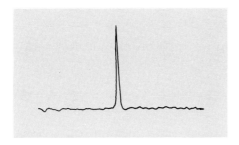

FIGURE 3-16. The magnetic resonance spectrum of dimethyl ether.

SUMMARY

A. Structures
  1. The tetrahedral angle: The angles between the carbon bonds of methane and related compounds are all 109° 28'. All four hydrogens of methane are equivalent, like the four corners of a tetrahedron.
  2. Conformers: An ordinary structural formula is not meant to suggest the shape of a molecule. Conformers are molecules that differ only in their shapes or conformations. Because molecules twist rather easily about their bonds, the differences between conformers are only temporary. In solution the conformation of a given molecule is constantly changing.
  3. Isomers are molecules having the same molecular formulas but differing from one another structurally. In contrast to the temporary differences between conformers, the differences between isomers are long lasting and can only be erased by breaking and making bonds.
  4. Isomers can be identified
      (a) By product counting (C_5H_{12} isomers).
      (b) By synthesis.
      (c) By characteristic reactions of functional groups (ethyl alcohol).
      (d) By light absorption.
B. Instrumental methods
  1. Light is an electromagnetic wave that carries energy in discrete amounts corresponding to photons.

2. Absorption of photons of ultraviolet or visible light moves electrons to orbitals of higher energy. Colored light is white light from which some wavelengths of the visible range are missing.
3. Absorption of photons of infrared light causes molecules and bonds to vibrate. Each kind of vibration requires the absorption of a particular wavelength of infrared light; this helps to identify the kinds of molecular structure present.
4. Radio waves and nuclear magnetic resonance. Hydrogen nuclei in a magnetic field absorb photons of radio-frequency radiation. The energy is used to reorient the proton magnets with respect to the field of the large external magnet. The magnetic field strengths at which the radio-frequency photons are absorbed depend on the position of the proton in the molecule and on the structure of the molecule.

EXERCISES

1. What are two main ways of determining the structure of a molecule?
2. Explain what a structural formula is meant to tell about a molecule.
3. What is wrong with each of the following structural formulas, if anything?

    (a)
    $$H-C \begin{matrix} \diagup H \\ -H \\ \diagdown H \end{matrix}$$

    (b)
    $$\begin{matrix} H \\ | \\ C-H-H \\ | \\ H \end{matrix}$$

    (c)
    $$\begin{matrix} H \\ \diagdown \\ \diagup \\ H \end{matrix} C=C \begin{matrix} H \\ | \\ -H \\ | \\ H \end{matrix}$$

4. Make a sketch showing the actual shape of a methane molecule.
5. Which of the following structural formulas are equivalent?

    (a)
    ```
        H H H H H
        | | | | |
    H—C—C—C—C—C—H
        | | | | |
        H H H H H
    ```

    (b)
    ```
        H H H
        | | |
    H—C—C—C—H
        | | |
        H H |
            H—C—H
                |
            H—C—H
                |
                H
    ```

    (c)
    ```
        H
        |
    H—C—H  H H H
        |  | | |
    H—C———C—C—C—H
        |  | | |
        H  H | H
             H—C—H
                |
                H
    ```

    (d)
    ```
        H     H     H
        |     |     |
    H—C———C———C—H
        |     |     |
              H
        |           |
    H—C—H     H—C—H
        |           |
        H           H
    ```

    (e)
    ```
        H H H H H
        | | | | |
    H—C—C—C—C—C—H
        | | | | |
        H H H | H
              H—C—H
                |
                H
    ```

6. Using only covalent bonds (no $-C^-$ or $-\overset{+}{O}H_2$, for example), draw structural formulas for the compounds listed below. Be sure that each isomer is shown just once, no equivalent structures should be given.
   (a) all the C_4H_{10} compounds
   (b) all the C_5H_{12} compounds
   (c) all the C_3H_8O compounds
   (d) all the compounds C_3H_8

7. Which of the following structural formulas are equivalent?

   (a)
   ```
        Cl H  H
        |  |  |
    H—C—C—C—H
        |  |  |
        H  H  H
   ```

   (b)
   ```
        H  H  H
        |  |  |
    H—C—C—C—H
        |  |  |
        Cl H  H
   ```

   (c)
   ```
        H  Cl H
        |  |  |
    H—C—C—C—H
        |  |  |
        H  H  H
   ```

   (d)
   ```
        H  H  H
        |  |  |
    H—C—C—C—H
        |  |  |
        H  H  Cl
   ```

8. How many monochlorinated compounds of molecular formula C_3H_7Cl can be made by chlorinating propane? Show their structures.

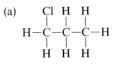
propane

9. How many monochlorinated compounds of molecular formula C_4H_9Cl can be made by chlorinating
   (a) butane, $CH_3CH_2CH_2CH_3$

   (b) isobutane,
   ```
                H
                |
       CH_3—C—CH_3
                |
                CH_3
   ```

   Show their structures.

10. Show the structures for the dichlorinated products, $C_4H_8Cl_2$, of both butanes.

11. Explain how the structures were assigned (from chemical reactions) to the isomeric compounds ethyl alcohol and dimethyl ether.

12. Predict, for the isomers below (a) the reactivity with sodium and (b) the boiling points (one higher, one lower).

   ```
        H  H  H                      H  H     H
        |  |  |                      |  |     |
    H—C—C—C—OH      and      H—C—C—O—C—H
        |  |  |                      |  |     |
        H  H  H                      H  H     H
   ```

13. Draw a diagram illustrating what is meant by the length of a wave. What is meant by the frequency of a wave?

14. What happens to the energy of the light when it is absorbed by a molecule in the case of (a) visible or ultraviolet light, (b) infrared light, (c) radio-frequency radiation in a magnetic field?

Chapter 4
Alkanes and Cycloalkanes

The last chapter discussed just a few compounds that were composed solely of the elements carbon and hydrogen. If almost any other two elements had been picked there would be few if any additional compounds to be considered. In the case of carbon and hydrogen, however, there are millions of compounds; many are already known and still others are easily synthesized in the laboratory whenever anyone cares to do so. Although the large number of C—H compounds may be regrettable from the point of view of one who is trying to learn as much as possible about them, it is the richness and complexity of the chemistry of carbon that makes the strange phenomenon of life possible.

In this chapter two classes of hydrocarbons will be taken up: the alkanes and the cycloalkanes.

Methane

The first member of the alkane series of hydrocarbons, which has the general formula C_nH_{2n+2}, is methane. As indicated by the fact that methane forms only one monochloro derivative, CH_3Cl, and only one dichloro derivative, CH_2Cl_2, all four hydrogens of methane are equivalent. This was explained in Chapter 3 on the basis of a tetrahedral model for methane in which the bonds, at angles of 109° 28′, extend from carbon at the center of the tetrahedron to the hydrogens at the corners. The tetrahedral model also explains the existence of only one dichloromethane, since any pair of corners of a tetrahedron are equivalent to any other pair. All four C—H bond lengths are also the same, namely 1.09×10^{-8} cm.

THE BOND ORBITALS OF METHANE

At this point it is desirable to examine briefly the reason why methane is tetrahedral. Covalent bonds are formed by the overlap of atomic orbitals, and if so, then **the orbitals of the central carbon atom of methane should also be tetrahedral**.

It is clear that none of the hydrogen-like atomic orbitals of Chapter 1 fit this requirement. The $2s$ orbital is spherical and does not point in any direction, and the $2p_x$, $2p_y$ and $2p_z$ orbitals point at angles of $90°$ to one another rather than at angles of $109°\ 28'$. The presence of the four hydrogens and the requirements for bond formation have brought about a change in the atomic orbitals.

The principal advantage of the tetrahedral set of carbon orbitals, and the reason that these are used in bond formation rather than the old set, is that a tetrahedral arrangement allows the four hydrogen atoms and four pairs of bonding electrons to be as far apart from each other as possible while still being close enough to the central carbon atom to form the C—H bonds.

Figure 4-1 is a diagram of this concept showing the four spherical $1s$ orbitals of the four hydrogen atoms overlapping with the four tetrahedral orbitals of the carbon atom. The overlap of a carbon tetrahedral atomic orbital with a hydrogen $1s$ atomic orbital forms a tetrahedrally oriented bonding orbital, or tetrahedral bond. Each bonding orbital is occupied by two electrons.

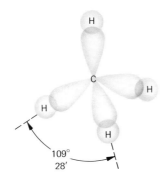

FIGURE 4-1. A carbon atom with four tetrahedral orbitals about to form bonds with four hydrogen atoms each with a $1s$ orbital.

The tetrahedral orbitals are also called sp^3 (read s p three) orbitals because it is possible to describe them mathematically as a combination, or **hybridization**, of an s orbital with three p orbitals. The combination of the s orbital and three p orbitals (four orbitals in all) turns out to be four equivalent tetrahedral or sp^3 hybrid orbitals. Because of the p orbitals that went into this combination, each sp^3 orbital has two lobes. Unlike the p orbital, however, each sp^3 orbital has one very large lobe and one very small one, as indicated in Figure 4-2. The larger of these lobes extends further out into space

FIGURE 4-2. The shape of an sp^3 orbital. In most pictures of sp^3 orbitals the smaller lobe is omitted for the sake of clarity.

than a p orbital does, and this contributes to better overlap with the hydrogen orbitals and a stronger bond.

Mathematically, the construction of a hybrid orbital is like the addition of velocities. When a man swims due west across a river that is flowing due south his actual motion is in a "hybrid" direction called southwest.

Most of the time it is convenient to ignore the detailed structure of the sp^3 orbitals and the resulting bonding orbitals and represent them merely by a line connecting the symbols for the two atoms, or in a model, by a stick connecting the two balls.

Methane, as should be expected from the fact that it is a covalent compound, has a low boiling point ($-184°C$). It is a colorless gas often found in coal mines, where it is called "fire damp," and causes explosions when the proportion of oxygen to methane is in the easily ignitable range.

$$CH_4 + 2O_2 \rightarrow CO_2 + 2H_2O + \text{heat} \tag{4-1}$$

Methane is also a common constituent of the marsh gas that bubbles up from the bottom of swamps and lakes, and marsh gas usually contains enough methane to be flammable. The reader can test this for himself on his next fishing trip as shown in Figure 4-3.

The methane in marsh gas is probably formed by the action of bacteria on plant debris. Methane is also found in natural gas (about 85% methane, 9% ethane), in the earth's atmosphere (0.00022%), and is probably a major constituent of the atmospheres of Jupiter and Saturn.

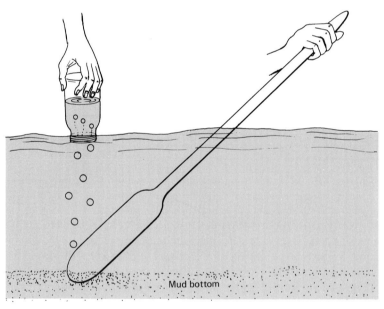

FIGURE 4-3. Stir the mud with an oar and collect the marsh gas by downward displacement of water in a bottle. A match will usually ignite the gas from the bottle. If the bottle is filled only about one-eighth full of marsh gas and the rest with air, a match will usually produce a small explosion.

Other Alkanes

The second member of the alkane series can be regarded as methane which has a methyl group (CH_3) in place of one of the hydrogen atoms.

```
      H                H  H
      |                |  |
   H—C—H            H—C——C—H
      |                |  |
      H                H  H
   methane            ethane
 a methyl           two methyl
group bonded to    groups bonded to
 hydrogen           each other
```

Because the bond connecting the two methyl groups is also constructed from tetrahedral atomic orbitals, the H—C—C angle is also 109° 28′, the same as the H—C—H angles. The C—C bond length is 1.54×10^{-8} cm.

If another CH_3 group is added in place of one of the hydrogens of ethane, the result is propane. Again, the bond angles (C—C—C) and (H—C—H) are all 109° 28′.

```
      H  H  H
      |  |  |
   H—C——C——C—H
      |  |  |
      H  H  H
       propane
```

The structural formulas of methane, ethane, propane, and the rest of the series of **normal alkanes** (that is, unbranched alkanes) as far as decane are given in Table 4-1. Here the structural formulas have been abbreviated. For example, $CH_3CH_2CH_3$ (propane) means a chain of three carbon atoms connected by single bonds with three hydrogen atoms on the first carbon, two on the next, and three on the last. This system is frequently used to save time in writing structural formulas, but it is always desirable to check to make sure that each carbon has four bonds.

TABLE 4-1. *The Series of Normal Alkanes to Decane*

	Name	Boiling point, °C
CH_4	methane	−161.5
CH_3CH_3	ethane	−88.6
$CH_3CH_2CH_3$	propane	−42.1
$CH_3CH_2CH_2CH_3$	butane	−0.5
$CH_3CH_2CH_2CH_2CH_3$	pentane	36.1
$CH_3CH_2CH_2CH_2CH_2CH_3$	hexane	68.7
$CH_3CH_2CH_2CH_2CH_2CH_2CH_3$	heptane	98.4
$CH_3CH_2CH_2CH_2CH_2CH_2CH_2CH_3$	octane	125.7
$CH_3CH_2CH_2CH_2CH_2CH_2CH_2CH_2CH_3$	nonane	150.8
$CH_3CH_2CH_2CH_2CH_2CH_2CH_2CH_2CH_2CH_3$	decane	174.0

From Table 4-1 it can be seen that the names pentane (five carbons), hexane (six carbons), heptane (seven carbons), octane (eight carbons), nonane (nine carbons) and decane (ten carbons) are based on the Latin names for the numbers five through ten. This system continues in the same way for still larger molecules.

The first four names, methane, ethane, propane, and butane however, are merely the result of historical accidents. By the time it was realized that there would eventually be an almost infinite series of alkanes, the names of these first four compounds had become too firmly established for them to be changed.

Methane and **methyl** come from the name methyl alcohol for the one-carbon compound CH_3OH. Methyl alcohol was at one time made by distilling wood chips and the name, coming from the Greek *methy* (wine) and *hyle* (wood) has a certain amount of logic behind it. However, methyl alcohol is poisonous and is not a constituent of drinkable wine, which contains the related compound ethyl alcohol.

Ethane and **ethyl** are derived from the Greek *aithein*, to blaze, via our word ether. "Ether" is the common name for diethyl ether $(CH_3CH_2)_2O$, a highly flammable liquid.

Propane and **propyl** are derived from **propionic acid**, a three-carbon compound whose name is derived from the Greek *prōtos* (first) and *piōn* (fat). Compounds related to propionic acid occur in fat.

Butane and **butyl** are derived from the four-carbon compound **butyric acid** which is found in rancid butter.

Refer again to Table 4-1 and note that as the size of the molecule increases so does the boiling point. Thus methane, ethane, and propane are gases; butane is a liquid only during a moderately cold winder; and pentane is a liquid except on an extremely hot summer day (97°F, a little below body temperature). Although a large molecule will usually have a higher boiling point than a small molecule, this is not always the case. For example H_2O (bp 100°C) is a smaller molecule than ethane (bp $-88.6°C$). The reason for its higher boiling point is that water molecules tend to form clusters as a result of hydrogen bonds.

A hydrogen bond is weaker than the covalent bonds holding the atoms together within a single molecule: it is merely the electrostatic attraction between the somewhat positive (δ^+) hydrogen nucleus on one water molecule and the electrons of the oxygen atom of another water molecule.

$$\begin{array}{c} H \\ | \\ :\!\ddot{O}\!-\!H \\ | \\ H \end{array}$$

$$\overset{\delta^-}{:\!\ddot{O}}\!-\!\overset{\delta^+}{H}\cdots\overset{\delta^-}{:\!\ddot{O}}\!-\!\overset{\delta^+}{H}\cdots:\!\ddot{O}\!-\!H\cdots:\!\ddot{O}\!-\!H$$

water

One reason that alkanes are not soluble in water is that water molecules prefer to stick to other water molecules rather than to alkane molecules. The C—H bonds are nonpolar, and the hydrogen is not sufficiently positive to hydrogen bond to a water molecule.

Various higher alkanes can be extracted from plants, for example, heptane from several kinds of pine, tetradecane (14 carbons) from chrysanthemums, and hexadecane (16 carbons) from roses. The natural protective waxy coating on pears contains a 29-carbon alkane.

The alkanes are relatively nontoxic, but may have narcotic properties at high concentrations. A pure gaseous alkane will asphyxiate, of course, merely because oxygen is excluded. The objection to the escape of alkanes from carburetors and automobile exhausts is not to the alkanes themselves but to the smog formed from their reaction with air which is promoted by sunlight.

BRANCHED ALKANES

As was noted in Chapter 3, the atoms of C_4H_{10} can be put together to form two isomers, the straight-chain compound butane and the branched-chain compound isobutane.

In the series of alkanes from CH_4 to C_nH_{2n+2}, where n can be any integer, the number of isomers is unity for CH_4 through C_3H_8 (propane) but increases rapidly thereafter. For example, C_9H_{20} has 35 isomers and $C_{40}H_{82}$ has 62,491,178,805,837. This means that nondescriptive names are inadequate to deal with the problem. What is needed is a **naming system** that will allow anyone to draw the structure that goes with any name even if he has never heard the name before.

The best way to learn the system is to look at some examples. Methane, ethane, propane, butane, pentane, and so on, are the unbranched normal alkanes, as before. "Isobutane" becomes 2-methylpropane, because it is propane with a methyl group replacing one of the two equivalent hydrogens on carbon atom 2.[1]

$$CH_3-\underset{\underset{CH_3}{|}}{\overset{\overset{H}{|}}{C}}-CH_3$$

The compounds 2-methylpentane and 3-methylpentane are isomers of hexane.

$$CH_3CH_2CH_2\underset{\underset{CH_3}{|}}{CH}CH_3 \qquad CH_3CH_2\underset{\underset{CH_3}{|}}{\overset{\overset{CH_3}{|}}{CH}}CH_2CH_3 \qquad \underset{\underset{CH_3}{|}}{\overset{\overset{CH_3}{|}}{CH_2}}CH_2CH_2CH_2$$

or   or   or

$$CH_3\underset{\underset{CH_3}{|}}{CH}CH_2CH_2CH_3 \qquad CH_3CH_2\underset{\underset{\underset{CH_3}{|}}{CH_2}}{\overset{\overset{CH_3}{|}}{CH}}CH_3 \qquad CH_3CH_2CH_2CH_2CH_2CH_3$$

2-methylpentane   3-methylpentane   hexane

Names for more complicated structures require more numbers, and may use larger substituents such as ethyl, isopropyl, and so forth.

$$CH_3-\underset{\underset{CH_3}{|}}{\overset{\overset{CH_3}{|}}{C}}-CH_2CH_3 \qquad CH_3-\underset{\underset{H}{|}}{\overset{\overset{CH_3}{|}}{C}}-CH_2-\underset{\underset{H}{|}}{\overset{\overset{CH_3}{|}}{C}}-CH_3$$

2,2-dimethylbutane   2,4-dimethylpentane

---

[1] Note that although one could easily draw structures if given the names 1-methylpropane or 3-methylpropane, these names are incorrect; butane is the correct name for the unbranched isomer.

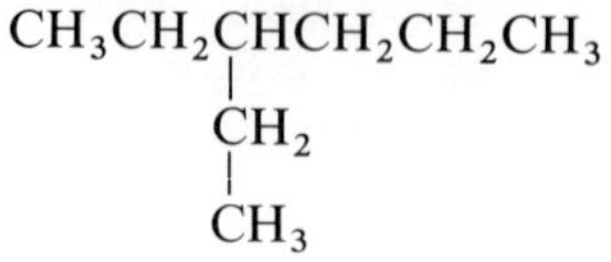

$$\begin{array}{c}\text{CH}_3\text{CH}_2\text{CHCH}_2\text{CH}_2\text{CH}_3\\|\\ \text{CH}_2\\|\\ \text{CH}_3\end{array}$$

3-ethylhexane

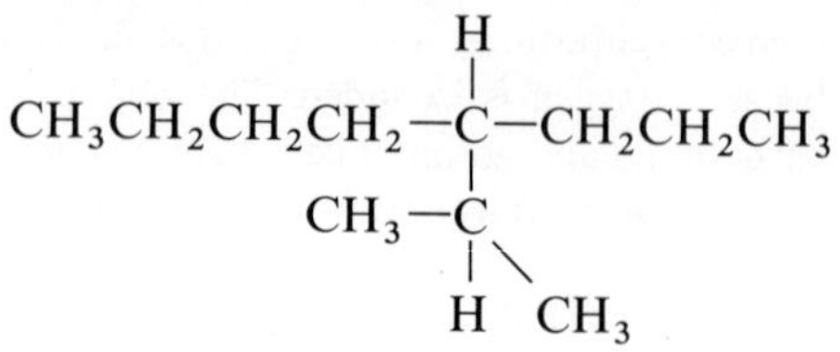

$$\begin{array}{c}\text{H}\\ |\\ \text{CH}_3\text{CH}_2\text{CH}_2\text{CH}_2-\text{C}-\text{CH}_2\text{CH}_2\text{CH}_3\\ |\\ \text{CH}_3-\text{C}\\ |\,\backslash\\ \text{H}\ \ \text{CH}_3\end{array}$$

4-isopropyloctane

Enough examples have now been given to illustrate the most important steps in naming an alkane.

1. Find the longest chain of carbon atoms. (Note that it need not be written horizontally and that it may go around a corner.) This establishes the main root of the name.
2. Everything not part of the longest chain is a substituent such as methyl, ethyl, and so on. (Note that the end methyl of a chain is part of the chain, not a substituent.)
3. Assign numbers to the substituents, starting with one end of the chain. Start the numbering with the end of the chain that results in the use of the lower numbers; for example 2-methylpentane rather than 4-methylpentane.
4. Put the substituents in the name in order of increasing size.
5. If two or more substituents of the same kind are present, use prefixes di, tri, tetra, and so forth.
6. Put in the numbers and punctuation.

The most common substituents used in naming alkanes are listed in Table 4-2.

TABLE 4-2. *Some Common Alkyl Substituents*

Name	Formula
Methyl	CH_3-
Ethyl	CH_3CH_2-
Propyl	$CH_3CH_2CH_2-$
Isopropyl (or *i*-propyl)	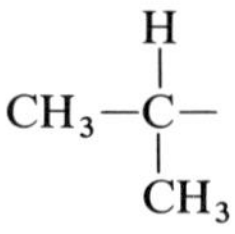 $CH_3-\overset{\overset{H}{\|}}{\underset{\underset{CH_3}{\|}}{C}}-$ or $CH_3CH(CH_3)-$
Butyl	$CH_3CH_2CH_2CH_2-$
Secondary butyl (or *sec*-butyl)	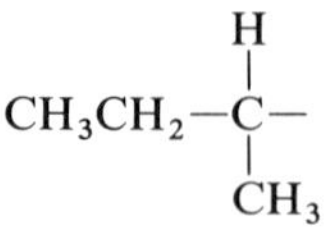 $CH_3CH_2-\overset{\overset{H}{\|}}{\underset{\underset{CH_3}{\|}}{C}}-$ or $CH_3CH_2CH(CH_3)-$
Tertiary butyl (or *t*-butyl)	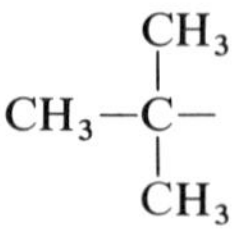 $CH_3-\overset{\overset{CH_3}{\|}}{\underset{\underset{CH_3}{\|}}{C}}-$ or $CH_3C(CH_3)_2-$

The systematic use of the rules in a complicated case is shown in Table 4-3.

TABLE 4-3. Name:

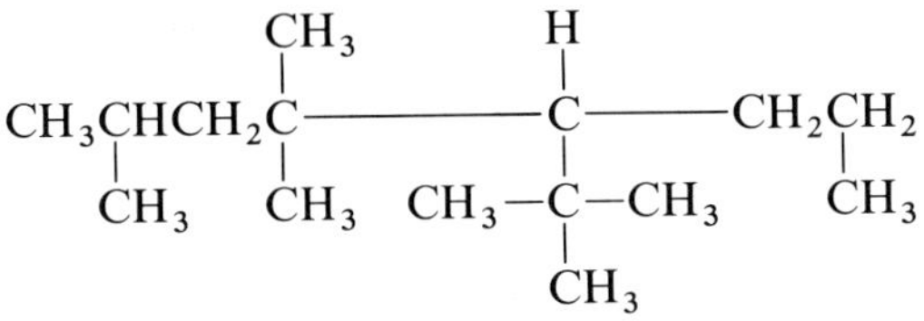

$$\text{CH}_3\text{CHCH}_2\overset{\overset{\text{CH}_3}{|}}{\underset{\underset{\text{CH}_3}{|}}{\text{C}}}\text{———}\overset{\overset{\text{H}}{|}}{\underset{\underset{\text{CH}_3\text{—C—CH}_3}{|}}{\text{C}}}\text{———CH}_2\text{CH}_2$$

with CH$_3$, CH$_3$, CH$_3$ substituents appropriately.

Step	Result
1. C—C—C—C—C—C—C—C	-octane
2. 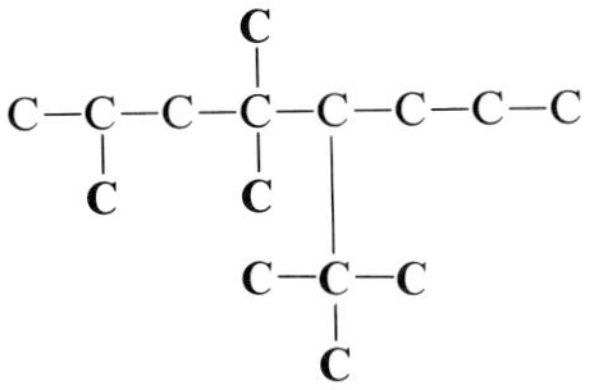 (structure with methyl branches and t-butyl group)	trimethyl ⎫ t-butyl ⎬ -octane
3. 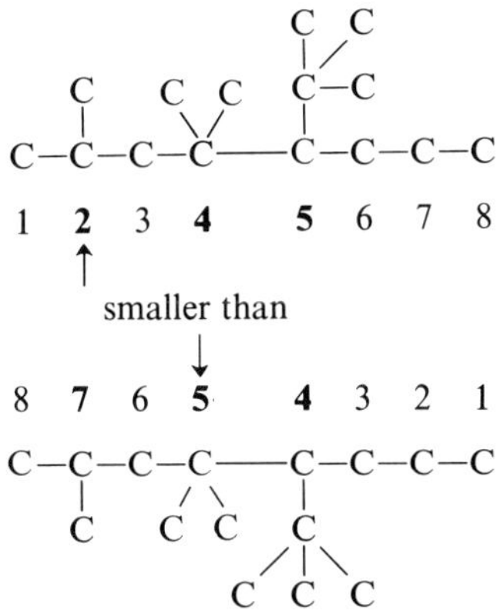 Numbering 1→8 gives methyls at 2,4,4; t-butyl at 5. Numbering 8→1 gives methyls at 5,5,7; t-butyl at 4. "smaller than" → first numbering chosen	methyls are 2,4,4; the t-butyl is 5 methyls are 5,5,7; the t-butyl is 4
4, 5. Add prefix tri, put in order of size 6. Add numbers and punctuate	trimethyl-t-butyloctane 2,4,4-trimethyl-5-t-butyloctane

SOURCE OF ALKANES

Most alkanes occur naturally either in the gas from oil fields or in the liquid crude oil. When a large quantity of one of the simpler alkanes is needed, it can be separated from the natural gas or petroleum by distillation or by gas-liquid chromatography. In distillation, materials of lower boiling point distill first and are removed and condensed back to the liquid state, then the next higher boilers, and so forth.

Purer samples are obtained by using a fractionating column which provides for an automatic redistillation. The distillate that condenses in the lower part of the column partly trickles back down into the pot and partly redistills and condenses at a higher part of the column. This portion of the material again divides itself into a more volatile part that goes further up the column and a less volatile part that trickles back down, and so on.

Gas-liquid chromatography is like distillation except that it separates substances according to their affinity for a liquid coating on a solid packing material rather than by boiling point. The packing material might be fire brick and the liquid coating a silicone[2] oil, for example. A carrier gas such as helium is passed through a column of the coated packing material. Compounds with only a weak affinity for the coating tend to wash right through with the helium, whereas those that tend to stick more firmly lag behind. Figure 4-4 is essentially a graph of the amount of material emerging from the column after various lengths of time. Ideally, each peak corresponds to just one compound coming off the column at a given time.

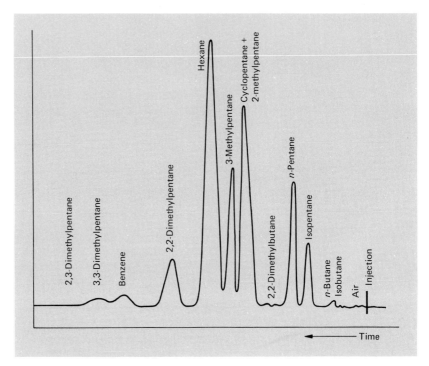

FIGURE 4-4. Chromatogram of a sample of low octane gasoline. [From Jean Tranchant (ed.), *Practical Manual of Gas Chromatography*, Elsevier, Amsterdam, 1969]

Occasionally, when the need arises for a very pure sample of a particular alkane or an alkane that would be difficult to separate from the mixture in petroleum, it is synthesized from compounds of other kinds (see Table 4-4). For example any Grignard reagent R—Mg—X will react with acid HX to give the corresponding compound RH. The letter R in R—Mg—X and in RH in the general

$$CH_3-Mg-Cl + HCl \rightarrow CH_4 + MgCl_2 \qquad (4\text{-}2)$$
methylmagnesium
chloride

$$R-Mg-X + HX \rightarrow RH + MgX_2 \qquad (4\text{-}3)$$

form, equation 4-3, stands for any alkyl group, for example methyl

[2] Carbon-silicon-oxygen compound, see Chapter 12.

(CH_3), ethyl (CH_3CH_2), and so forth. A more detailed discussion of such reactions, as well as how one gets the reagent in the first place, will be deferred until later.

TABLE 4-4. *Examples of Reactions Leading to Alkanes*

Reaction	Equation
Grignard synthesis	$CH_3CH_2-Mg-Cl + HCl \rightarrow CH_3CH_3 + MgCl_2$
Hydrogenation of an alkene	$CH_3CH=CHCH_3 + H_2 \xrightarrow[\text{(catalyst)}]{\text{Pt}} CH_3CH_2CH_2CH_3$
Hydrogenation of an alkyne	$HC\equiv CCH_3 + 2H_2 \xrightarrow{\text{Pt}} CH_3CH_2CH_3$
Wurtz reaction	$2CH_3-Br + 2Na \rightarrow CH_3CH_3 + 2NaBr$
Reduction of an alkyl iodide	$CH_3CH_2I + HI \xrightarrow{\text{heat}} CH_3CH_3 + I_2$
Kolbe electrolysis	$2CH_3CH_2\overset{O}{\underset{\|}{C}}-O^-\,Na^+ + 2H_2O \xrightarrow{\text{electrolysis}} \underbrace{CH_3CH_2CH_2CH_3 + 2CO_2}_{\text{at the anode}} + \underbrace{2NaOH + H_2}_{\text{at the cathode}}$

Comparatively few alkanes are found in living organisms. This is probably because of the low reactivity of alkanes; most life processes require a functional group of some sort. However, the exceptions can be very interesting. Certain ants, for example, discharge odors to communicate with other ants of the colony when they are alarmed. The pheromones[3] responsible for the odor have been separated by gas-liquid chromatography. They include the alkanes undecane and tridecane, as well as several compounds with hydroxyl or $>C=O$ functional groups.

$$CH_3CH_2CH_2CH_2CH_2CH_2CH_2CH_2CH_2CH_2CH_3$$
undecane

$$CH_3CH_2CH_2CH_2CH_2CH_2CH_2CH_2CH_2CH_2CH_2CH_2CH_3$$
tridecane

Another example of the use of an alkane by living organisms is a sex pheromone of the tiger moth.

$$CH_3-\underset{\underset{CH_3}{|}}{\overset{\overset{H}{|}}{C}}-CH_2CH_2CH_2CH_2CH_2CH_2CH_2CH_2CH_2CH_2CH_2CH_2CH_2CH_3$$

2-methylheptadecane

a sex pheromone of tiger moths

This alkane, which was obtained by extracting the abdominal tips of fifty female moths with a solvent (methylene chloride, CH_2Cl_2)

---

[3] Chemicals used for communication between members of the same species are called pheromones.

stimulates the male moths to instant activity. Nature's use for the sex pheromone is to attract the male to the female, but it can be used by man to attract male moths to a trap. Because the attractive power of sex pheromones is usually quite specific, it is possible to use them to construct traps for destructive insects without harming useful insects.

Cycloalkanes

The compound **cyclohexane** has two hydrogen atoms less than hexane. A model of hexane can be converted into a model of cyclohexane by removing hydrogen atoms from carbons 1 and 6 and connecting these carbons together to form a ring. The cyclohexane ring is still somewhat flexible, although not quite so flexible as hexane itself. There are two possible shapes, or conformations, one called the chair form and the other called the boat form (Figure 4-5). Of the two forms of cyclohexane, the chair form is by far the

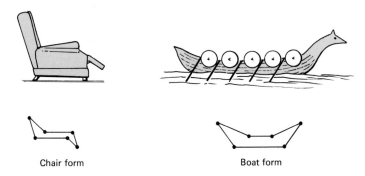

Chair form   Boat form

FIGURE 4-5. The chair and boat forms of cyclohexane. The hydrogen atoms have been omitted for the sake of clarity.

$$\begin{array}{cc} & CH_3 \\ H_2C & CH_3 \\ | & | \\ H_2C & CH_2 \\ & CH_2 \end{array} \qquad \begin{array}{cc} & CH_2 \\ H_2C & CH_2 \\ | & | \\ H_2C & CH_2 \\ & CH_2 \end{array}$$

hexane   cyclohexane
bp 68.7°C   bp 80.8°C

more stable and at any given instant almost all of the molecules in a sample of cyclohexane will be in the chair form. Figure 4-6 shows the chair form with the hydrogen atoms included.

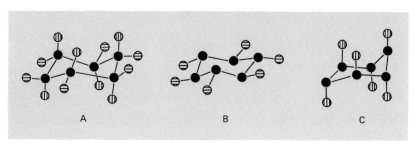

FIGURE 4-6. Chair form of cyclohexane. **(A)** Cyclohexane. **(B)** Equatorial hydrogens only. **(C)** Axial hydrogens only.

There appear to be two kinds of hydrogen in the chair form of cyclohexane. One set, the **equatorial** hydrogens, is arranged in a ring around the circumference of the carbon skeleton. The other set, the **axial** hydrogens, is pointed in a vertical direction, three on top of the ring of carbon atoms and three beneath it.

At room temperature the flexing of the cyclohexane ring occurs so rapidly that the sets of equatorial and axial hydrogens are indistinguishable. Thus a nuclear magnetic resonance spectrometer (Chapter 3) sees only one kind of hydrogen and the spectrum has only a single peak. At $-100°C$, however, the interchange of hydrogens between equatorial and axial positions is slowed enough so that the spectrometer sees two separate peaks, one for the six equatorial hydrogens and one for the six axial hydrogens.

From models showing the actual sizes of the atoms it can be seen that the three axial hydrogens on the same side (top or bottom) of the cyclohexane ring are quite close together and rather crowded. This crowding is even more pronounced if one of the hydrogen atoms is replaced by a bulkier group such as CH_3. Thus the **methylcyclohexane** molecule spends most of its time in the conformation that has the CH_3 group in an uncrowded equatorial position rather than in an axial position (Figure 4-7).

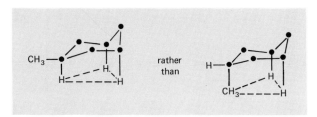

FIGURE 4-7. The conformation with the CH_3 in an equatorial position is preferred for methylcyclohexane.

OTHER CYCLOALKANES

Larger and smaller cycloalkanes such as cyclopentane (five carbons), cycloheptane (seven carbons) and so on, also exist. Most of them are very much like the alkanes in chemical reactivity and will not be discussed separately. The very smallest rings such as cyclopropane, however, have quite different properties and will be discussed in Chapter 5.

A ring of 120 carbon atoms has been made in the laboratory.[4]

Substituted cycloalkanes are named by a system like that used for branched alkanes. For example, 1,3,5-trimethylcyclohexane has

---

[4] There is a good chance that a few of the many molecules in the sample of C_{120} ring compound had knots in them or were interlinked with other molecules.

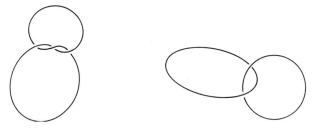

Alkanes and Cycloalkanes

methyl groups on every third carbon atom.

$$\begin{array}{c}
 \quad \quad \quad CH_3 \\
 H_2C — CH \\
CH_3 — CH \quad \quad CH_2 \\
 H_2C — CH \\
 \quad \quad \quad CH_3
\end{array}$$

1,3,5-trimethylcyclohexane

Adamantane (see structure in Figure 4-8) is a molecule related to the chair conformation of cyclohexane. Although this molecule has only four more carbons than cyclohexane, careful inspection of the figure will reveal four cyclohexane rings, all exactly equivalent,

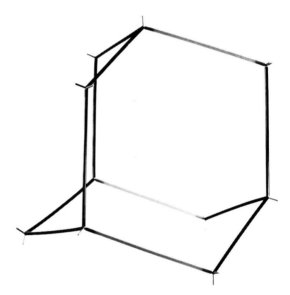

FIGURE 4-8. The structural relationship between cyclohexane and adamantane. For the sake of clarity only the C—C bonds are shown.

which form the walls of a cagelike structure. By adding more carbon atoms, the walls of the adamantane cage can each be used as walls of adjoining cages, as in the models shown in Figures 4-9 and 4-10. As the structure is made bigger, two interesting things become apparent. One is that the proportion of hydrogen to carbon is going down. (A speck of dust coated with paint is mostly paint; a rock coated with paint is mostly rock.) Carried far enough this will give a substance that has too little hydrogen to be detected and which is essentially a form of the element carbon.

Diamonds are in effect enormous multicage molecules, so big that we can actually see them. Return again to Figure 4-10 where it can be seen that the carbon atoms seem to be in parallel layers extending in several directions. When a diamond is cut by a diamond cutter, he cleaves the molecule along planes that separate one layer of carbon atoms from the next. Because bonds are broken, this is a chemical reaction not just a physical process.

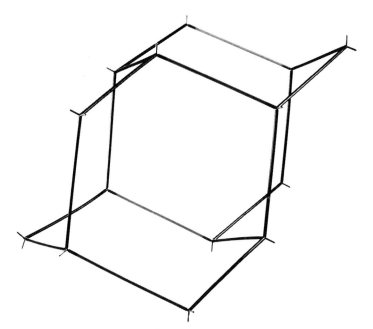

FIGURE 4-9.

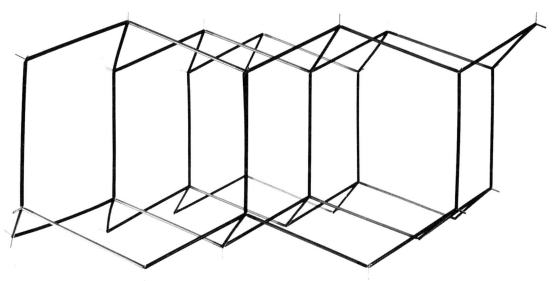

FIGURE 4-10.

The covalent bonds are also responsible for the hardness of the diamond. The softness of crystals of other covalent substances is due to the fact that there are no covalent bonds *between* the molecules making up the crystal.

The name adamantane is nonsystematic; it is derived from the relationship between adamantane and diamonds, adamant (hard) being an old word for diamond. The systematic names for compounds with more than one ring tend to be complicated; adamantane is tricyclo-$[3,3,1,1^{3,7}]$-decane. Adamantane is found in Czechoslovakian petroleum. A derivative of adamantane with a $-NH_2$ group in place of one of the tertiary ($\geqslant$C—H) hydrogen atoms has some antiviral activity.

Reactions of Alkanes

CHLORINATION

The chlorination of alkanes is catalyzed by light and is a free radical chain reaction. It will be recalled from the last chapter that the absorption of visible light by a substance moves electrons from orbitals of lower energy to orbitals of higher energy. In the case of chlorine gas, which appears green because of its absorption of light near the red end of the spectrum, the promoted electrons are no longer able to hold the molecules together and free chlorine atoms are produced.

$$Cl_2 \xrightarrow{light} 2\, Cl\cdot \qquad (4\text{-}4)$$

Free chlorine atoms are very reactive and will react with alkanes by removing a hydrogen atom.

$$Cl\cdot + CH_4 \rightarrow HCl + H-\overset{\overset{\displaystyle H}{|}}{\underset{\underset{\displaystyle H}{|}}{C}}\cdot \qquad (4\text{-}5)$$

The resulting **free radical** (a methyl in the case of methane) reacts with a chlorine molecule to produce the chlorination product and another free chlorine atom.

$$H-\overset{\overset{\displaystyle H}{|}}{\underset{\underset{\displaystyle H}{|}}{C}}\cdot + Cl_2 \rightarrow H-\overset{\overset{\displaystyle H}{|}}{\underset{\underset{\displaystyle H}{|}}{C}}-Cl + Cl\cdot \qquad (4\text{-}6)$$

methyl chloride
bp $-24°C$

The new chlorine atom goes on to react with another molecule of alkane, producing a new alkyl radical, which reacts with another molecule of chlorine to give still another chlorine atom, and so on. Note that it takes only one photon of light to cause the reaction of many molecules by this chain process.

Although the *actual* reactions are those shown in equations 4-4 through 4-6, the net result is that shown by equation 4-7.

$$CH_4 + Cl_2 \rightarrow CH_3Cl + HCl \qquad (4\text{-}7)$$

Of course methyl chloride (CH_3Cl) still has C—H bonds, and these will also react with chlorine atoms. Hence the mixture of products of the reactions of methane and chlorine will include **methylene chloride** (CH_2Cl_2, bp 40°C), **chloroform** ($CHCl_3$, bp 61–62°C) and **carbon tetrachloride** (CCl_4, bp 76.8°C). This mixture can be separated by distillation or by gas-liquid chromatography.

Methyl chloride boils at $-24°C$ and is used as a local anaesthetic; it acts by cooling the skin. Any evaporating liquid has to absorb a certain amount of heat in order to evaporate and, because a jet of methyl chloride evaporates very rapidly, it also cools its surroundings very rapidly.

Methylene chloride or dichloromethane (CH_2Cl_2) is used as a solvent and paint remover.

Chloroform was used as an anaesthetic in an operation for Queen Victoria in 1847. It is not a very safe anaesthetic, because the side effects of large doses include hypotension, respiratory depression, and death. Back in the Victorian era, before the novelty wore off, there were chloroform binges at which the guests sat around a bowl of slowly evaporating chloroform until they toppled over; the Victorian era, like ours, was one of several periods during which drug abuse was a problem. *Plus ça change, plus c'est la même chose.*

Chloroform is usually stored in tightly capped brown bottles and has about 1% ethyl alcohol added to it to discourage the formation of phosgene, a highly toxic gas used as a chemical warfare agent in World War I.

$$\text{Cl}-\overset{\overset{\displaystyle O}{\|}}{\text{C}}-\text{Cl}$$
phosgene

Carbon tetrachloride has been used as an anthelmintic against hookworm and tapeworm. It is toxic, however, and also forms phosgene. For this reason it is no longer used in fire extinguishers.

COMBUSTION

Alkanes burn in oxygen or in air to form H_2O, CO_2, and CO. The relative amounts of CO_2 and CO depend on how much oxygen is present. Under some conditions carbon in the form of soot

$$CH_4 + 2\,O_2 \rightarrow 2\,H_2O + CO_2 \qquad (4\text{-}8a)$$

$$2\,CH_4 + 3\,O_2 \rightarrow 4\,H_2O + 2\,CO \qquad (4\text{-}8b)$$

is produced, as in a candle flame, for example.

The combustion reaction is a chain reaction, like chlorination. Without going into details, we note that it has to be initiated, usually by a spark, and that its initiation is quite sensitive to the relative amounts of alkane and air. The explosive limits for methane and air mixtures, for example, are 5.53% methane, below which nothing happens, and 14% methane, above which the mixture merely burns quietly. The charge of fuel vapor and air drawn into the cylinder of an internal combustion engine must be within the explosive limits if the engine is to fire properly—as anyone will recall who has ever tried to start a "flooded" engine.[5]

The rate of the combustion reaction in an engine can vary all the way from a comparatively gentle explosion to a detonation, depending on the pressure and on the fuel. The ping of a knocking gasoline engine is a fuel detonation caused either by too high a compression or the wrong kind of fuel. Generally speaking, branched alkanes have less tendency to knock than normal alkanes. Substances like tetraethyllead in gasoline also reduce knocking by interrupting some of the reaction chains.

$$CH_3CH_2-\underset{\underset{\displaystyle CH_2CH_3}{|}}{\overset{\overset{\displaystyle CH_2CH_3}{|}}{Pb}}-CH_2CH_3$$
tetraethyllead

However, lead compounds are toxic and the lead from automobile

---

[5] A flooded engine is one in which the ratio of fuel to air is too high.

exhausts probably constitutes a health hazard. The ancient Romans also suffered from lead poisoning, but in their case it came from the use of lead water pipes. For the same reason, it is dangerous to use a pewter pitcher to store drinking water.

The highly branched alkane 2,2,4-trimethylpentane, commonly known as "isooctane" and the unbranched hydrocarbon heptane are used to establish a scale of antiknock quality for gasolines. Isooctane is rated as 100, and heptane is rated as 0. A fuel with the same behavior as a mixture of 90% isooctane and 10% heptane has an octane rating of 90, and so on.

Other reactions of alkanes are shown in Table 4-5.

TABLE 4-5. *Examples of Other Alkane Reactions*

Reaction	Equation
Nitration by nitric acid	$CH_4 + HO-NO_2 \xrightarrow{400°C} CH_3NO_2 + H_2O$ (nitromethane)
Isomerization	$CH_3CH_2CH_2CH_2CH_3 \xrightarrow[100°C]{AlCl_3} CH_3CHCH_2CH_3$ with CH_3 branch
Dehydrogenation	$CH_3CH_3 \xrightarrow[\text{high temperatures}]{Pt} CH_2=CH_2 + H_2$
Cracking	$CH_3CH_2CH_3 \xrightarrow[\text{high temperature}]{catalyst} CH_4 + CH_2=CH_2$

SUMMARY

1. Tetrahedral orbitals: The bonding orbitals of carbon in alkanes are tetrahedral or sp^3.
2. Alkanes, C_nH_{2n+2}
   (a) As the size of the alkane molecule increases so does the boiling point.
   (b) Alkanes can be obtained by distillation of petroleum or by synthesis from Grignard reagents.
   (c) Alkanes can be chlorinated by exposure to chlorine and light.
   (d) Combustion of alkanes: Smooth reaction in a gasoline engine is favored by branches in the alkane skeleton.
   (e) Nomenclature: The large number of possible isomers makes it necessary to use a systematic nomenclature. The names are based on the longest chain, use the lowest possible numbers to locate substituents, and mention the substituents in order of increasing size.
3. Cycloalkanes, C_nH_{2n}
   (a) Reactions, except for those of the small rings (Chapter 5), are like those of the alkanes.
   (b) Cyclohexane:
      (i) The chair conformation is preferred.
      (ii) The hydrogens can be divided into two groups, axial and equatorial.
      (iii) Conformational changes interchange the positions of axial and equatorial hydrogens.
   (c) Adamantane: The basis for the carbon skeleton of diamond.

EXERCISES

1. Give the names of the first ten unbranched alkanes.

2. List three compounds, by structural formula, *in the order of increasing boiling point.*

3. Which of the following structural formulas represent the same compounds?

(a)
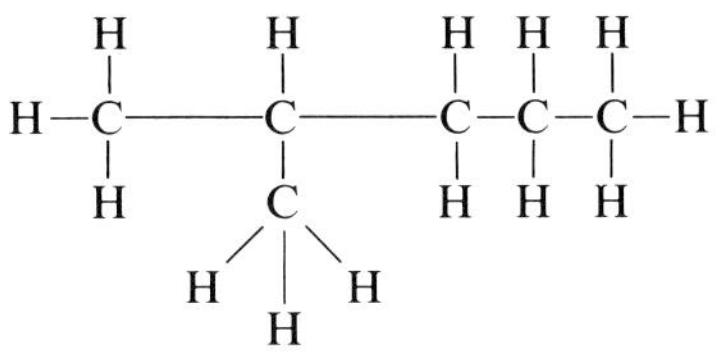

(b)
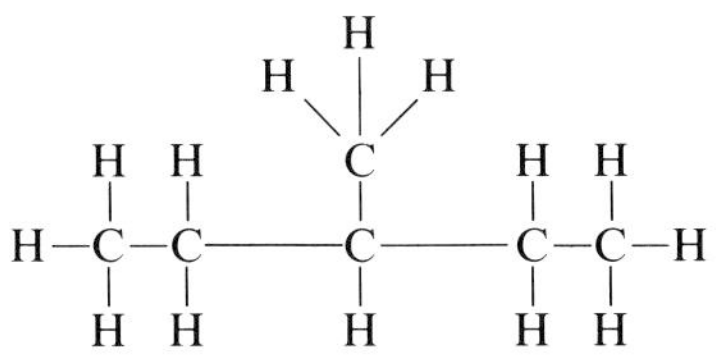

(c)
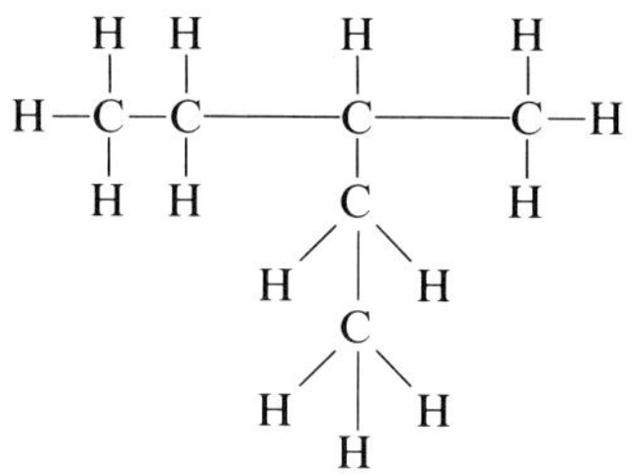

(d)
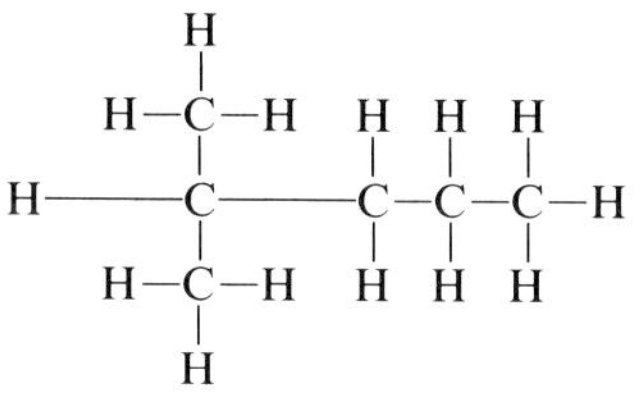

(e)
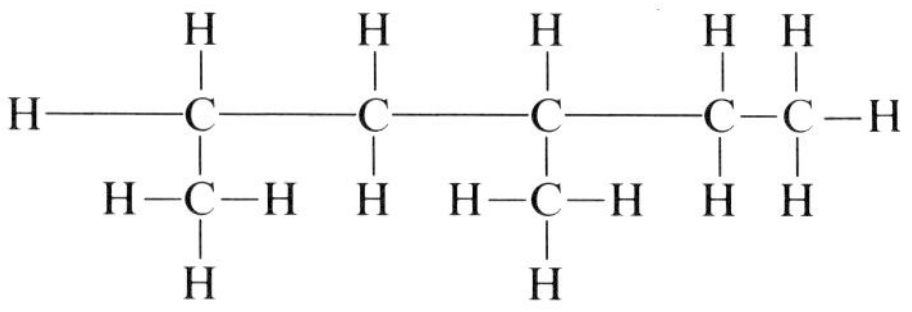

(f)
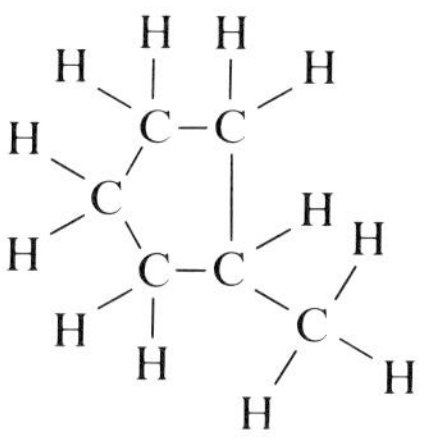

(g)
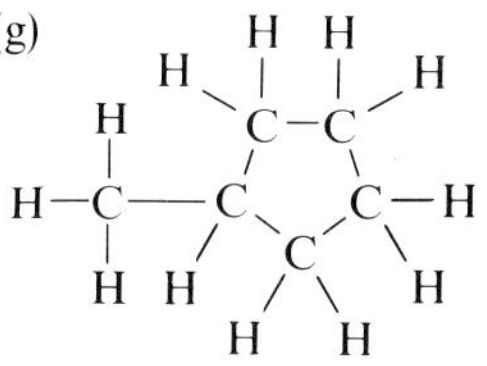

4. Write structural formulas for the following:
(a) 2,2-dimethylheptane
(b) 2,3-dimethyloctane
(c) 2,9-dimethyldecane

(d) 2,3,4,5,6-pentamethylheptane
(e) 1,1,2,2,3,3,4,4,5,5,6,6-dodecamethylcyclohexane
(f) 1,2-dimethylcyclononane
(g) 3,4-diethylhexane

5. What is incorrect about the name 2-ethylpentane?

6. Draw structural formulas for all of the isomeric C_8H_{18} compounds whose names end in -pentane. There are six of them.

7. Draw a structure for 4-isopropyl-5-*t*-butyloctane.

8. Draw structural formulas for all of the isomeric C_7H_{16} compounds whose names end in -butane.

9. Describe two methods for separating a mixture of alkanes.

10. Write an equation for the preparation of methane from a magnesium-containing compound.

11. Write an equation for the preparation of ethane from a magnesium-containing compound.

12. Draw a picture of the chair conformation of cyclohexane.

13. Would *t*-butylcyclohexane prefer to be in the conformation with the *t*-butyl group in an equatorial position or in an axial position? Make an intelligent guess, but explain your choice.

14. Draw a structural formula for adamantane.

15. If adamantane reacts with chlorine to replace just one hydrogen atom in each molecule, how many isomeric monochloroadamantanes will be formed?

16. If pentane reacts with chlorine to form monochloropentane, how many isomers will be formed?

17. Give a use for one of the products of the reaction of methane with chlorine.

18. Give three examples of the occurrence of alkanes in nature.

19. Is there a smallest possible diamond? If so, how big is it?

20. Explain how infrared light can be used to detect the presence of C—H bonds in a molecule. (See Chapter 3.)

21. A compound C_5H_{12} reacts with chlorine in the light

$$C_5H_{12} + Cl_2 \xrightarrow{light} C_5H_{11}Cl + HCl$$

Although the reaction mixture contains several products, *only one of them* has the molecular formula $C_5H_{11}Cl$. What is the original C_5H_{12} compound and what is the structure of the single $C_5H_{11}Cl$ product?

Chapter 5
Small Rings, Alkenes, and Alkynes

Unsaturated compounds are compounds which can easily react with hydrogen to form new C—H bonds. The unsaturated hydrocarbons discussed in this chapter have less hydrogen (per carbon atom) than the alkanes, or saturated hydrocarbons. Besides being readily convertible to alkanes by addition of hydrogen, unsaturated compounds also give other new reactions.

The Small Rings

Cyclopropane, for example, reacts with hydrogen, under the conditions written on the arrow of equation 5-1, to give propane.

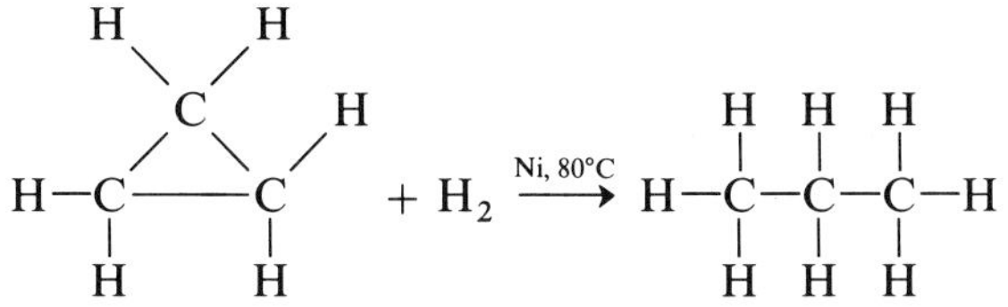

$$\underset{\substack{\text{cyclopropane}\\ \text{bp}\ -33°C}}{\text{C}_3\text{H}_6} + \text{H}_2 \xrightarrow{\text{Ni, 80°C}} \underset{\substack{\text{propane}\\ \text{bp}\ -42°C}}{\text{CH}_3\text{CH}_2\text{CH}_3} \qquad (5\text{-}1)$$

The reaction involves breaking the H—H bond (by the catalyst, metallic nickel) and breaking a C—C bond, then making two new C—H bonds.

It might reasonably be expected that the same catalyst and reaction conditions would break other C—C bonds as well, but this is not generally the case. For example, propane is not cleaved further to give methane and ethane; that is, reaction 5-2 does not take place.

$$CH_3CH_2CH_3 + H_2 \xrightarrow{Ni, 80°C} \!\!\!\!\!\!\!\!\!\!\!\! \times\!\!\times\!\!\times\!\!\times\!\!\!\!\! \to CH_4 + CH_3CH_3 \qquad (5\text{-}2)$$

In fact, we cannot bring about reaction 5-2 even with a much more active catalyst, such as platinum, or with a moderately higher reaction temperature.

The C—C bonds of cyclopropane are therefore somehow different from the C—C bonds of the alkanes and larger cycloalkanes, none of which undergo this cleavage by hydrogenation.

When one attempts to make a model of cyclopropane using tetrahedral bond angles, a probable reason for the weakness of the bonds is seen immediately: **the orbitals are not pointing in the right direction for maximum overlap**. In ethane the carbon sp^3 orbitals overlap along their long axes as shown in Figure 5-1A, whereas in cyclopropane they overlap at a less favorable angle, as shown in Figure 5-1B and the region of overlap is much less.

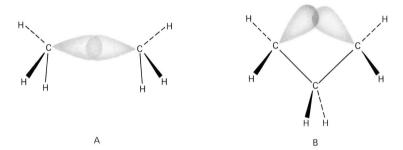

FIGURE 5-1. Overlap of tetrahedral carbon orbitals in (A) ethane and (B) cyclopropane.

Cyclobutane will give the hydrogenation reaction, but only at a higher temperature, as indicated by the conditions for reaction 5-3.

$$\begin{array}{c} CH_2-CH_2 \\ |\qquad\ \ | \\ CH_2-CH_2 \end{array} + H_2 \xrightarrow{Ni, 200°C} CH_3CH_2CH_2CH_3 \qquad (5\text{-}3)$$

cyclobutane   butane
bp 13°C   bp 0°C

Apparently the C—C bonds of cyclobutane are weaker than those of an ordinary alkane (such as butane), but stronger than those of cyclopropane. Again, a model of cyclobutane (Figure 5-2) shows that the orbitals are not quite pointing in the best direction for maximum overlap, although the discrepancy is not so great as in the case of cyclopropane. Hence, cyclobutane can be hydrogenated, but it is more stable, less reactive, and requires more drastic reaction conditions than does cyclopropane.

Cyclopentane and cyclohexane and larger rings are still more stable, and the chemistry of these compounds is essentially the same as that of the alkanes. The reactive bonds of cyclopropane and cyclobutane are said to be **strained**, by analogy with a bolt or rivet that breaks easily after it has been distorted.

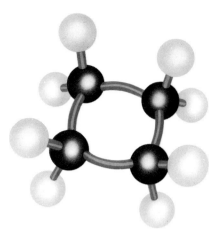

FIGURE 5-2. A model of cyclobutane. Note that flexible springs are used to allow distortion of the bond angles from 109° 28' to 90°.

Cyclopropane can also be cleaved by bromine (Br_2) and by hydrogen iodide (HI), whereas cyclobutane is inert to these reagents.[1]

$$\underset{\substack{\text{cyclopropane}\\ \text{bp } -33°C}}{\overset{CH_2}{\underset{CH_2 - CH_2}{\diagdown\diagup}}} + \underset{\substack{\text{bromine}\\ \text{dark red liquid;}\\ \text{bp } 59°C}}{Br_2} \xrightarrow[\text{in the dark}]{\text{room temperature}} \underset{\substack{\text{1,3-dibromopropane}\\ \text{colorless liquid;}\\ \text{bp } 167°C}}{\underset{Br \qquad \quad Br}{CH_2-CH_2-CH_2}} \qquad (5\text{-}4)$$

$$\underset{CH_2 - CH_2}{\overset{CH_2}{\diagdown\diagup}} + HI \xrightarrow{\text{room temperature}} \underset{\substack{\text{1-iodopropane or } n\text{-propyl iodide}\\ \text{bp } 102.4°C}}{CH_3-CH_2-CH_2-I} \qquad (5\text{-}5)$$

Cyclopropane has been used as a general inhalation anesthetic since 1929. Like methane, it is flammable, and mixtures with air or oxygen may explode if exposed to a flame or spark. This is a hazard in the operating room, although certain precautions are usually taken.

Three-membered rings are sometimes found in naturally occurring compounds and compounds used in medicines.

A
thujone
colorless liquid present in the essential oils of many plants; LD_{100} i.p. in rats, 240 mg/kg; human toxicity, ingestion causes convulsions

B

C
6-β-hydroxy-3,5-cyclopregnane-20-one, Neurosterone, or cyclopregnol
colorless solid; mp 181°C; used as psychotropic agent

---

[1] Note that cyclobutane, like the alkanes in general, will react with bromine *in the light* to give substitution of Br for H. The ring is not cleaved, however, and no reaction occurs in the dark.

(B) and (C) are abbreviated structural formulas in which each line-end or corner stands for a carbon atom. Each carbon atom is assumed to have enough attached hydrogen atoms to bring its valence up to four.

The information given here about thujone and cyclopregnol was taken from the *Merck Index* (8th ed.). The expression "LD_{100} i.p." refers to the fact that a dosage of 240 mg/kg of body weight given to rats intraperitoneally, caused the death of 100% of the rats.

The naming of pharmaceutical compounds is in a rather unfortunate state, most drugs having at least three different names: These are:

1. The proprietary, or trademark name, for example Neurosterone. In this example the proprietary name suggests to the physician that the drug may have some action on the nervous system, and to the chemist that it is related to a class of compounds known as the steroids. However, many proprietary names convey no information whatsoever, and a different manufacturer would of course have to use a different proprietary name.
2. The "trivial" name need not be capitalized. Trivial names have the advantage of brevity, but there is no systematic way of telling what the structure is. Cyclopregnol and adamantane are examples of trivial names.
3. The systematic name, as used in *Chemical Abstracts*, is the only one that might enable a chemist to write the structure of the compound. Unfortunately, the systematic name for a complicated molecule is also likely to be complicated, although we try to keep the system as simple as possible.

Alkenes

Alkenes differ from alkanes and cycloalkanes by the presence of **double bonds** between carbon atoms. The simplest member of the series of alkenes is the gas ethylene and the next member is propylene.

$$\begin{array}{cc} H \\ \diagdown \diagup H \\ C=C \\ \diagup \diagdown \\ H H \end{array} \begin{array}{cc} H \\ \diagdown \diagup H \\ C=C \\ \diagup \diagdown \\ CH_3 H \end{array}$$

ethylene propylene

colorless gas; bp $-48°C$
bp $-103.9°C$

Trigonal Carbon Atoms and the π Bond

A trigonal carbon atom is one, like the carbons of the double bonds in ethylene or propylene, that is directly connected only to three other atoms. Another example is the central atom of a carbonium ion (Chapter 1). A common feature of trigonal carbon atoms is a bond angle of about 120° and a planar rather than tetrahedral arrangement of the bonds, as shown in Figure 5-3.

The bonds to trigonal carbon atoms are formed by a set of hybridized atomic orbitals. There are three of them, all in the same plane, and at angles of 120°, as shown in Figure 5-4A. Called sp^2 orbitals, they are constructed by mixing the original carbon $2s$ orbital and the $2p_x$ and $2p_y$ orbitals, leaving out the $2p_z$ orbital. The left-over $2p_z$ orbital is perpendicular to the plane of the other three orbitals, as shown in Figure 5-4B.

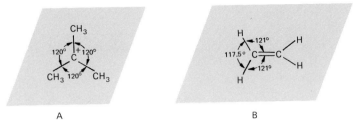

FIGURE 5-3. The geometry of trigonal carbon atoms. (A) Trimethylcarbonium ion. (B) Ethylene.

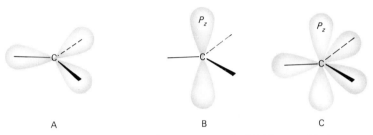

FIGURE 5-4. (A) The three sp^2 carbon orbitals. (B) The p_z orbital, perpendicular to the other three. (C) The three sp^2 orbitals and the p_z orbital.

If we ignore the p_z orbitals temporarily, ethylene is constructed from two carbon atoms and four hydrogens, using the hydrogen 1s orbitals and the trigonal carbon sp^2 orbitals. A C—C sigma (σ) bond is formed by the overlap of two of the sp^2 orbitals, and the C—H bonds are formed by the overlap of the hydrogen 1s and the remaining carbon sp^2 orbitals. This is shown in Figure 5-5. So far we have used all but two of the valence electrons and all but two of the carbon orbitals, one p_z orbital on each carbon. Figure 5-6 shows the bonds made so far and the p_z orbitals.

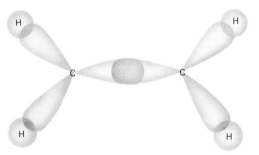

FIGURE 5-5. The sigma (σ) bonding orbitals of the ethylene molecule formed by overlap of sp^2 carbon atomic orbitals.

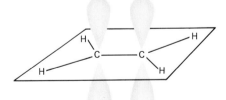

FIGURE 5-6. The carbon p_z orbitals are perpendicular to the plane containing the σ bonds.

Small Rings, Alkenes, and Alkynes

Next, we notice that although the p_z orbitals cannot overlap *lengthwise*, they can overlap *sideways* provided the C—C distance is shortened a little. Figure 5-7 shows this overlap which forms a π bond[2] from the two p_z orbitals. The π bond contains the remaining two electrons of ethylene.

FIGURE 5-7. The sidewise overlap of two p_z orbitals on adjacent carbon atoms forms a π bonding orbital.

CONSEQUENCES OF THE π BOND

In order for the two p_z orbitals to overlap sidewise, as in Figure 5-7, they have to be pointing in parallel directions. Since they are perpendicular to the σ bonds, the latter are all forced to lie in one plane, as shown. Unlike the single bond in ethane, the double bond in ethylene resists twisting and there is no free rotation of the CH$_2$ groups (Figure 5-8). A second consequence of the π bond is that the two carbon atoms have to be closer together. The C—C distance is 1.54×10^{-8} cm in ethane but only 1.34×10^{-8} cm in ethylene.

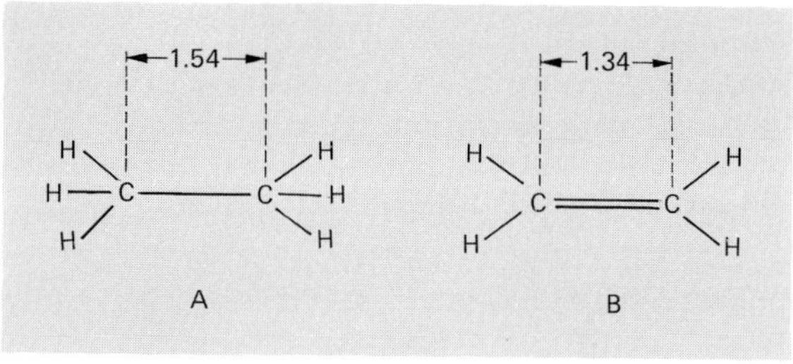

FIGURE 5-8. **(A)** Ethane, longer bond and freely rotating. **(B)** Ethylene, shorter bond and planar.

Ethylene is synthesized in the leaves of many species of tropical trees (for example, *Euonymus japonica*). Exposure of the tree to an ethylene atmosphere increases the rate at which the leaves die and fall off. The ethylene naturally present in the leaf may be part of the mechanism by which old leaves are removed to make way for new ones. Ethylene is also used commercially to bring out the ripe color of citrus fruits that have been picked green.

Tetrachloroethylene is a widely used drycleaning fluid. It is not flammable.

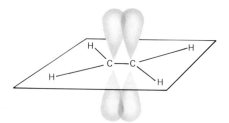

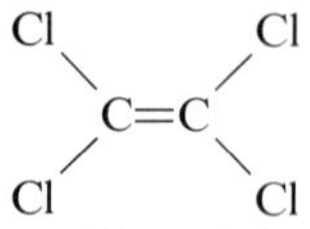

tetrachloroethylene
bp 121.02°C

[2] The letter π in the Greek alphabet corresponds to *p* in our alphabet and σ (sigma) corresponds to *s*.

The Butenes and Cis-trans Isomerism

DOUBLE BOND POSITION ISOMERISM

As we go up the series of alk**ene**s, ethyl**ene**, propyl**ene** (or prop**ene**), but**ene**, pent**ene**, hex**ene**, and so on, we note that ethylene and propylene have no isomers. In the case of propylene, the double bond might be between atoms 1 and 2 or 2 and 3, but these positions are really equivalent (Figure 5-9). Thus "2-propene" is just 1-propene

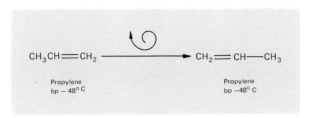

FIGURE 5-9. Propylene does not have double bond position isomers.

turned end-for-end; it is not necessary to break any bonds. At butene and the higher alkenes it is necessary to indicate the position of the double bond, because 1-butene and 2-butene are different compounds with different boiling points. There is no such com-

$$CH_2=CHCH_2CH_3 \qquad CH_3CH=CHCH_3$$
1-butene   2-butene
bp $-6°C$ bp $>0°C$

pound as 3-butene, since this would be just 1-butene turned end-for-end. There are also only two pentenes that differ only in the position of the double bond.

$$CH_2=CHCH_2CH_2CH_3 \qquad CH_3CH=CHCH_2CH_3 \qquad CH_3CH_2CH=CHCH_3$$
1-pentene   2-pentene   2-pentene

There are three hexene isomers differing only in the position of the double bond.

$$CH_2=CHCH_2CH_2CH_2CH_3 \qquad CH_3CH=CHCH_2CH_2CH_3$$
1-hexene   2-hexene

$$CH_3CH_2CH=CHCH_2CH_3$$
3-hexene

CIS-TRANS ISOMERISM

When the structure of 2-butene is drawn in detail, we find that there are two such compounds. The name 2-butene is not good enough (unless one is talking about a mixture), and it is necessary to specify whether the compound is *cis*-2-butene or *trans*-2-butene (Figure 5-10).

The prefix **cis** comes from the Latin for "on the same side" and **trans** means across, that is, on opposite sides of the double bond. Notice carefully what is meant by "side." For example, the compound 2-methyl-1-propene has two methyl groups on the same *end* rather than on the same *side* of the double bond (Figure 5-11). This compound does not have cis and trans isomers.

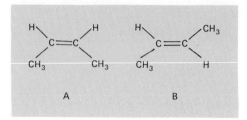

FIGURE 5-10. Cis-trans isomers. **(A)** *cis*-2-Butene, bp +4°C. **(B)** *trans*-2-Butene, bp +1°C.

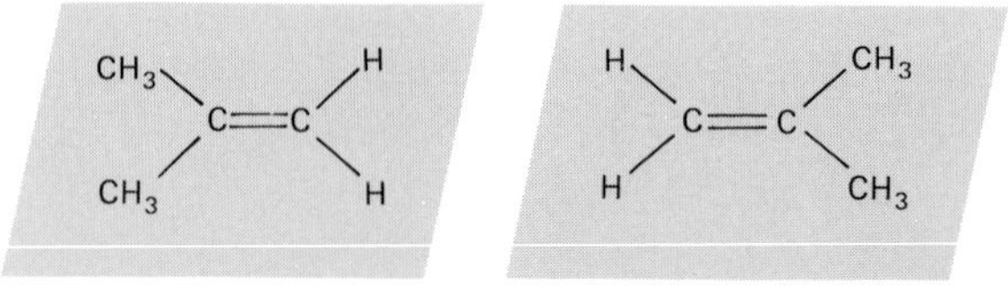

FIGURE 5-11. Two molecules of 2-methyl-1-propene, or **isobutylene** (trivial name).

If we return our attention to the pentenes and examine Figure 5-12, it can be seen that 1-pentene does not have cis-trans isomers, but "2-pentene" has to be either *cis*-2-pentene or *trans*-2-pentene.

Cis-trans isomerism can also occur in substituted cycloalkanes (Figure 5-13).

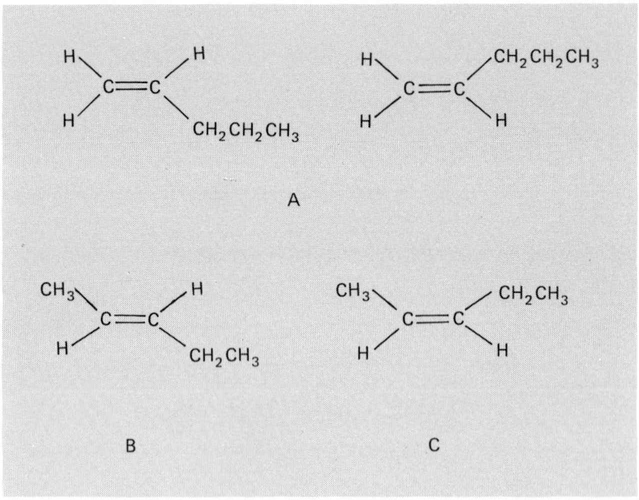

FIGURE 5-12. Cis-trans isomerism in the pentenes. **(A)** 1-Pentene, bp 30°C. **(B)** *trans*-2-Pentene, bp 36.4°C. **(C)** *cis*-2-Pentene, by 37.1°C.

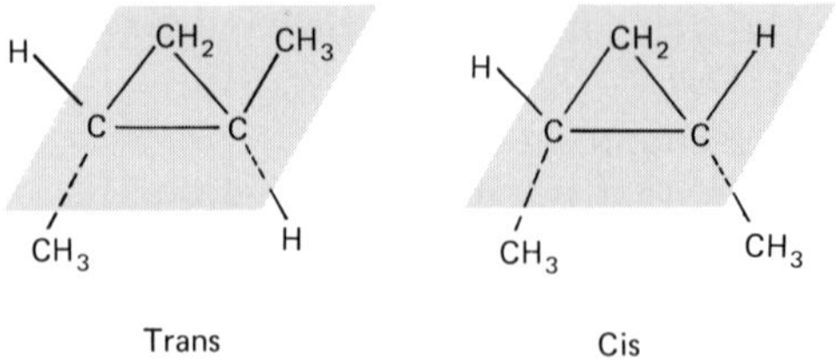

FIGURE 5-13. *Cis*- and *trans*-1,2-dimethylcyclopropane.

The Role of the π Electrons in Reactions

The most important reactions of alkenes involve conversion of the double bond to a single bond; these reactions are given by any alkene. For example, ethylene, propylene, *cis*-2-butene, and almost *any* alkene will add HCl. This is essentially a reaction of the π electrons of the C=C functional group, and the atoms or groups attached to the four corners play only a minor role in the reaction.

$$\diagup\!\!\!\!\mathrm{C}\!=\!\mathrm{C}\diagdown + \mathrm{HCl} \rightarrow \diagup\!\!\!\!\mathrm{C}\!-\!\mathrm{C}\diagdown\!\!\!\overset{\mathrm{H}}{\underset{}{}} \quad (5\text{-}6)$$
$$\mathrm{Cl}$$

The reason for the considerable reactivity of the C=C double bond, is that the electron pair of the π orbital is easily accessible to an attacking reagent and is not very firmly bound to the atomic nuclei. In fact, the reactions of ethylene are in some respects like the reactions of other electron-rich compounds, such as ammonia, for example.

$$\mathrm{H\!-\!\underset{H}{\overset{H}{N}}\!:\;\;H^+Cl^-} \rightarrow \mathrm{H\!-\!\underset{H}{\overset{H}{\overset{\pm}{N}}}\!-\!H\;Cl^-} \quad (5\text{-}7)$$

$$\underset{HH}{\overset{HH}{\diagdown\mathrm{C}\diagup}}\!\!\!\underset{}{\overset{}{\|}}\;H^+Cl^- \rightarrow \underset{HH}{\overset{HH}{\diagdown\mathrm{C}\diagup}}\!\!\overset{+}{}\;\;Cl^- \quad (5\text{-}8)$$

In general, π electrons (and also unshared pairs of electrons in the outer shell) are subject to attack by electron-poor reagents known as **electrophiles** (Greek: electron-lovers). In the same terminology, electron-rich reagents such as ethylene or ammonia are known as **nucleophiles** (Greek: nucleus-lovers). A large proportion of organic reactions are reactions in which an electrophile, with a vacant orbital capable of accepting electrons, forms a bond with a nucleophile, using an electron pair supplied by the latter. Equations 5-7 and 5-8 are typical examples, the proton (H^+) being the electrophile in both cases.

The charged carbon atom (carbonium ion) formed in reaction 5-8 still has the planar, sp^2 geometry of the original double bond, whereas the other carbon atom has become an ordinary tetrahedrally bonded CH_3 group (Figure 5-14).

The carbonium ion is a powerful electrophile and, in the presence of the nucleophilic Cl^-, it has only a very brief existence. The next step in the reaction, also shown in Figure 5-14, converts the carbonium ion–chloride ion pair into a covalently bonded molecule, ethyl chloride, in which all of the bonds are tetrahedrally oriented. The overall reaction is represented by equation 5-9.

$$CH_2\!=\!CH_2 + HCl \rightarrow CH_3CH_2Cl \quad (5\text{-}9)$$

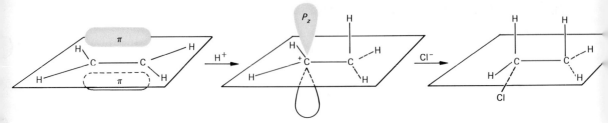

FIGURE 5-14.

Another example of a reaction of alkenes that can be explained best in terms of electrophilic and nucleophilic reagents is the addition of bromine. In its reactions the bromine molecule (Br_2) acts as though it were Br^+Br^- because its covalent electron-pair bond is easily polarized.

$$:\!\ddot{Br}\!:\!\ddot{Br}\!: \quad \text{acts like} \quad :\!\ddot{Br}^+\!:\!\ddot{Br}\!:^-$$

The nucleophile in the addition of bromine is Br^-, and the electrophile is Br^+.

$$(5\text{-}10)$$

A few reactions of the double bond involve atoms or radicals as the reagent, instead of electrophiles and nucleophiles. An atom or a **free radical** (R·) has one unpaired electron and needs only one more to make an electron-pair bond. Examples to be discussed later are the addition of hydrogen and vinyl polymerization. In these

$$(5\text{-}11)$$

$$(5\text{-}12)$$

reactions the intermediate is an sp^2 hybridized free radical rather than an sp^2 hybridized carbonium ion.

Reactions of Alkenes

ADDITION OF HALOGENS

The halogens chlorine and bromine add to alkenes in the way discussed in the previous section. In the case of bromine in particular, the disappearance of the typical color of the bromine as it is used up by the reaction is a convenient test for the presence of a double bond in a molecule. Thus

$$\underset{\substack{\text{ethylene}\\ \text{colorless gas;}\\ \text{bp}-120}}{\overset{H}{\underset{H}{>}}C=C\overset{H}{\underset{H}{<}}} + \underset{\substack{\text{bromine}\\ \text{dark red}\\ \text{liquid}}}{Br_2} \rightarrow \underset{\substack{\text{1,2-dibromoethane or}\\ \text{ethylene bromide}\\ \text{colorless liquid;}\\ \text{bp 131.6}}}{\overset{H}{\underset{H}{>}}\underset{Br}{\overset{|}{C}}-\underset{Br}{\overset{|}{C}}\overset{H}{\underset{H}{<}}} \quad (5\text{-}13)$$

$$\underset{\substack{\text{propylene}\\ \text{colorless gas}}}{\overset{CH_3}{\underset{H}{>}}C=C\overset{H}{\underset{H}{<}}} + \underset{\substack{\text{bromine}\\ \text{dark red}\\ \text{liquid}}}{Br_2} \rightarrow \underset{\substack{\text{1,2-dibromopropane}\\ \text{colorless liquid;}\\ \text{bp 146.6}}}{\overset{CH_3}{\underset{H}{>}}\underset{Br}{\overset{|}{C}}-\underset{Br}{\overset{|}{C}}\overset{H}{\underset{H}{<}}} \quad (5\text{-}14)$$

$$\underset{\substack{\text{cis-2-butene}\\ \text{colorless liquid}}}{\overset{CH_3}{\underset{H}{>}}C=C\overset{CH_3}{\underset{H}{<}}} + \underset{\substack{\text{bromine}\\ \text{dark red}\\ \text{liquid}}}{Br_2} \rightarrow \underset{\substack{\text{2,3-dibromobutane}\\ \text{colorless oil}}}{\overset{CH_3}{\underset{H}{>}}\underset{Br}{\overset{|}{C}}-\underset{Br}{\overset{|}{C}}\overset{CH_3}{\underset{H}{<}}} \quad (5\text{-}15)$$

And in general

$$\underset{\substack{\text{any alkene}\\ \text{usually colorless}}}{\overset{R}{\underset{R}{>}}C=C\overset{R}{\underset{R}{<}}} + \underset{\substack{\text{bromine}\\ \text{dark red}\\ \text{liquid}}}{Br_2} \rightarrow \underset{\substack{\text{dibromo derivative}\\ \text{of the alkene}}}{\overset{R}{\underset{R}{>}}\underset{Br}{\overset{|}{C}}-\underset{Br}{\overset{|}{C}}\overset{R}{\underset{R}{<}}} \quad (5\text{-}16)$$

Chlorination of alkenes is just like the addition of bromine. Ethylene chloride, or 1,2-dichloroethane, is a colorless liquid, heavier than water, and boiling at 83°C. Unlike completely chlorinated alkanes, such as CCl_4, ethylene chloride will burn, although it burns with a smoky flame. It is used industrially as a solvent for oils, resins, and rubber and in the manufacture of cellulose acetate (Chapter 13). It is toxic, and deaths have occurred due to liver or kidney damage after large doses.

HYDROGENATION

Alkenes absorb hydrogen gas in the presence of a metal catalyst to give the corresponding alkane.

$$CH_3-CH=CH_2 + H_2 \xrightarrow{\text{Pt or Ni}} CH_3-CH_2-CH_3 \quad (5\text{-}17)$$

Note that isomeric alkenes, either cis-trans isomers or double bond position isomers, can give the same alkane.

$$\underset{H}{\overset{CH_3}{\diagup}}C=C\underset{CH_3}{\overset{H}{\diagdown}} \quad \underset{H}{\overset{CH_3}{\diagup}}C=C\underset{H}{\overset{CH_3}{\diagdown}} \quad CH_2=CH-CH_2-CH_3$$

$$\xrightarrow[\text{Ni}]{H_2} \quad \downarrow \overset{H_2}{\underset{Ni}{}} \quad \xleftarrow[\text{Ni}]{H_2}$$

$$CH_3CH_2CH_2CH_3 \tag{5-18}$$

Hydrogenation is used often commercially and in the laboratory to get rid of unwanted double bonds. For example, the unsaturated fats (Chapter 10), which tend to be oils, can be converted to higher melting substances of butter or waxlike consistency by hydrogenation.

Hydrogenation is believed to involve hydrogen atoms adsorbed on the surface of the metal catalyst. These add simultaneously to the side of the alkene facing the catalyst.

$$R-\underset{\underset{\cdot H \; H \cdot}{|}}{\overset{\overset{R}{|}}{C}}\overset{R}{\underset{|}{C}}-R \rightarrow R-\underset{\underset{H}{|}}{\overset{\overset{R}{|}}{C}}-\underset{\underset{H}{|}}{\overset{\overset{R}{|}}{C}}-R \tag{5-19}$$

(on catalyst)

ADDITION OF HYDROGEN HALIDES

Alkenes will add dry[3] HBr, HCl, or HI rapidly and without a catalyst to give haloalkanes. 2-Butene, for example, with HBr gives 2-bromobutane (equation 5-20). The reactions of other alkenes with

$$CH_3CH=CHCH_3 \xrightarrow{HBr} CH_3CH_2\underset{\underset{Br}{|}}{C}HCH_3 \tag{5-20}$$

2-butene
(cis or trans)

2-bromobutane
bp 91.3°C

the halogen acids are similar, except for one complication. If the two ends of the double bond are unlike, the HX (where X represents a halogen) can add in either of two ways. For example, 1-butene and HCl gives both 1-chlorobutane and 2-chlorobutane (equations 5-21).

$$\begin{array}{c} H-Cl \\ + \\ CH_2=CHCH_2CH_3 \\ + \\ Cl-H \end{array} \diagup\!\!\!\diagdown \begin{array}{c} CH_3\underset{\underset{}{|}}{\overset{Cl}{C}}HCH_2CH_3 \\ \text{major product} \\ \\ CH_2CH_2CH_2CH_3 \\ \underset{Cl}{|} \\ \text{minor product} \end{array} \tag{5-21}$$

Markovnikov's rule, for predicting the major product, is that **the carbon already having the most hydrogen gets the hydrogen.**

---

[3] Dilute acid tends to catalyze the addition of the water which serves as solvent instead of the acid itself.

The explanation for Markovnikov's rule is simply that a carbonium ion with alkyl substituents directly attached to it is more easily formed and more stable than one without.

$$\underset{\text{more stable}}{\overset{+}{\underset{H}{C}}H_2-\overset{+}{C}HCH_2CH_3} \quad \underset{\text{less stable}}{CH_2-\overset{H}{\underset{|}{C}}HCH_2CH_3}$$

The proton is attached first, forming the more stable of the two possible carbonium ions, and the halide ion is then attached to the positively charged carbon as in Figure 5-14.

ADDITION OF WATER

Water, in dilute sulfuric acid, adds to the double bonds of alkenes. A proton from the sulfuric acid acts as the nucleophile and a water molecule acts as the electrophile.

$$CH_2=CH_2 + H_2O \xrightarrow[\substack{\text{dilute } H_2SO_4 \\ \text{room temperature}}]{H^+} CH_3CH_2OH \quad (5\text{-}22)$$

The sulfuric acid, which has to be fairly dilute, merely provides protons which act as a catalyst; that is, a proton is borrowed from the sulfuric acid early in the reaction, but is replaced later on.

$$CH_2=CH_2 + H^+ \rightleftharpoons [CH_3CH_2^+] \overset{H_2O:}{\rightleftharpoons} \left[CH_3CH_2-\overset{+}{O}\diagdown^{H}_{H}\right]$$

$$\left[CH_3CH_2\overset{+}{-}O\diagdown^{H}_{H}\right] \rightarrow CH_3CH_2OH + H^+ \quad (5\text{-}23)$$

Note that no protons are used up, since the last step of reaction 5-23 produces a proton for every one added to ethylene in the first step.

TABLE 5-1. *Other Reactions of Alkenes*

Reaction	Equation
Addition of sulfuric acid	$CH_2=CH_2 + H^{+\,-}OSO_2OH \rightarrow CH_3-CH_2-OSO_2OH$ ethyl hydrogen sulfate
Addition of hypochlorous acid (as $Cl^{+\,-}OH$)	$CH_2=CH_2 + ClOH \rightarrow \underset{\underset{Cl}{\|}}{CH_2}-\overset{\overset{OH}{\|}}{CH_2}$ ethylene chlorohydrin
Alkylation	$CH_3-\underset{\underset{CH_3}{\|}}{C}=CH_2 + H-\underset{\underset{CH_3}{\|}}{\overset{\overset{CH_3}{\|}}{C}}-CH_3 \xrightarrow{H_2SO_4} CH_3-\underset{\underset{CH_3}{\|}}{\overset{\overset{H}{\|}}{C}}-CH_2-\underset{\underset{CH_3}{\|}}{\overset{\overset{CH_3}{\|}}{C}}-CH_3$
Oxidation	$CH_3-CH=CH_2 \xrightarrow[H_2O]{KMnO_4} CH_3-\overset{\overset{OH}{\|}}{CH}-\overset{\overset{OH}{\|}}{CH_2}$

Preparation of Alkenes

Alkenes are usually prepared by what we like to call **elimination** reactions, because some small molecule such as Cl_2, HCl, or H_2O is split off or eliminated from the larger molecule. In most cases an elimination reaction can be considered to be the reverse of a corresponding addition reaction.

DEHYDRATION

The elimination of water from any of a class of compounds called alcohols (Chapter 7) is one of the most useful ways of making alkenes. Ethylene, for example, can be made by the dehydration of ethyl alcohol (equation 5-24).

$$\underset{\text{ethyl alcohol}}{\underset{H \quad OH}{\underset{|\quad\;\;\;|}{CH_2-CH_2}}} \xrightarrow[170°C]{95\%\ H_2SO_4} CH_2=CH_2 + H_2O \qquad (5\text{-}24)$$

This reaction is the reverse of the hydration of alkenes (equation 5-22) that takes place with dilute sulfuric acid at room temperature. In both reaction 5-22 and 5-24 protons from the sulfuric acid are acting as catalysts. In general, whether a reaction is a simple one catalyzed by an acid or a complicated biological one catalyzed by an enzyme, the catalyst must always catalyze the reaction in *both* directions. The direction in which the reaction goes is controlled by changing the other reaction conditions. To hydrate an alkene, we use dilute acid (because it has a higher concentration of water) and a lower temperature (because higher temperature favors the reverse reaction). To dehydrate an alcohol, we use hot, 95% sulfuric acid.

It may be recalled that when sulfuric acid and water are mixed, the mixture becomes hot. Evidently a reaction of some sort takes place (equation 5-25).

$$\underset{H}{\overset{H}{\diagdown}}\ddot{O}: + H_2SO_4 \rightleftharpoons \underset{H}{\overset{H}{\diagdown}}\overset{+}{\underset{H}{\diagup}}\ddot{O} + HSO_4^- \qquad (5\text{-}25)$$

Alcohols are like water, and so it is not unexpected that ethyl alcohol, CH_3CH_2-O-H, also reacts with strong acids (equation 5-26).

$$CH_3CH_2\ddot{\underset{H}{\overset{\diagup}{O}}} + H^+ \rightleftharpoons CH_3CH_2\overset{+}{\underset{H}{\overset{H}{\underset{\diagdown}{\overset{\diagup}{O}}}}} \qquad (5\text{-}26)$$

ethyl hydronium ion

The reaction that leads to the formation of ethylene is actually a reaction of the ethyl hydronium ion rather than of the alcohol itself.

$$\underset{\underset{H\;\;H}{\diagup\;\;\diagdown}}{\underset{\ddot{O}}{\overset{|}{\underset{H}{\underset{|}{CH_2-CH_2}}}}}\overset{\overset{H}{\diagdown}\overset{+}{\underset{\diagup}{O}}\text{--}H}{} \rightarrow \underset{\underset{H\;\;H}{\diagup\;\;\diagdown}}{\underset{\overset{+}{\underset{|}{\ddot{O}}}}{\underset{|}{\underset{H}{CH_2=CH_2}}}} \quad \underset{}{\overset{H}{\diagdown}\ddot{O}\text{--}H} \qquad (5\text{-}27)$$

From equation 5-27, it can be seen that the $\overset{\overset{H}{\underset{+}{|}}}{-O-H}$ part structure (which used to be the $-\ddot{\underset{\cdot\cdot}{O}}H$ group) departs with the bonding electrons and becomes an ordinary water molecule $\overset{H}{\underset{\diagdown}{}}\ddot{O}-H$. At the same time, another molecule pulls a proton off the other carbon atom, leaving two electrons behind to form the double bond. The proton that was removed and the attacking water molecule form a hydronium ion, H_3O^+. Note that a proton from the acid H_2SO_4 (or $HSO_4^- + H_3O^+$) is just borrowed, not used up.

We can indicate what happens to the electrons during this reaction by means of curved arrows (Figure 5-15). Note that one pair of bonding electrons goes with the departing $H-\overset{+}{O}-H$, neutralizing the charge. The electrons of the C-H bond are used to form the double bond. The pair of unshared electrons on the attacking H_2O molecule is used to form the new H-O bond in H_3O^+.

FIGURE 5-15. What happens to the electrons.

DEHYDROHALOGENATION

The elimination of a hydrogen halide such as HCl, HBr, or HI is called dehydrohalogenation. This reaction is produced by the attack of a strong base on an alkyl halide or haloalkane.

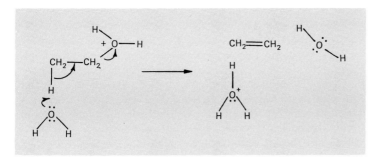

$$\text{CH}_3-\underset{\underset{Cl}{|}}{\overset{\overset{H}{|}}{C}}-\underset{\underset{H}{|}}{\overset{\overset{H}{|}}{C}}-\text{CH}_3 \xrightarrow{K^+ {}^-OH \text{ in alcohol}} K^+Cl^- + CH_3CH=CHCH_3 \quad \text{major product}$$

$$\xrightarrow{K^+ {}^-OH \text{ in alcohol}} K^+Cl^- + CH_2=CHCH_2CH_3 \quad \text{minor product} \quad (5\text{-}28)$$

In reaction 5-28, the base HO^- removes a proton from one carbon atom while the chlorine departs from the adjacent carbon atom as Cl^-. The proton leaves its bonding electrons and the Cl^- takes its bonding electrons with it, as shown in Figure 5-16.

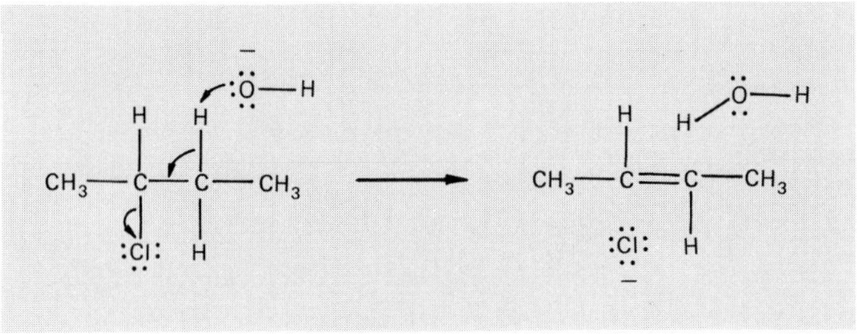

FIGURE 5-16.

In the addition of HCl to alkenes, we saw that the hydrogen goes on the carbon already having the greater number of hydrogens. The rule for the elimination of HX from an alkyl halide is the opposite of Markovnikov's rule: **take the hydrogen from the carbon having the fewest hydrogens.**

Double Bonds in Larger Molecules

The chemistry of living organisms involves a great many compounds in which a double bond is just one feature of a larger molecule. We will give just two examples at this point.

IPS CONFUSUS, THE BARK BEETLE[4]

Bark beetles are of considerable concern to ecologists because they destroy about six times as much timber each year as is lost to forest fires. During his initial attack on the bark, the male *Ips confusus* leaves a substance that attracts females and other males to the tree and eventually they multiply and kill the tree. The same substance could presumably be used to attract the beetles to a trap instead, thus saving the trees. The identification of the attracting substance, so that it can be produced for this purpose, is a typical example of the sort of task performed by organic or biological chemists.

First, several kilograms of the mixture of wood and feces left on the tree by the unmated male beetles was collected, and the active attractants were extracted by a solvent. The attractants turned out to be a mixture of three different compounds; however we will discuss only one of them that was separated from the mixture by chromatography.

The first step in determining the structure of the molecule was simply to measure its molecular weight (the sum of the atomic weights of all the atoms), which turned out to be 154. Next, an infrared spectrometer was used. The infrared light absorbed by the molecule at a wavelength of 2960 nm[5] showed

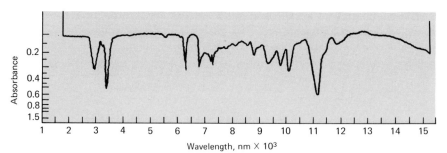

FIGURE 5-17. The infrared spectrum of an attractant substance. [After R. M. Silverstein, *J. Chem. Educ.*, **45**:12 (1968)]

that the molecule contained an O—H bond. Similarly, absorption bands in other parts of the infrared spectrum showed that the molecule contained a part structure

$$\diagdown_{\diagup}C=CH_2$$

and probably also a part structure

$$CH_3-\underset{H}{\overset{|}{C}}-CH_3$$

[4] See R. M. Silverstein, *Journal of Chemical Education*, **45**, 794 (1968).
[5] 3400 cm^{-1}.

Recourse to still another instrument, which measured the absorption of ultraviolet light in the neighborhood of 226 nm, indicated that the molecule contained the part structure

$$C=C-C=C$$

A nuclear magnetic resonance spectrum (see Chapter 3) permitted the above part structure to be identified as

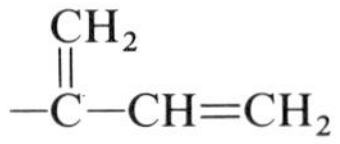

$$\begin{array}{c} CH_2 \\ \| \\ -C-CH=CH_2 \end{array}$$

At this stage the information could be summarized as follows

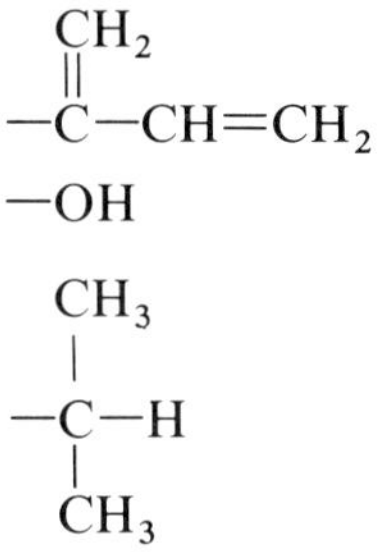

$$\begin{array}{c} CH_2 \\ \| \\ -C-CH=CH_2 \end{array}$$
—OH
$$\begin{array}{c} CH_3 \\ | \\ -C-H \\ | \\ CH_3 \end{array}$$

plus enough other atoms to bring the molecular weight up to 154.

Conversion of the molecule by chemical reaction to a **derivative** in which the OH group was replaced by an acetate group, and then an examination of the mass spectrum of the derivative, made it possible to deduce that the molecular formula of the original compound was $C_{10}H_{18}O$. Further experiments with the nuclear magnetic resonance absorption spectrometer showed that there had to be two

$$-CH_2-$$

part structures as well, one of them adjacent to the isopropyl group, the other close to a double bond and to the OH. Putting all of this information together suggests the structure shown below.[6]

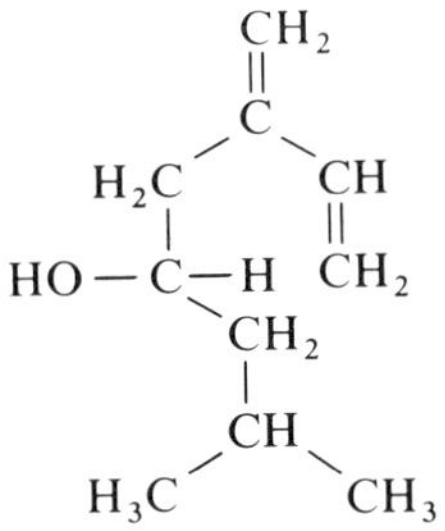

$$\begin{array}{c} CH_2 \\ \| \\ C \\ H_2C \diagup \quad \diagdown CH \\ | \qquad \qquad \| \\ HO-C-H \quad CH_2 \qquad C_{10}H_{18}O \\ \diagdown CH_2 \\ | \\ CH \\ H_3C \diagup \quad \diagdown CH_3 \end{array}$$

Final confirmation of this structure could be obtained by synthesizing a sample in the laboratory and comparing the synthetic and the natural sample. These should have identical physical and chemical properties if the structure is correct.

THE PIGMENTS OF THE EYE

In order for the eye (Figure 5-18) to be able to detect light and therefore to see, some compound or mixture of compounds must first absorb the light. The alkenes discussed so far in this chapter absorb only ultraviolet or infrared light; because we cannot see by light from this part of the spectrum, the pigments of the eye cannot be just simple alkenes, although they do contain double bonds.

[6] The substance was also found to rotate the plane of polarized light, see Chapter 9.

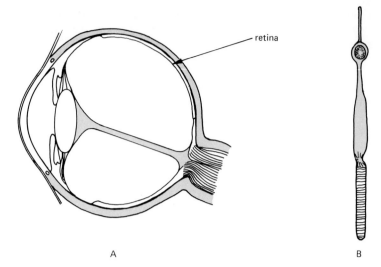

FIGURE 5-18. **(A)** Cross section of the eye showing the retina. **(B)** A rod cell.

Double bonds that do not have another double bond within two carbon atoms of their part of the molecule are called isolated double bonds and absorb only in the deep ultraviolet part of the spectrum. Double bonds in the 1,3 positions relative to each other, are called conjugated double bonds (because their electrons interact somewhat).

$$-CH_2-\underbrace{\overset{1}{C}H=\overset{2}{C}H-\overset{3}{C}H=CH}_{\text{conjugated}}-CH_2-\underbrace{\overset{6}{C}H=CH}_{\text{isolated}}-$$

As the number of conjugated double bonds is increased, the molecule begins to absorb ultraviolet light that is closer to the visible range, and finally *in* the visible range. An example of a compound that has a visible color is vitamin A (Figure 5-19). One of the pigments in the eye, **retinal**, is closely related to vitamin A_1; the biggest difference is that the terminal CH_2OH group of the vitamin has been oxidized enzymatically to an **aldehyde** group,[7] $-\overset{\overset{O}{\|}}{C}-H$ (usually written simply as CHO). The reason that vitamin A helps us to see better in dim light is that the vitamin serves as a source of retinal.

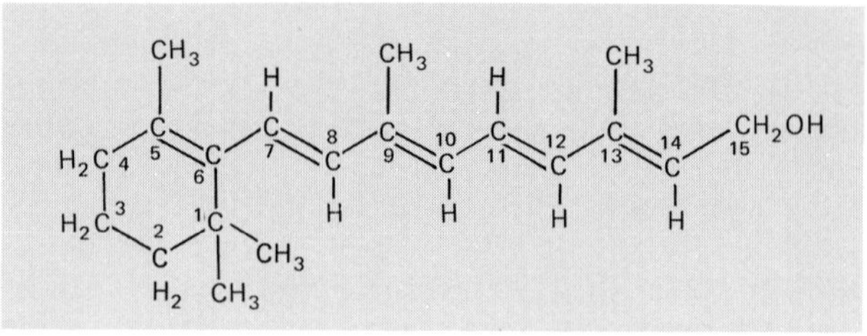

FIGURE 5-19. Vitamin A_1 (or retinol$_1$). This substance is yellow and is found in yellow vegetables. Note the all-trans structure of the chain of conjugated double bonds.

---

[7] The aldehyde group is also attached to a protein molecule (Chapter 14). The subscripts in vitamin A_1, A_2, and so on, refer to minor differences in chemical structure. Note that many nutritionists prefer to call all vitamins by their chemical names.

Although vitamin A has the all-trans structure in its double bonds (Figure 5-19), the corresponding retinal has one double bond that can be either cis or trans. The form of retinal that is effective in vision has a cis configuration of the double bond at position 11 (Figure 5-20). During the visual process this 11-cis retinal is isomerized by light and is converted to the all trans form. Bright light temporarily destroys our ability to see in dim light because time must be allowed for the formation of a new supply of the 11-*cis*-retinal.

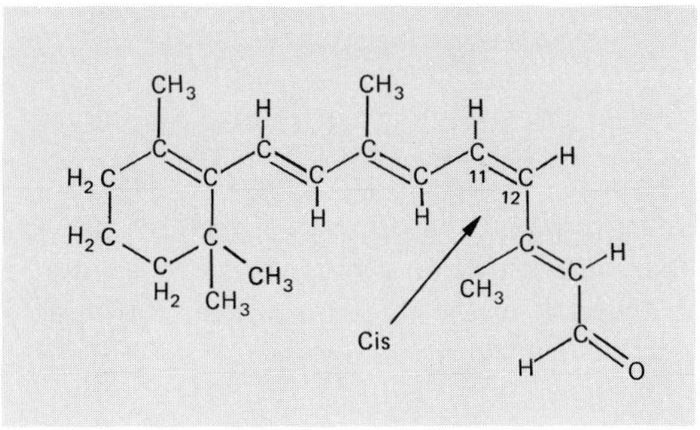

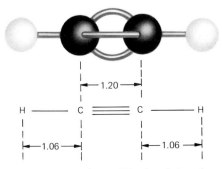

FIGURE 5-20. Exposure to light causes the all-trans form retinal$_1$ to isomerize to the 11-cis form. Although double bonds do not ordinarily undergo rotation to convert cis to trans (or vice versa) very readily, absorption of light energy promotes some of the electrons to new orbitals that do permit the rotation.

Alkynes

The alkynes are a series of unsaturated hydrocarbons containing triple bonds between carbon atoms. The two carbon atoms of the triple bond and the two atoms attached to them all lie in a straight line (Figure 5-21). That is to say, the bond angle is 180° rather than

FIGURE 5-21. Acetylene or ethyne. The bond lengths are in units of 10^{-8} cm.

the approximately 120° angle of the alkenes or the 109° 28′ angle of methane. Because the molecule is linear, there is of course no cis-trans isomerism in alkynes. The alkynes are named like alkenes but with a **yne** ending in plane of the *ene* ending.

The Triple Bond

The two carbon atoms at the ends of a triple bond are connected to just two other atoms. As in the case of ethylene, some of the bonding electrons are used to form σ bonds by the end-to-end overlap of orbitals and others are used to form π bonds by side-to-side overlap. We will first consider the σ bonds and the orbitals used to form them.

Combining one carbon 2s orbital and one $2p_x$ orbital gives two sp hybrid orbitals (Figure 5-22). The sp hybrid orbitals are both on

FIGURE 5-22. Two sp orbitals.

the x-axis, but they point in opposite directions. Overlap between the sp orbitals of two carbon atoms gives the "sigma" part of the triple bond. Overlap of the remaining two sp orbitals with the 1s orbitals of two hydrogen atoms gives the two C—H bonds (Figure 5-23).

FIGURE 5-23. Formation of the "sigma" part of the triple bond and two C—H bonds.

This leaves two p_z and two p_y orbitals, one of each on each carbon. The p_z orbital of one carbon atom overlaps sideways with the p_z orbital of the other carbon atom, and the p_y orbitals of each carbon also overlap. This is illustrated in Figure 5-24. Actually, the π_z and π_y bonding orbitals themselves overlap to accommodate a symmetrical, **cylindrical** sheath of electrons around the C—C bond axis.

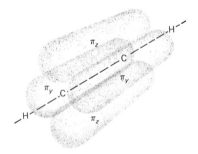

FIGURE 5-24. The p_z and p_z and the p_y and p_y carbon orbitals overlap sideways to form the π_z and π_y parts of the triple bond.

Note that if a reaction uncouples the electrons of either the π_z or the π_y orbital, the one that remains constitutes the π part of an ordinary double bond. For example, by using a "poisoned" catalyst it is possible to add just two hydrogens to acetylene, giving ethylene (equation 5-29). The addition of hydrogen to the triple bond to give alkenes (with a poisoned catalyst) or alkanes (equation 5-30, for example) is a general one.

$$H-C\equiv C-H \xrightarrow[\text{catalyst}]{\text{H}_2 \text{ poisoned}} \begin{array}{c} H \\ \diagdown \\ C=C \\ \diagup \\ H \end{array} \begin{array}{c} H \\ \diagup \\ \\ \diagdown \\ H \end{array} \quad (5\text{-}29)$$

$$\xrightarrow[\text{active catalyst}]{\text{H}_2} \left[\begin{array}{c} H \\ \diagdown \\ C=C \\ \diagup \\ H \end{array} \begin{array}{c} H \\ \diagup \\ \\ \diagdown \\ H \end{array} \right] \xrightarrow[\text{catalyst}]{\text{H}_2} CH_3-CH_3 \quad (5\text{-}30)$$

Reactions of Alkynes

Alkynes undergo various addition reactions such as the hydrogenation reaction just discussed. They also add halogen, and if an excess of halogen is present the intermediate halo-substituted alkene continues to react, as in equation 5-31.

$$H-C\equiv C-H + Cl_2 \rightarrow \begin{array}{c} Cl \\ \diagdown \\ C=C \\ \diagup \\ H \end{array} \begin{array}{c} H \\ \diagup \\ \\ \diagdown \\ Cl \end{array} \xrightarrow{Cl_2} \begin{array}{c} Cl\;\;Cl \\ |\;\;\;\;| \\ H-C-C-H \\ |\;\;\;\;| \\ Cl\;\;Cl \end{array} \quad (5\text{-}31)$$

acetylene
colorless gas; bp $-83.6°C$
(sublimes from the solid state)

trans-1,2-dichloroethylene
bp 48.4°C

1,1,2,2-tetrachloroethane
bp 146.3°C

WATER

The addition of water to an alkyne is catalyzed by mercury salts, usually $HgSO_4$ in dilute H_2SO_4 solvent.

$$H-C\equiv C-H + H_2O \xrightarrow[H_2SO_4]{Hg^{2+}} \left[\begin{array}{c} H\;\;OH \\ |\;\;\;\;| \\ H-C=C-H \end{array} \right] \quad (5\text{-}32)$$

The product that would be expected by analogy with the behavior of alkenes is shown in brackets in equation 5-32. However, this substance, called an **enol**, is rapidly converted to an isomer containing a C=O rather than a C=C double bond[8] (equation 5-33).

$$\left[\begin{array}{c} H\;\;OH \\ |\;\;\;\;| \\ H-C=C-H \end{array} \right] \rightarrow \begin{array}{c} H\;\;O \\ |\;\;\;\;\| \\ H-C-C-H \\ | \\ H \end{array} \quad (5\text{-}33)$$

acetaldehyde
bp 21°C

Hence for most practical purposes it can be considered that the hydration of acetylene gives acetaldehyde, as indicated by equation 5-34.

$$H-C\equiv CH + H_2O \xrightarrow[\text{dil. }H_2SO_4]{Hg^{2+}} CH_3-\overset{\overset{\displaystyle O}{\|}}{C}-H \quad (5\text{-}34)$$

acetaldehyde

---

[8] A very small percentage of the enol form is still present in the product at equilibrium.

$$CH_3-C\equiv C-CH_3 \xrightarrow[\text{dil. H}_2\text{SO}_4]{\text{Hg}^{2+}} CH_3-CH_2-\underset{\underset{\text{2-butanone}}{}}{\overset{\overset{O}{\|}}{C}}-CH_3 \quad (5\text{-}35)$$

dimethylacetylene
or 2-butyne
bp 27.2°C   bp 79.6°C

HYDROGEN HALIDES

The addition of hydrogen halides can be made to go in two stages, either stopping at the haloalkene or going on to give a dihaloalkane. Note that the **unsymmetrical** dihaloalkane is formed (equation 5-36), as predicted by Markovnikov's rule.

$$H-C\equiv C-H + HCl \rightarrow \underset{H}{\overset{H}{C}}=\underset{Cl}{\overset{H}{C}} \xrightarrow{HCl} CH_3-\underset{Cl}{\overset{H}{\underset{|}{\overset{|}{C}}}}-Cl \quad (5\text{-}36)$$

1,1-dichloroethane
bp 57.3°C

The Terminal ≡C—H

Just as hydrogen bonded to oxygen in water or in ethyl alcohol has different chemical properties from hydrogen bonded to sp^3 carbon, hydrogen bonded to an sp carbon also has different properties.

$$\underset{\underset{\text{carbon}}{sp^3}}{CH_4} + \underset{\text{sodium}}{Na} \rightarrow \text{no reaction} \quad (5\text{-}37)$$

$$H-O-H + Na \rightarrow H_2 + Na^+\ {}^-OH \quad (5\text{-}38)$$

$$H-C\equiv C-H + Na \xrightarrow[\text{as solvent}]{\text{liquid ammonia, NH}_3,} H_2 + Na^+\ {}^-C\equiv CH \quad (5\text{-}39)$$

As is shown by equation 5-39, the hydrogen in acetylene is acidic enough to be replaced by sodium, forming an ionic compound, sodium acetylide. We can also make sodium acetylide by using a strong base, as in equation 5-40.

$$H-C\equiv C-H + \underset{\text{sodamide}}{NaNH_2} \rightarrow NH_3 + Na^+\ {}^-C\equiv C-H \quad (5\text{-}40)$$

The hydrogen-replacing reaction of equation 5-39 or 5-40 also occurs with other 1-alkynes (for example, equation 5-41). It is a

$$CH_3CH_2C\equiv C-H + NaNH_2 \rightarrow NH_3 + CH_3CH_2C\equiv C^-Na^+$$

$$(5\text{-}41)$$

reaction of hydrogen on an sp hybridized carbon, or triply bonded carbon, so it is **not** given by alkynes with their triple bonds anywhere but at the end of the chain.

$$\underset{\substack{\text{2-butyne, has no}\\\equiv C-H\text{ part structure}}}{CH_3-C\equiv C-CH_3} + NaNH_2 \rightarrow \text{no reaction} \quad (5\text{-}42)$$

A convenient way of distinguishing between compounds like 1-butyne and 2-butyne is to use a similar salt-forming reaction with a heavy metal which gives an easily recognizable white precipitate (equations 5-43 and 5-44). These precipitates of silver or copper

$$CH_3C{\equiv}CCH_3 \xrightarrow[\text{in alcohol or aqueous ammonia}]{AgNO_3} \text{no reaction, solution remains clear} \quad (5\text{-}43)$$

$$CH_3CH_2C{\equiv}C{-}H \xrightarrow[\text{in alcohol or aqueous ammonia}]{AgNO_3} CH_3CH_2C{\equiv}C^-Ag^+ \quad (5\text{-}44)$$
$$\text{heavy white precipitate}$$

acetylides are destroyed by warming with nitric acid (HNO_3) while still wet. If allowed to dry, they are likely to explode.

Heavy metal salts of acetylene itself have been used as the primary charge that is detonated by the fall of the hammer on a primer or percussion cap. The heat or shock wave from the primer then ignites the powder or sets off the main explosive. For use with high explosives such as dynamite (Chapter 7), the primer, which is easily set off by percussion, is stored separately from the high explosive, which is ordinarily a substance sufficiently insensitive to shock to allow it to be safely handled.

Preparation of Alkynes

Two general methods for making alkynes will be considered and one special method for acetylene only.

FROM ALKENES

The first of the general methods is like the dehydrohalogenation method for making an alkene, and in fact it starts with an alkene.

$$CH_3{-}\underset{\underset{H}{|}}{C}{=}\underset{\underset{H}{|}}{C}{-}CH_2CH_3 + Cl_2 \rightarrow CH_3{-}\underset{\underset{Cl}{|}}{\overset{\overset{H}{|}}{C}}{-}\underset{\underset{Cl}{|}}{\overset{\overset{H}{|}}{C}}{-}CH_2CH_3 \quad (5\text{-}45)$$

$$CH_3{-}\underset{\underset{Cl}{|}}{\overset{\overset{H}{|}}{C}}{-}\underset{\underset{Cl}{|}}{\overset{\overset{H}{|}}{C}}{-}CH_2CH_3 + 2KOH \rightarrow 2KCl + CH_3{-}C{\equiv}C{-}CH_2CH_3 \quad (5\text{-}46)$$

The sequence of reactions 5-45 and 5-46 is quite general and will work with just about any alkene provided that it has hydrogen on both ends of the double bond.

FROM ACETYLIDES

Acetylene or a terminal alkyne, in other words any alkyne with a ≡C—H part structure, can be converted into its sodium salt by reaction 5-41. The sodium salt can then be converted to a larger, nonterminal alkyne by means of the **halide ion displacement reaction**, equation 5-47.

$$R{-}C{\equiv}C^-Na^+ + CH_3CH_2{-}Cl \rightarrow NaCl + R{-}C{\equiv}C{-}CH_2CH_3 \quad (5\text{-}47)$$

where R = any alkyl group. This process will work with most

primary alkyl chlorides ($-CH_2-Cl$ part structure) and secondary alkyl chlorides ($-\underset{\underset{C}{|}}{CH}-Cl$ part structures), but not with tertiary alkyl chlorides ($C-\underset{\underset{C}{|}}{\overset{\overset{C}{|}}{C}}-Cl$ part structures). The tertiary alkyl chlorides tend to eliminate HCl and give the corresponding alkene instead of reaction 5-47.

Reaction 5-47 is a displacement of the nucleophile Cl^- by a stronger nucleophile, the carbon anion (**carbanion**). The ionic electrons of the carbanion form a new bond to the carbon of the alkyl chloride while the covalent bond electrons of the $C-Cl$ bond become ionic electrons as the Cl^- is formed. This is shown diagrammatically in Figure 5-25.

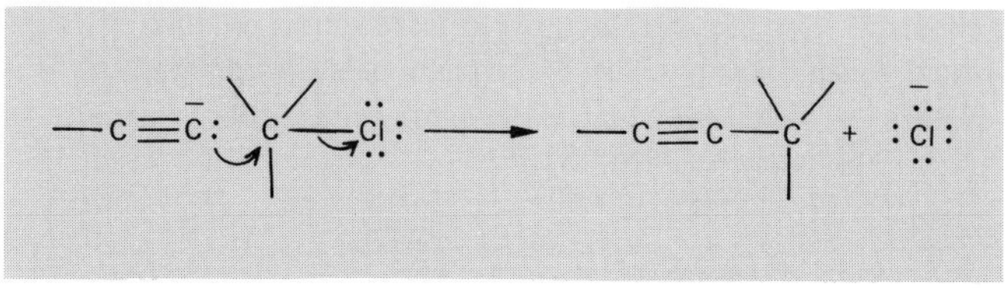

FIGURE 5-25. Flow of electrons in the displacement reaction.

Preparation of Acetylene

We include this special method of making this particular compound because acetylene is an important industrial starting material for a large variety of useful substances, including synthetic rubber (Chapter 12). The inorganic compound $Ca^{2+}C_2^{2-}$, calcium carbide, is essentially a salt of acetylene in which the divalent ion Ca^{2+} has replaced both of the protons. Salts of acetylene (acetylides) regenerate the parent acetylene when treated with acids. The acid need not be a strong one; even water is strong enough.

$$CaC_2 + 2H_2O \rightarrow Ca(OH)_2 + H-C\equiv C-H \quad (5\text{-}48)$$
calcium carbide, (white solid)   slaked lime   acetylene

The calcium carbide is easily made by heating lime (CaO) with carbon (coke or charcoal).

$$CaO + C \xrightarrow{2000°C} CaC_2 + CO \quad (5\text{-}49)$$

Even a very primitive economy, for example, Robinson Crusoe on his island, should be able to make calcium carbide and hence acetylene for lighting, percussion caps for muzzle loading rifles, and perhaps even synthetic rubber to repair shoes and caulk boats. Crusoe would have to start with limestone or coral from the reef that fringes the island. Limestone is calcium carbonate, $CaCO_3$. Heating limestone decomposes it to CaO (also useful in making mortar for bricklaying) and CO_2.

$$CaCO_3 \xrightarrow{heat} CO_2 + CaO \quad (5\text{-}50)$$
the coral reef   lime

Reaction 5-49 requires a container, but Crusoe could have made that out of clay pottery. The high temperature needed should be obtainable with a hot fire, encouraged by a bellows.

Acetylene is used as a fuel in oxyacetylene torches for welding and cutting steel. The oxygen and acetylene are kept in separate gas cylinders, the oxygen as a highly compressed gas and the acetylene as a solution in acetone, also under pressure. The reason for the solvent acetone is that compressed acetylene without the solvent is likely to explode.

FIGURE 5-26. A carbide lamp, very handy during power shortages. [After 1908 Catalog, Sears, Roebuck & Co.]

SUMMARY

A. Small rings (three or four carbon atoms)
 1. Strained bonds: Their strained bonds make them unusually reactive towards such reagents as Br_2 or HI.
B. Alkenes
 1. Reactivity: The reactivity and planar structure can best be explained in terms of a π bond.
 2. Isomers: The rigid planar structure also accounts for the existence of cis and trans isomers.
 3. Reactions: Alkenes undergo addition reactions which convert the double bond to a single bond.
    (a) Br_2 and H_2.
    (b) HCl and water. Markovnikov's rule.
 4. Preparation: The reactions used to make alkenes are *elimination* reactions that resemble the addition reactions in reverse.
    (a) Dehydration.
    (b) Dehydrohalogenation with alcoholic KOH
    (c) The proton is taken from the carbon with the fewest protons.
 5. Some alkenes in nature—the bark beetle and the pigment of the eye.
C. Alkynes
 1. Reactivity: The triple bond, as predicted by the π orbital model, acts like two double bonds superimposed.
    (a) Addition of one molecule of halogen, and so on, gives a substituted alkene. Addition of a second molecule gives a substituted alkane.
 2. Shape: Because of sp hybridization, the alkyne functional group —C≡C— is linear.
    (a) There are no cis and trans isomers.
 3. Acidity of terminal alkynes, —C≡C—H: The end hydrogen in this structure is weakly acidic.
    (a) Reaction with $NaNH_2$.
    (b) Formation of heavy metal salts.

4. **Preparation**: Methods of preparation are elimination reactions like those used for alkenes. Special reactions are
   (a) The displacement reaction with sodium acetylides.
   (b) The formation of acetylene from CaC_2.

EXERCISES

1. Write equations for the reaction of cyclopropane with (a) hydrogen (Ni, 80°) and (b) bromine (in the dark).

2. Do the same for 1,2,3-trimethylcyclopropane.

3. Do the same for methylcyclopropane. Can more than one product be obtained from these reactions?

4. Explain why cyclopropane, but not cyclobutane, reacts with bromine.

5. Complete the equation:

$$\square\!-\!CH_2\!-\!\triangle \xrightarrow{H_2,\ Ni,\ 80°C}$$

The H_2 over the arrow means that you can use as much as you want. Show *two* possible products.

6. Explain why the CH_2 groups in ethylene cannot rotate.

7. Using the names for the isomeric hexenes (page 73) as a model, name the isomeric octenes. (Ignore cis-trans isomerism in this question.)

8. Explain why each of the following names is incorrect and give a correct name.
   (a) 3-butene
   (b) 2-ethyl-1-propene
   (c) 4-pentyne
   (d) 3,4-dimethyl-4-pentene

9. Look at the two structures given for 1-pentene in Figure 5-12. If double bonds cannot be twisted, why aren't the two structures two different compounds?

10. Which of the following compounds can have cis-trans isomers?

    (a) 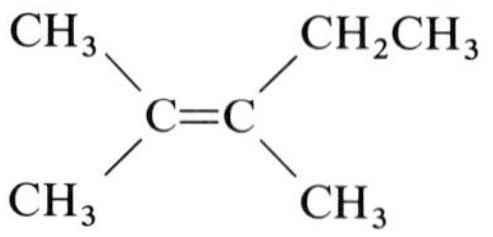 CH_3 and CH_2CH_3 on one carbon; CH_3 and CH_3 on the other, with C=C

    (b) 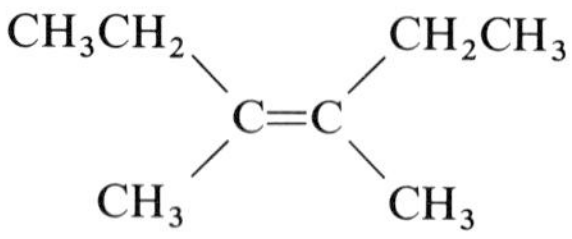 CH_3CH_2 and CH_2CH_3 on one carbon; CH_3 and CH_3 on the other, with C=C

    (c) Cl and Cl on one carbon; Br and Br on the other, with C=C

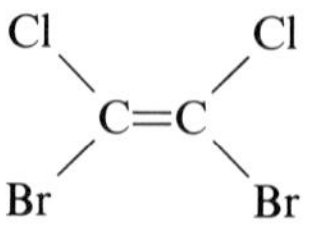

    (d) 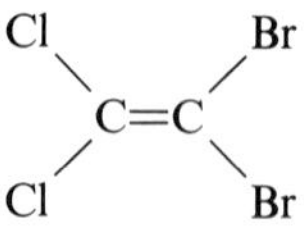 Cl and Br on one carbon; Cl and Br on the other, with C=C

(e) 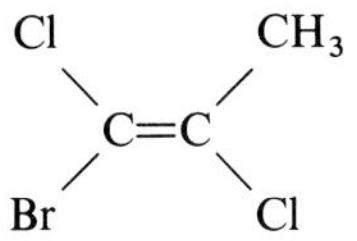
$$\begin{array}{c}Cl\\ \diagdown\\ Br\end{array}C=C\begin{array}{c}CH_3\\ \diagup\\ Cl\end{array}$$

11. Comment briefly on the possibility of existence of the compound *trans*-cyclopropene.

12. Write structural formulas for all twelve possible isomers of molecular formula C_5H_{10}.

13. Two bottles have had their labels drop off. One contains cyclohexane and the other contains cyclohexene. Both are colorless. Invent an easy experiment, based on a color change, that would allow one to tell which bottle contains which compound.

14. Write equations illustrating three reactions of alkenes, using 1-hexene as an example.

15. A compound C_6H_{10} will add only *1* mole of bromine, giving $C_6H_{10}Br_2$. Draw one of the several possible structural formulas for the compound C_6H_{10} that would predict this result.

16. Write structural formulas for the following:
    (a) cyclopropane
    (b) 1,1-dimethylcyclopropane
    (c) *cis*-1,2-diethylcyclobutane
    (d) *cis*-2-pentene
    (e) 3-heptyne
    (f) *trans*-3-octene

17. Complete the following reactions

   (a) 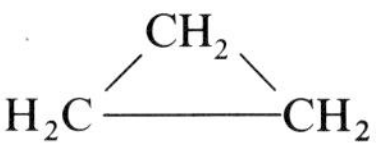 $\begin{array}{c}\quad CH_2\\ \diagup\;\;\diagdown\\ H_2C\!-\!\!-\!\!-\!CH_2\end{array}$ + HI →

   (b) $CH_3CH=CHCH_3$ + HCl →
   (c) $CH_3CH_2CH=CH_2$ + HCl →
   (d) cyclohexene + Br_2 →
   (e) propyne + 2 HCl →
   (f) $H-C\equiv C-H$ + H_2O $\xrightarrow{Hg^{2+},\,H_2SO_4}$
   (g) *trans*-2-butene + H_2 $\xrightarrow{Ni}$
   (h) *cis*-2-butene + H_2 $\xrightarrow{Ni}$
   (i) $CH_2=CH_2$ + H_2O $\xrightarrow{cold\ dilute\ H_2SO_4}$

18. Write equations showing how to prepare:
    (a) 2-chlorobutane from an alkene
    (b) 2,2-dichloropropane from an alkyne
    (c) 1-butyne from acetylene and an alkyl halide

19. Complete the following reactions

   (a) CH_3CH_2OH $\xrightarrow{95\%\ H_2SO_4,\,170°C}$
   (b) CH_3CH_2Cl + KOH $\xrightarrow{alcohol}$
   (c) 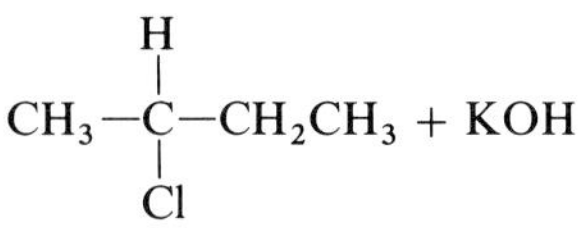 $CH_3-\underset{\underset{Cl}{|}}{\overset{\overset{H}{|}}{C}}-CH_2CH_3$ + KOH $\xrightarrow{alcohol}$

(d) H—C≡C—CH$_3$ + NaNH$_2$ →
(e) H—C≡C$^-$Na$^+$ + CH$_3$Cl →

20. How could one distinguish between two liquids, one a 1-alkyne and the other a 2-alkyne?

21. Match each of the following bond angles with the most nearly correct value.

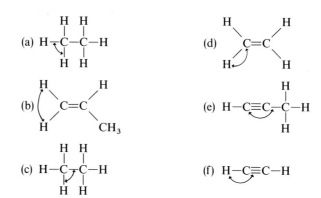

Angles: 109° 28′, 120°, 180°.

Chapter 6
Aromatic Hydrocarbons

The Bonds in Benzene

If cyclohexane is heated to a high temperature over a catalyst such as platinum, it loses six hydrogen atoms and forms a compound C_6H_6 known as benzene.

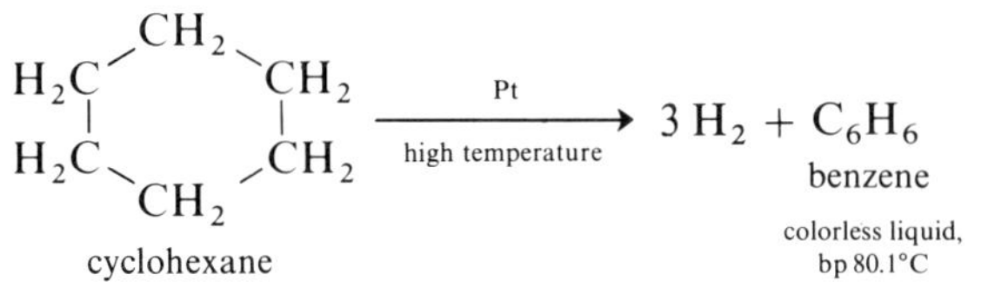

$$\text{cyclohexane} \xrightarrow[\text{high temperature}]{\text{Pt}} 3H_2 + C_6H_6 \text{ (benzene, colorless liquid, bp 80.1°C)} \quad (6\text{-}1)$$

The fact that six atoms of hydrogen have been removed suggests that the product ought to have three double bonds. It is easy to construct a *model* of a compound with three double bonds in a six-membered ring. If such a compound existed we would call it 1,3,5-cyclohexatriene. It would be a slightly lopsided hexagon, as illustrated in Figure 6-1, since double bonds are shorter than single bonds. It would presumably give various reactions, for example, the addition of bromine, that are typical of compounds with carbon-carbon double bonds.

However, the compound produced by reaction 6-1 *and by numerous other reactions that might have been expected to give 1,3,5-cyclohexatriene* is always benzene.[1] Benzene does not react very much like an alkene. Furthermore, it can be shown by x-ray

---
[1] One example is the formation of benzene from three molecules of acetylene,

$$3\ HC\equiv CH \xrightarrow[\text{iron}]{\text{hot}} C_6H_6$$

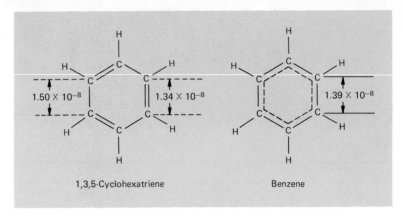

FIGURE 6-1. The hypothetical 1,3,5-cyclohexatriene would be a slightly lopsided hexagon because double bonds are shorter than single bonds. The actual molecule, benzene, is a regular hexagon. All six sides are 1.39×10^{-8} cm.

diffraction methods that the C—C bonds of benzene all have the same length. We conclude, then, that benzene is something new and not 1,3,5-cyclohexatriene after all.

The easiest explanation or electronic model for benzene starts with the six carbon atoms sp^2 hybridized, as though ready to form double bonds. One of the three sp^2 orbitals of each carbon atom is used to attach a hydrogen atom, and the other two make bonds with the neighboring carbon atoms. This leaves a ring of six equally spaced C—H units and six as yet unused p_z atomic orbitals, one on each of the carbon atoms. Figure 6-2 shows the planar, hexagonal structure of σ bonds formed by the overlap of the sp^2 orbitals.

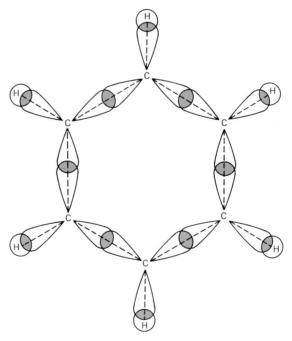

FIGURE 6-2. The σ bond skeleton of benzene.

The σ bonds shown in Figure 6-2 account for all but six of the electrons and use all the orbitals but the six p_z orbitals that were not used in constructing the sp^2 hybrids. Each carbon atom has one p_z orbital, and these orbitals are all perpendicular to the plane of the benzene ring and parallel to each other. Furthermore, the distances between them are all the same.

Next, the six p_z orbitals of Figure 6-3 are allowed to overlap sideways to form π bonding orbitals of some kind. If each p_z orbital overlapped only with *one* of its two neighbor p_z orbitals, the molecule

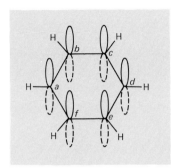

FIGURE 6-3. The six as yet unused p_z orbitals.

would have three double bonds as shown in Figure 6-4. But since the C—C bond lengths are all the same, each p_z orbital overlaps equally with *both* of its neighboring p_z orbitals. As a result of this overlap, all six p_z orbitals combine to form **benzene π molecular orbitals**. Unlike the double bond orbitals that would otherwise have been formed, the benzene orbitals extend over the whole ring. The π orbitals of benzene are not fundamentally very different from the π orbital of ethylene except that they extend over six carbon atoms instead of just two.

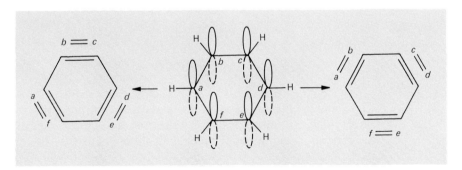

FIGURE 6-4. Two ways of pairing the p_z orbitals and their electrons to make double vonds.

Figure 6-5 shows one of the resulting orbitals. The electrons in the benzene π molecular orbitals form a cloud with two doughnut-shaped parts, one above and one below the ring. It turns out that the orbitals extending all around the ring give a more stable molecule than cyclohexatriene would have been. If cyclohexatriene could be prepared it would turn into benzene spontaneously.

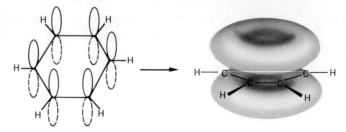

FIGURE 6-5. Overlap of the carbon p_z orbitals to form a benzene molecular orbital.

Because the C—C bonds of benzene are all alike, modern formulas represent benzene and its derivatives by hexagons with a circle (the six π electrons) inside. In this chapter we will usually show the carbon and hydrogen atoms explicitly to avoid possible misunderstandings.

benzene or benzene

On some occasions it is still convenient to represent benzene as though it were a hexagon with double and single bonds. When this is done, it is to be understood that the "single" and "double" bonds are really all alike. One way of emphasizing this is to draw not one but two such structures, differing in the positions of the "single" and "double" bonds and connected by a double headed arrow.

Resonance structures of benzene

Such structures, alike in the shape of the molecule, and appearing to represent different positions for the electrons, are called resonance structures.[2] It should be emphasized that, although the electrons of one resonance structure can be shifted and the structures can be interconverted as indicated by the curved arrows in the following formulas, the apparently different resonance structures are merely convenient fictions. No such movement of the electrons actually occurs. **All of the C—C bonds of benzene are alike.**

[2] A similar use of the double headed arrow will be found on page 105.

When coal is heated to 1000–1300°C out of contact with air, the complicated molecules derived from the organic plant material from which the coal was formed are broken down. The products are benzene, almost all the other hydrocarbons to be discussed in this chapter, hydrogen, methane, numerous compounds containing nitrogen, oxygen, or sulfur, coke (essentially a form of carbon), and coal tar.

Benzene is used in the laboratory and industrially as a solvent and in paint removers. It is somewhat toxic (bone marrow damage) and can be absorbed through the skin. For this reason, most products designed for use in the home contain the less toxic compound toluene (methylbenzene) instead of benzene.

toluene
bp 110.6°C

Benzene and related compounds are called **aromatic** hydrocarbons because the earliest known derivatives were fragrant substances from plants.

Benzine, spelled with an i instead of an e, is a common name for a low-boiling mixture of alkanes and should not be confused with benzene.

Reactions of Benzene

COMBUSTION AND OXIDATION

Benzene and other hydrocarbons containing benzene rings burn readily but with a very sooty flame. Benzene reacts much less readily with oxidizing agents than alkenes. Ethylene, for example, will decolorize potassium permanganate at room temperature whereas benzene does not react at all.

$$3\,CH_2{=}CH_2 + 2\,KMnO_4 + 4\,H_2O \rightarrow 3\,\underset{}{CH_2(OH){-}CH_2(OH)} + 2\,MnO_2 + 2\,KOH$$

potassium permanganate (purple solution) → colorless solution with black MnO_2 precipitate

(6-2)

$$C_6H_6 + KMnO_4 \xrightarrow{H_2O} \text{no reaction, solution remains purple}$$

(6-3)

ADDITION REACTIONS

Benzene undergoes addition reactions much less readily than alkenes do, but the products of the reaction are those that would have been expected from a cyclohexatriene. Hydrogen, for example will add to benzene under somewhat drastic conditions (equation 6-4) and three molecules of hydrogen are taken up as though three double bonds were being saturated. Reaction 6-4 is essentially the reverse of reaction 6-1. The reaction is made to go in this direction by using hydrogen at a high pressure.

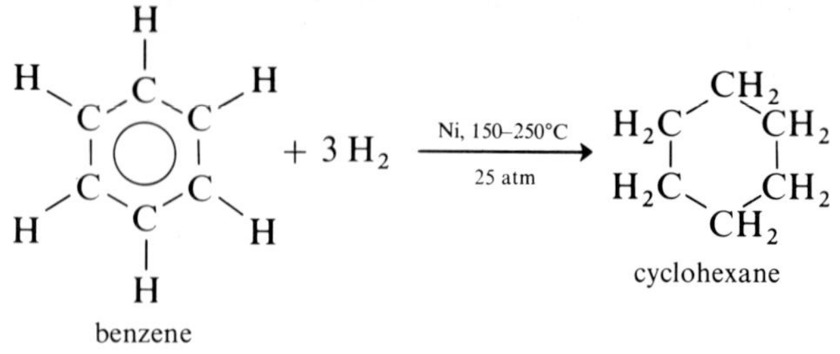

$$\text{benzene} + 3\,H_2 \xrightarrow[25\text{ atm}]{\text{Ni, 150–250°C}} \text{cyclohexane} \quad (6\text{-}4)$$

The addition of bromine or chlorine to benzene is also more difficult than the corresponding reactions of alkenes. The reaction will go only if some extra energy is provided in the form of sunlight.

$$CH_2=CH-CH_3 + Br_2 \xrightarrow{\text{fast}} \underset{\underset{Br}{|}}{CH_2}-\underset{\underset{Br}{|}}{CH}-CH_3 \quad (6\text{-}5)$$

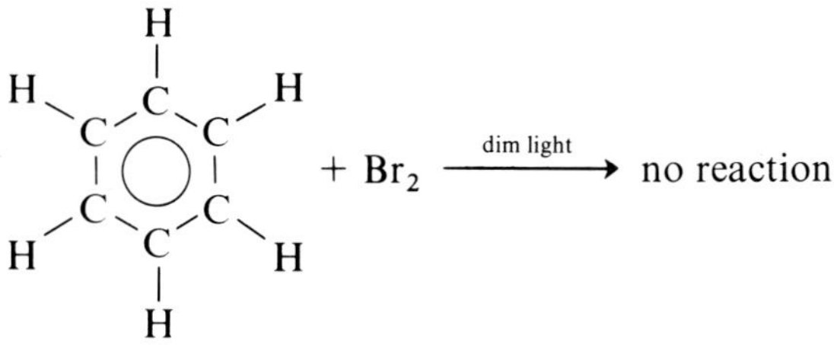

$$\text{benzene} + Br_2 \xrightarrow{\text{dim light}} \text{no reaction} \quad (6\text{-}6)$$

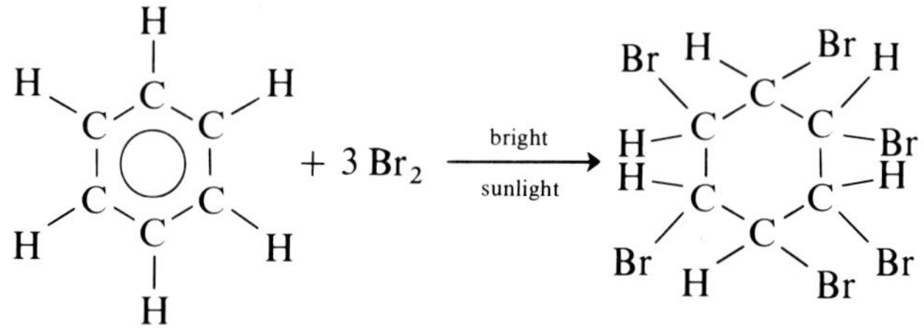

$$\text{benzene} + 3\,Br_2 \xrightarrow{\text{bright sunlight}} \text{hexabromocyclohexane} \quad (6\text{-}7)$$

The relative unreactivity of benzene towards oxidizing agents and towards addition of halogen or hydrogen illustrates the greater stability of aromatic rings as compared to double bonds.

SUBSTITUTION REACTIONS

When ethylene and chlorine are allowed to react at room temperature the product is 1,2-dichloroethane, the result of an addition reaction. At very high temperature, however, the product is chloroethylene (also called vinyl chloride). Reaction 6-8 is a substitution

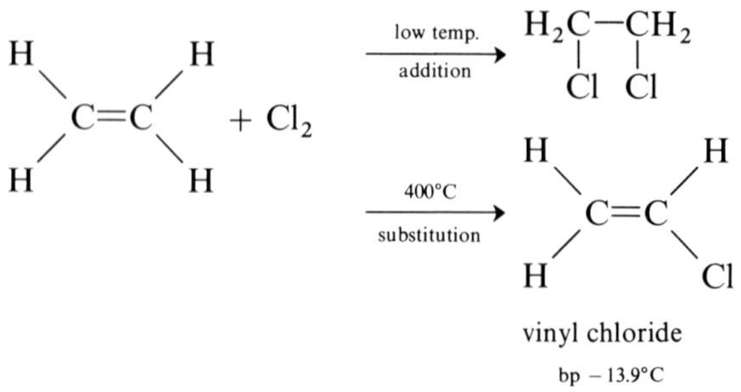

$$H_2C=CH_2 + Cl_2 \xrightarrow[\text{addition}]{\text{low temp.}} \underset{\underset{Cl}{|}}{H_2C}-\underset{\underset{Cl}{|}}{CH_2}$$

$$\xrightarrow[\text{substitution}]{400°C} \underset{\text{vinyl chloride} \atop \text{bp }-13.9°C}{H_2C=CHCl} \quad (6\text{-}8)$$

rather than an addition reaction since it leaves the double bond intact. Because aromatic rings are more stable than double bonds, aromatic compounds have a strong tendency to give reactions that leave the aromatic ring intact. Thus, substitution reactions take place readily with aromatic compounds and, as we have seen in the last section, addition reactions take place only with difficulty.

CHLORINE AND BROMINE SUBSTITUTION

Benzene reacts with bromine in the presence of a small amount of $FeBr_3$ catalyst to give bromobenzene and HBr (equation 6-9). The role of the $FeBr_3$ catalyst is to convert the mildly electrophilic reagent Br_2 into the much more strongly electrophilic reagent $Br^+FeBr_4^-$.

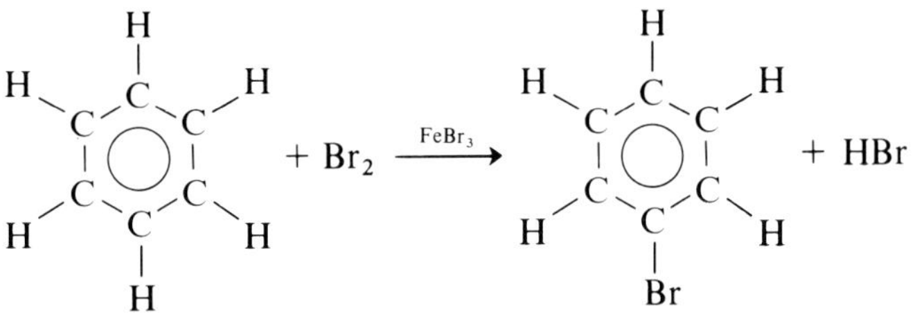

$$\text{benzene} + Br_2 \xrightarrow{FeBr_3} \text{bromobenzene} + HBr \quad (6\text{-}9)$$

bromobenzene
bp 155–156°C

$$Br-Br \quad \overset{Br}{\underset{Br}{Fe}}-Br \rightarrow Br^+ \; Br-\overset{Br}{\underset{Br}{Fe}}-Br \quad (6\text{-}9a)$$

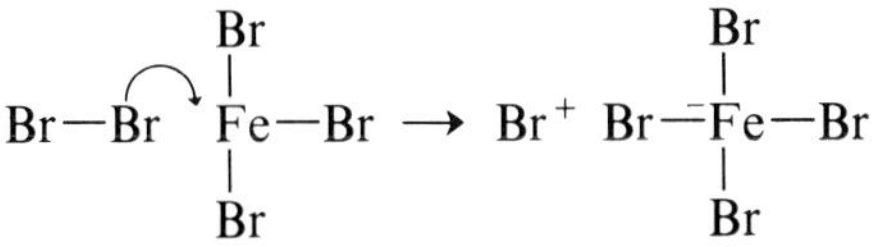

The Br^+ adds to the benzene ring in much the same way as to a double bond, but the next step is loss of H^+. This reaction leaves the ring aromatic or benzenelike, whereas addition of Br^- would have destroyed the aromatic structure.

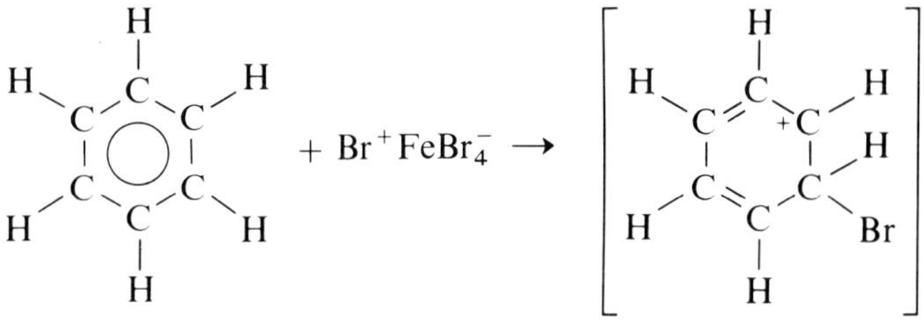

$$\text{benzene} + Br^+FeBr_4^- \rightarrow [\text{arenium ion}] \quad (6\text{-}9b)$$

$$[\text{arenium ion}] \rightarrow \text{bromobenzene (Kekulé 1)} \;\; \text{or} \;\; \text{bromobenzene (Kekulé 2)} + H^+ \quad (6\text{-}9c)$$

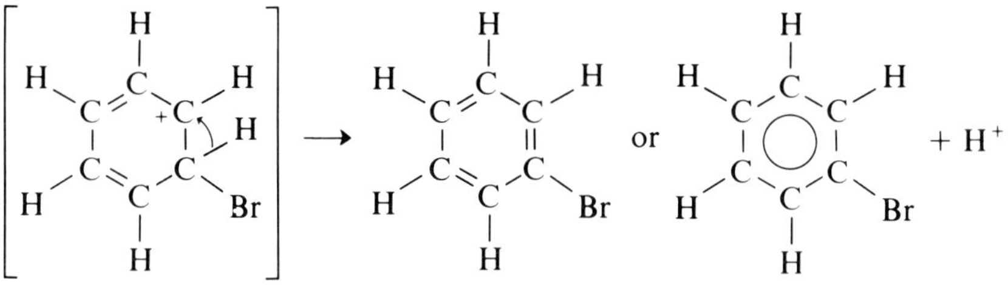

The reaction with chlorine (catalyzed by FeCl₃) is similar.

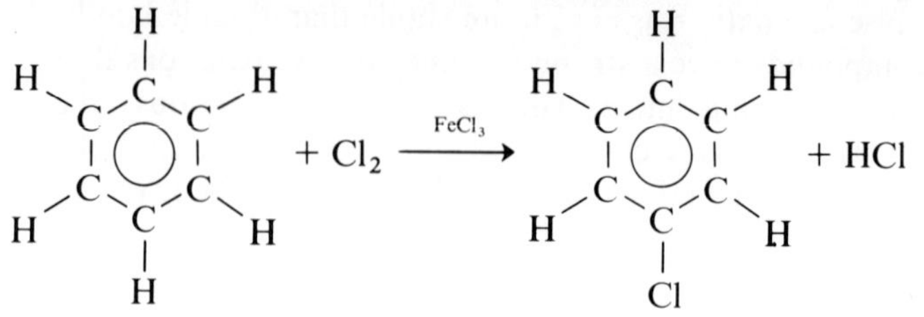

$$C_6H_6 + Cl_2 \xrightarrow{FeCl_3} C_6H_5Cl + HCl \quad (6\text{-}10)$$

Note that because all six hydrogens of benzene are alike, there is only one monochlorobenzene and only one monobromobenzene.

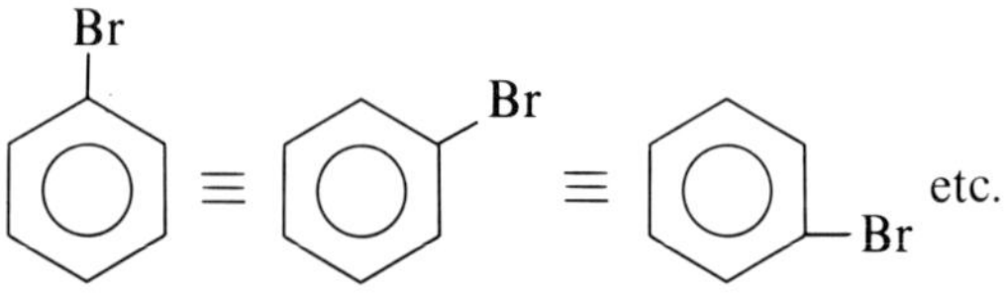

$$\text{C}_6\text{H}_5\text{Br} \equiv \text{C}_6\text{H}_5\text{Br} \equiv \text{C}_6\text{H}_5\text{Br} \quad \text{etc.}$$

OTHER AROMATIC SUBSTITUTION REACTIONS

A few other useful aromatic substitution reactions are nitration (equation 6-11), sulfonation (equation 6-12), and alkylation (equation 6-13).

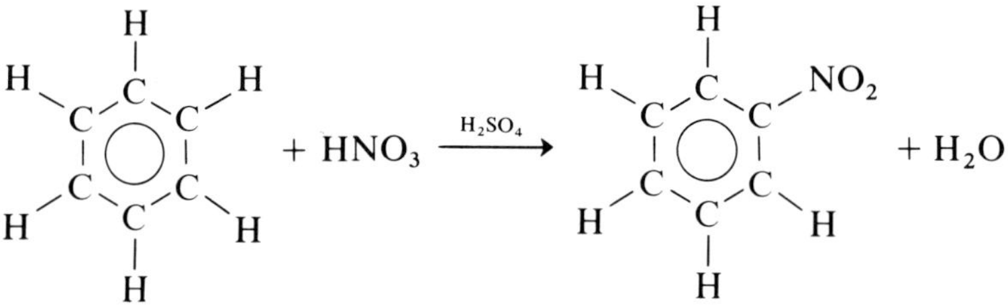

$$C_6H_6 + HNO_3 \xrightarrow{H_2SO_4} C_6H_5NO_2 + H_2O \quad (6\text{-}11)$$

nitrobenzene
bp 211°C

Nitrobenzene is an oily liquid with an almondlike odor, boiling at 211°C. Earlier generations associated the odor of nitrobenzene with shoe polish, in which it was formerly used as a solvent to soften the wax. It is no longer used for that purpose because it is quite toxic, both as the vapor and on the skin.

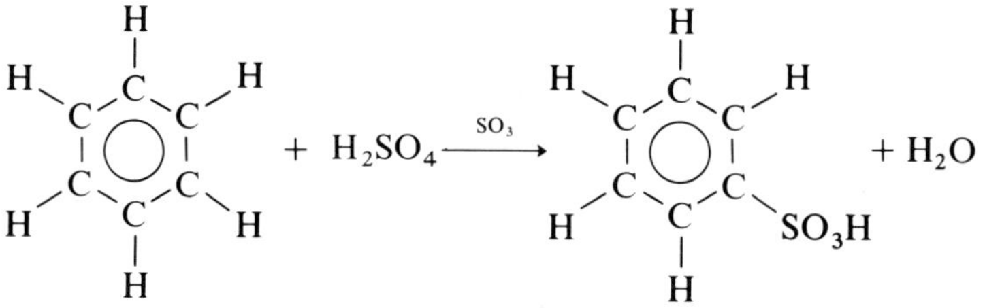

$$C_6H_6 + H_2SO_4 \xrightarrow{SO_3} C_6H_5SO_3H + H_2O \quad (6\text{-}12)$$

benzenesulfonic acid

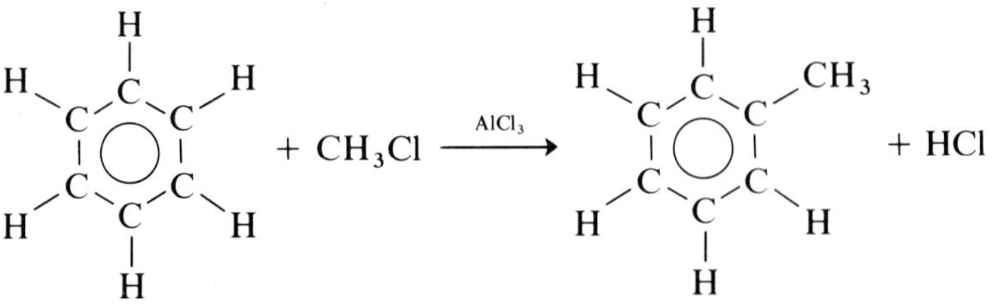

$$C_6H_6 + CH_3Cl \xrightarrow{AlCl_3} C_6H_5CH_3 + HCl \quad (6\text{-}13)$$

toluene
bp 110.6°C

The nitration, sulfonation, and alkylation of benzene are also examples of electrophilic displacement reactions in which a strong electrophile displaces hydrogen ion (a weaker electrophile) from the electron-rich benzene ring. Thus in nitration,

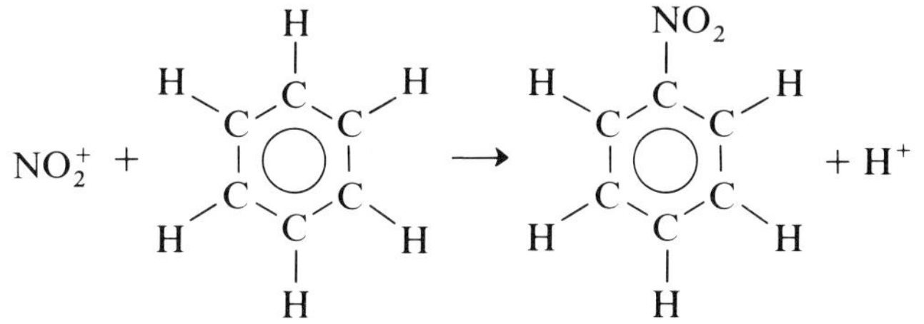

$$NO_2^+ + C_6H_6 \longrightarrow C_6H_5NO_2 + H^+$$

The purpose of the catalyst H_2SO_4 is to convert the less reactive HNO_3 into the more electrophilic and, hence, more reactive NO_2^+.

$$O_2N{-}OH + H_2SO_4 \longrightarrow NO_2^+ + H_2O + HSO_4^-$$

The rest of the mechanism of displacement of H^+ by NO_2^+ is like that for bromination (see Chlorine and Bromine Substitution in this chapter).

Alkybenzenes

The simplest alkylbenzene is methylbenzene, or toluene. Toluene can be made either from benzene and methyl chloride (equation 6-13) or by dehydrogenation of the heptane fraction of petroleum (equation 6-14).

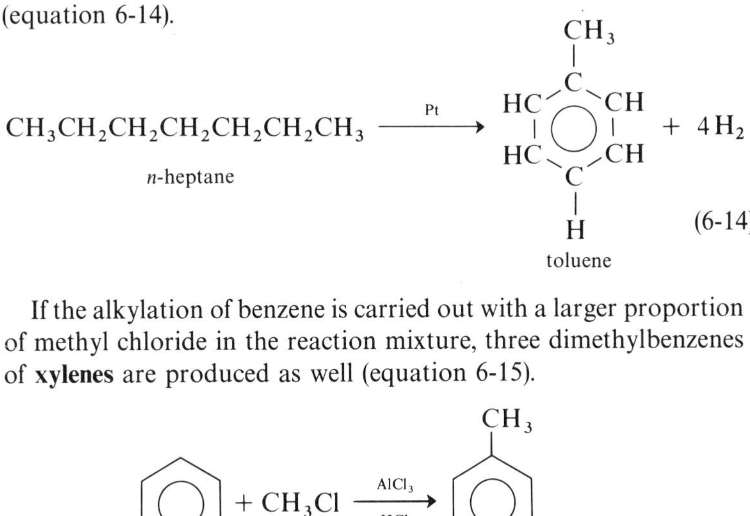

$$CH_3CH_2CH_2CH_2CH_2CH_2CH_3 \xrightarrow{Pt} \text{toluene} + 4H_2 \quad (6\text{-}14)$$

n-heptane

If the alkylation of benzene is carried out with a larger proportion of methyl chloride in the reaction mixture, three dimethylbenzenes of **xylenes** are produced as well (equation 6-15).

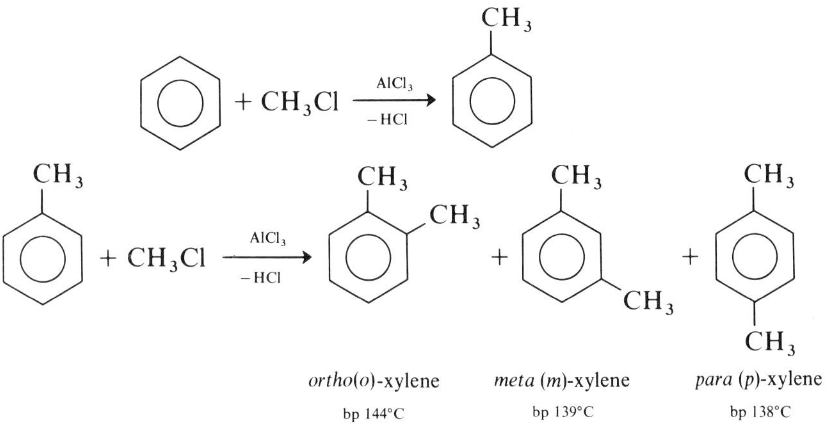

$$C_6H_6 + CH_3Cl \xrightarrow[-HCl]{AlCl_3} C_6H_5CH_3$$

$$C_6H_5CH_3 + CH_3Cl \xrightarrow[-HCl]{AlCl_3} \text{ortho}(o)\text{-xylene} + \text{meta}(m)\text{-xylene} + \text{para}(p)\text{-xylene}$$

ortho(*o*)-xylene
bp 144°C

meta (*m*)-xylene
bp 139°C

para (*p*)-xylene
bp 138°C

(6-15) Aromatic Hydrocarbons

The three dimethylbenzenes can be called 1,2-, 1,3-, and 1,4-dimethylbenzenes, and this nomenclature is frequently used. However, chemists also find it convenient to use the terms **ortho**, **meta**, and **para** in discussion of isomers of this type. The major product of reaction 6-15 is a mixture of *ortho*-xylene (*o*-xylene) and *para*-xylene (*p*-xylene) when the reaction is run at 0°C, but the product is mainly *meta*-xylene (*m*-xylene) when the reaction is run at 80°C. A mixture of all three xylenes can be distilled from coal tar. Although the individual pure xylenes are expensive, the mixture, which is known in commerce simply as xylene, is cheap. It is used as a solvent and for sterilizing catgut.

The fact that only one ortho-xylene exists was one of the early pieces of evidence against a 1,3,5-cyclohexatriene structure for benzene.

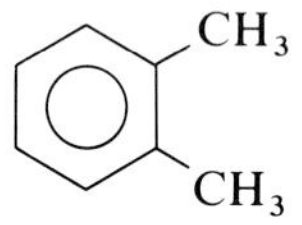

o-xylene 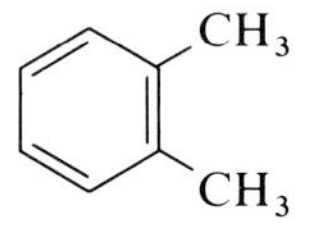   "1,2-dimethyl-1,3,5-cyclohexatriene" 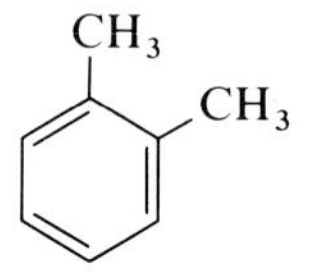   "2,3-dimethyl-1,3,5-cyclohexatriene"

Like benzene, the alkylbenzenes can be halogenated, nitrated, and sulfonated.

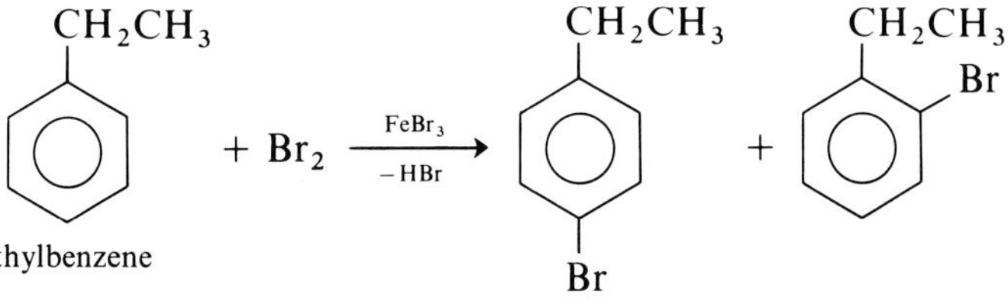

ethylbenzene + Br_2 $\xrightarrow[-HBr]{FeBr_3}$ [para-bromoethylbenzene] + [ortho-bromoethylbenzene]   (6-16)[3]

The products are mainly the para and ortho isomers, so the alkyl group is called an ortho-para director.

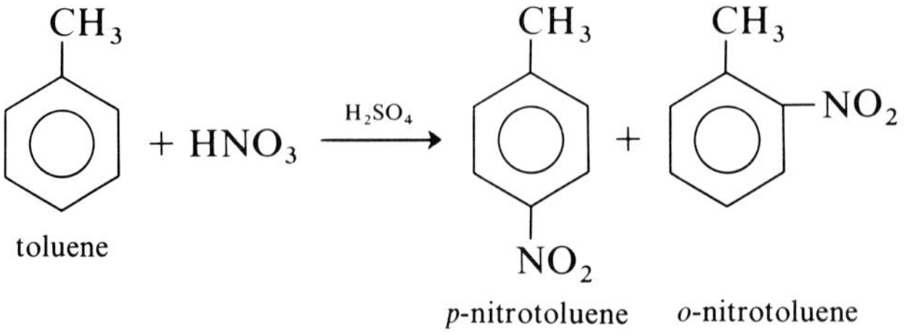

toluene + HNO_3 $\xrightarrow{H_2SO_4}$ *p*-nitrotoluene + *o*-nitrotoluene   (6-17)

p-nitrotoluene
solid; mp 51.3°C

o-nitrotoluene
oil; bp 222.3°C

---

[3] If $FeBr_3$ is rigorously excluded and the reaction is carried out in sunlight, the substitution will take place on the alkyl group rather than on the ring. This reaction is essentially the same as the halogenation of alkanes.

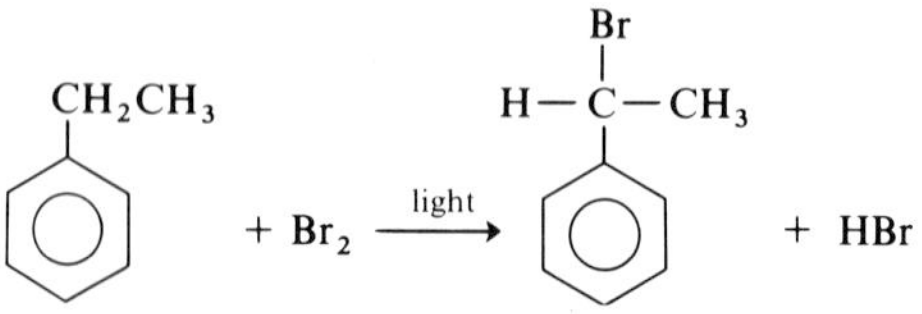

ethylbenzene + Br_2 $\xrightarrow{light}$ [1-bromo-1-phenylethane] + HBr

Not only the ortho-para substitution of toluene but also the direction taken by many other reactions can be explained by the following simple principle: **alkyl substituents release electrons, and a carbonium ion $-\overset{|}{\underset{|}{C}}{}^+$ with alkyl substituents is more easily formed than one without such substituents**. This principle was used in Chapter 5 to explain the direction of addition of unsymmetrical reagents to alkenes (Markovnikov's rule).

To explain the ortho-para substitution of toluene it is best to start with the double bond version of the structural formula. A reagent such as NO_2^+ (from HNO_3 and H_2SO_4) can add either in the ortho or para positions, giving a stable methyl-substituted ion, or in the meta position, giving a less stable, unsubstituted carbonium ion.

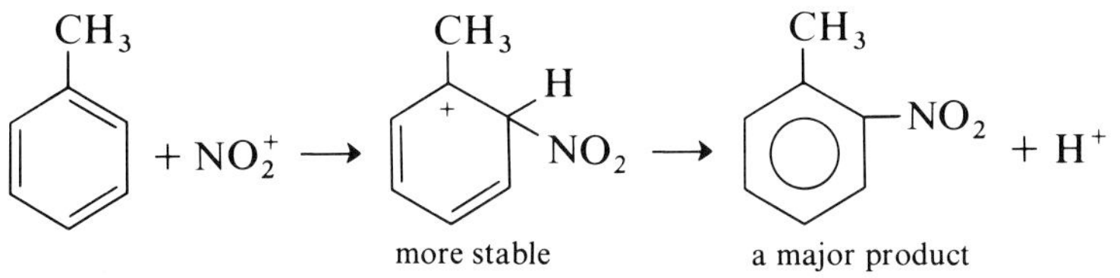

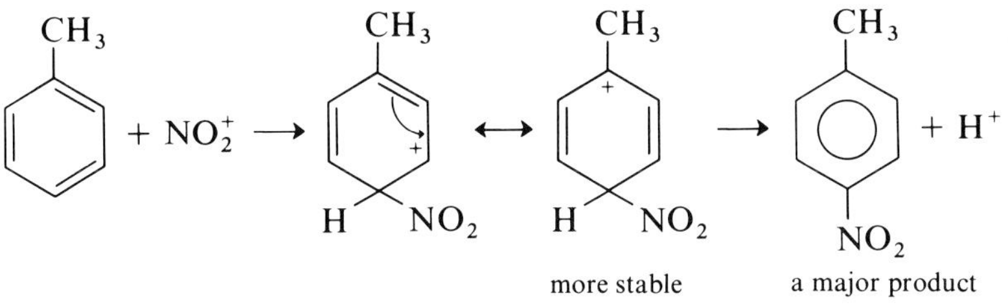

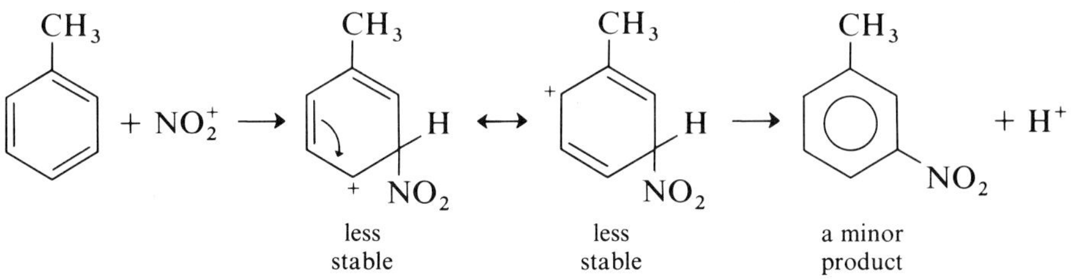

On prolonged reaction of toluene with HNO_3 and H_2SO_4, three NO_2 groups are introduced, occupying both ortho positions and the para position.

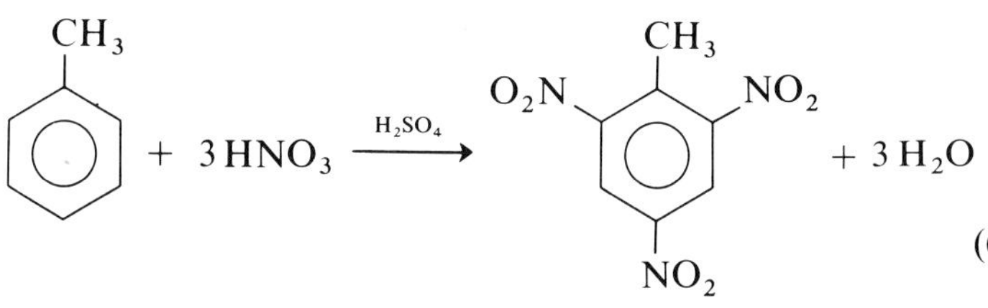

$$\text{toluene} + 3\,HNO_3 \xrightarrow{H_2SO_4} \text{2,4,6-trinitrotoluene (TNT)} + 3\,H_2O \quad (6\text{-}18)$$

mp 81°C; explodes at 240°C

Aromatic Hydrocarbons

TNT is a high explosive used both for military purposes and for blasting. Because it can be melted safely, it is easily cast into blocks or in the form of lead-sheathed wires (Primacord). When Primacord is detonated at one end, the explosion travels along the wire and can be used to detonate other explosive charges in quick succession if a turn of the Primacord is wrapped around each charge. It has also been used to dig ditches. The Primacord is buried in a shallow groove in the soil, detonated, and the result is instant ditch.

It should be noted that in the formation of trinitrotoluene from toluene, the second nitro group entered positions ortho or para to the methyl group, but meta to the first nitro group. That is, a nitro substituent already on the aromatic ring acts in just the opposite way from a methyl substituent. The nitro substituent is a **meta director**. Thus in the nitration of nitrobenzene, the main product is *m*-dinitrobenzene.

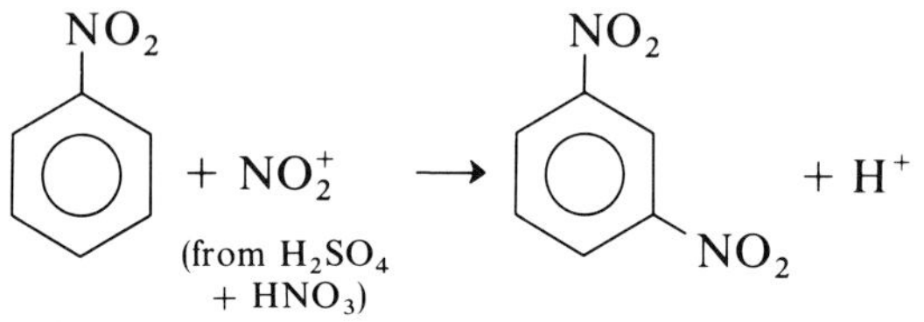

$$\text{C}_6\text{H}_5\text{NO}_2 + \text{NO}_2^+ \longrightarrow \text{C}_6\text{H}_4(\text{NO}_2)_2 + \text{H}^+$$

(from H_2SO_4 + HNO_3)

The explanation for this result is essentially the reverse of the explanation for the ortho-para directive effect of methyl substituents: $-NO_2$ makes a positive charge on an adjacent carbon atom less stable.[4] When the three possible adducts of NO_2^+ to nitrobenzene are examined, it is seen that the ortho and para adducts are *less* stable than the meta adduct.

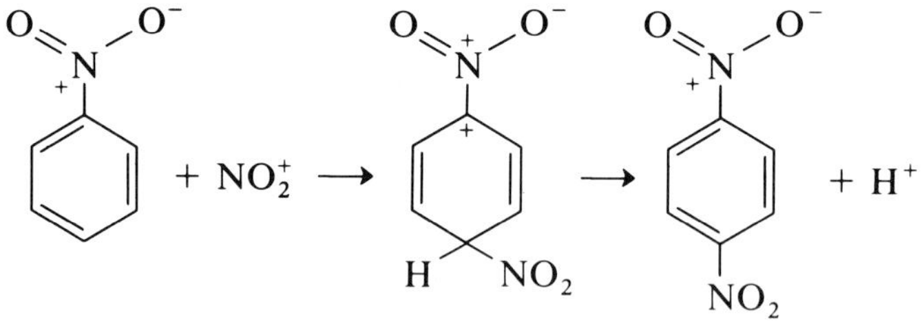

less stable ion / minor product

less stable ion / minor product

[4] The reason for this effect is the positive charge on the nitrogen atom of the nitro group, which can be depicted as

$$\text{C}_6\text{H}_5\text{NO}_2 + \text{NO}_2^+ \longrightarrow [\text{intermediate}] \longleftrightarrow [\text{intermediate}] \longrightarrow \text{m-dinitrobenzene} + \text{H}^+$$

more stable ion · major product

TNT in blocks can be used to impose patterns on steel, a process that depends on the shaped charge or Munroe effect. The artist carves the pattern in the face of a TNT block, lays it face down on a heavy steel plate, and detonates the TNT. Contrary to what you might reasonably have supposed, the incised part of the pattern in the steel is opposite the incised part of the pattern in the TNT. That is, the explosive that is *farther away* from the metal indents the metal more than the explosive that is in direct contact with it.

It is also possible to reproduce finely detailed patterns of delicate objects such as flowers or lace or ferns on a steel surface by pressing the object between the steel plate and a flat block of TNT, then detonating the block. TNT may be dangerous to your health because repeated contact with the skin can cause dermatitis.

The liberation of chemical energy that takes place in an explosion is a process not very different from the liberation of chemical energy in the process of combustion. Historically, the first explosive was black powder, which is just a mixture of combustible material (carbon and sulfur) with an oxidizing agent, KNO_3. Black powder burns rapidly even in the absence of air because the KNO_3 acts as the oxidant and air is not needed. Explosives like TNT are more efficient because the oxidant (the NO_2 groups) is not only already mixed with the fuel but is actually part of the same molecule.

Of course, the conversion of chemical energy into other forms of energy by producing an explosion is a very crude, primitive process when compared to what goes on in living systems such as a contracting muscle, or the bioelectric systems of nerve tissue and the electrical organs of the electric eel.

The fragment C_6H_5-, corresponding to benzene minus one hydrogen atom, is called the **phenyl** group.[5] Examples of the use of this nomenclature are the compounds **diphenylmethane** and **biphenyl**.

diphenylmethane, colorless solid; mp 25.9°C

biphenyl, colorless solid; mp 69–71°C

Biphenyl is used in molten form as a heat transfer agent when, for one reason or another, it is not feasible to use water. It has also been used as a fungistat to

---

[5] From a Greek root meaning "I bear light." The name *phene* was originally suggested for benzene because benzene was found in illuminating gas, where it increases the luminosity of the flame. Phene did not stick, however, except as phenyl for the substituent group.

prevent the growth of fungus on oranges during shipment. Diphenylmethane has an orangelike odor.

The compound **d**ichloro**d**iphenyl**t**richloroethane, known as DDT, has been used as an insecticide since 1942.

$$Cl-C_6H_4-CH(CCl_3)-C_6H_4-Cl$$

DDT

Although DDT has undoubtedly saved thousands of lives that would otherwise have been lost to insect-borne diseases such as malaria, its use has led to some unforeseen ecological effects. It turns out that small organisms that act as food for fish tend to concentrate the DDT that gets into rivers and oceans. The fish in turn concentrate the DDT still more and, when the fish are eaten by predators such as the osprey or the pelican, the concentration of DDT in the fatty tissue of the predator reaches alarming proportions. In the osprey, DDT is believed to make the egg shells too brittle for the young to survive. In our own body fat we do not know of any bad effects, so far.

FIGURE 6-6. A pelican.

The part structure $C_6H_5CH_2-$ is called the benzyl substituent.

benzyl chloride

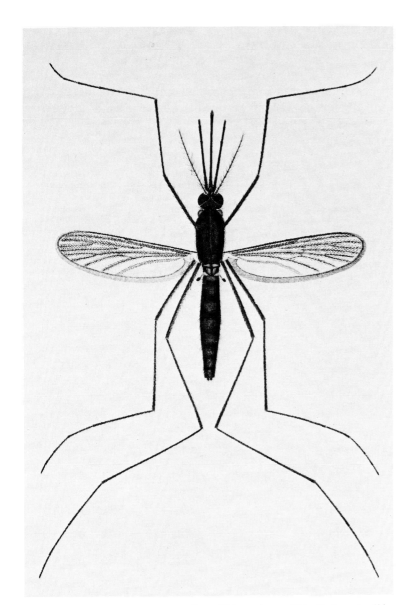

FIGURE 6-7. *Anopheles atropos* (from "Mosquitoes of North America," by S. J. Carpenter and W. J. La Casse). [Originally published by the University of California Press; reprinted by permission of The Regents of the University of California]

Aromatic Hydrocarbons

Polynuclear Hydrocarbons

NAPHTHALENE

In ethylene we have a molecule in which two electrons occupy a π orbital extending over two carbon atoms; in benzene we have a larger system in which six electrons occupy π orbitals extending over six carbon atoms. The compound naphthalene, $C_{10}H_8$, has the same sort of bonds as benzene except that there are ten π electrons in orbitals extending over a ten-carbon framework. Like benzene, it has all its atoms in the same plane.

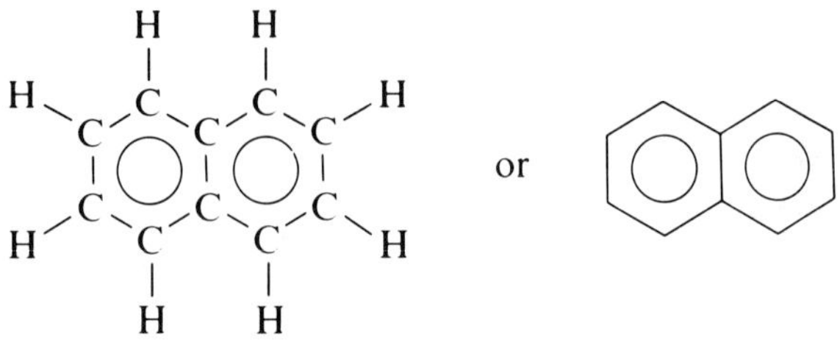

naphthalene
white crystals; mp 80.2

Note that two of the carbon atoms form part of both six membered rings and that *these carbon atoms have no hydrogen*. The reason for emphasizing this is that abbreviated formulas are often seen that do not show the atoms specifically. However, the valence of each carbon atom is always 4: three σ bonds to three adjacent atoms and one electron contributed to the π-electron cloud. Because the middle two carbons are each bonded to three other carbon atoms and each contributes one electron to the π cloud, these carbons do not have a valence left to bind a hydrogen atom as well.

The reactions of naphthalene are similar to those of benzene. For example, it can be hydrogenated (equation 6-19). Note that

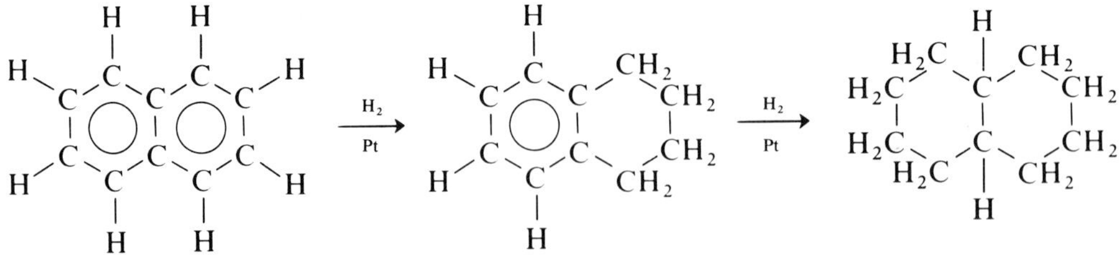

naphthalene → (H₂, Pt) → tetrahydronaphthalene or tetralin, bp 207°C → (H₂, Pt) → decahydronaphthalene or decalin, bp 194.6°C

(6-19)

naphthalene has two distinguishable positions, alpha (α) and beta (β), in which hydrogen may be replaced by a substituent. Chlorination, for example, gives a mixture of α and β monochlorination products (equation 6-20). All four α positions are alike and all four β positions are alike, so there is only one α-chloronaphthalene and one β-chloronaphthalene.

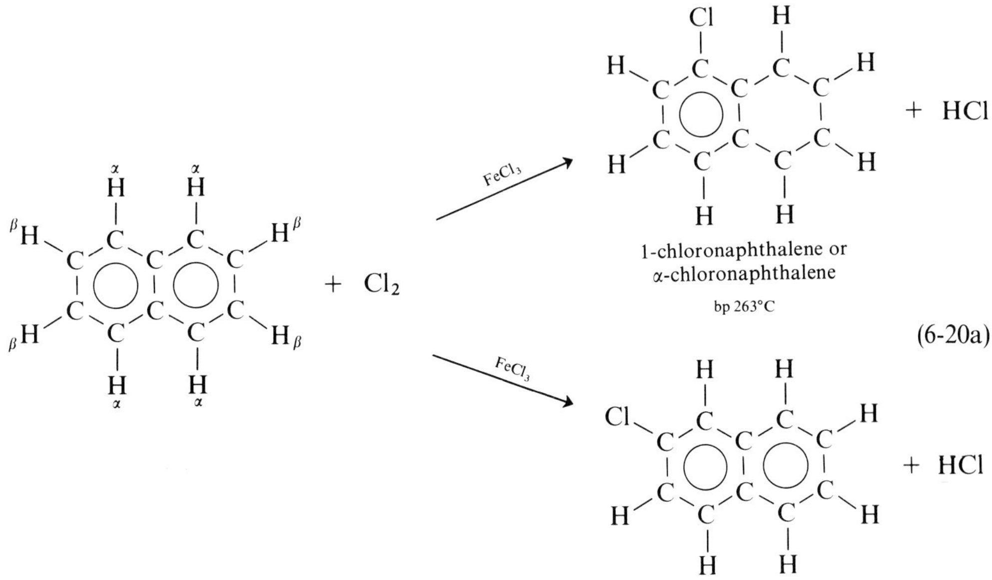

1-chloronaphthalene or α-chloronaphthalene

bp 263°C

(6-20a)

2-chloronaphthalene or β-chloronaphthalene

mp 55°C

(6-20b)

Naphthalenes with more than one substituent are named by a numbering system.

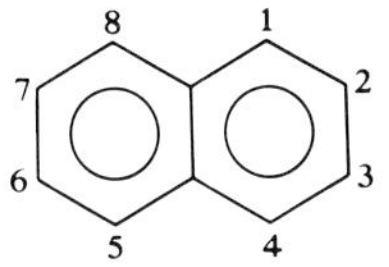

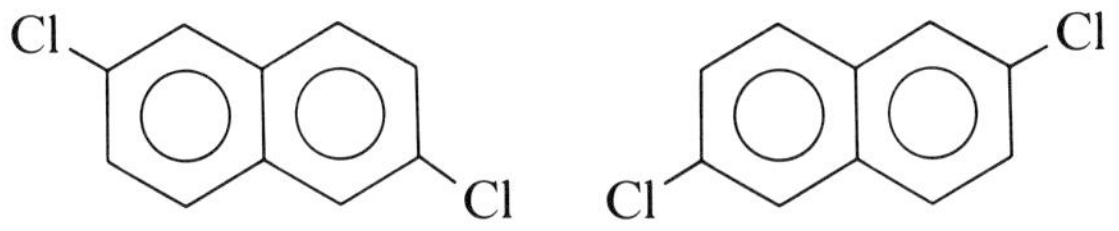

2,6-dichloronaphthalene

Naphthalene, in the form of compacted balls of the crystals, is used as a moth repellant and has a characteristic odor. It occurs in coal tar to the extent of about 11% and is the most abundant single constituent. Tetralin and decalin are used as solvents in floor waxes and shoe polish. Tetralin is used in commercial degreasing agents for cleaning tools and engines, and gives these products a characteristic pungent odor.

ANTHRACENE AND HIGHER MEMBERS

A large number of polynuclear aromatic hydrocarbons are known, but only a few samples will be discussed here. If we put one additional ring on the end of a naphthalene molecule we obtain the three-ring compound **anthracene**, which is also found in coal tar.

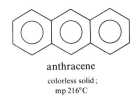

anthracene

colorless solid; mp 216°C

Pure anthracene crystals, as seen by an ordinary incandescent light, appear to be white but, in sunlight or in ultraviolet light in the dark, they can be seen to give off a beautiful blue glow called fluorescence.

Fluorescent compounds absorb light of short wavelength and high energy (as in sunlight or ultraviolet light) and store the energy for a very short time as electronic energy. Ordinary molecules convert this electronic excitation into heat, but fluorescent molecules reradiate it as light of a different color. Anthracene absorbs ultraviolet light, which is beyond the visible part of the spectrum, and reradiates it as blue light.

Phenanthrene is an isomer of anthracene. Unless phenanthrene is specially purified it is likely to contain traces of anthracene, which can be detected by the fluorescence of the crystals.

phenanthrene
colorless; mp 101°C

The angular sequence of three six membered rings exemplified by phenanthrene is also found in a varied group of compounds called steroids. Steroids such as cholesterol (page 133) and testosterone (page 143) are substances of considerable importance in living organisms.

FIGURE 6-8. Architectural use of the steroid ring system which is common in natural products.

When benzene or a polynuclear aromatic hydrocarbon absorbs light, the energy is used to promote one of the π electrons to a hitherto unused π orbital of higher energy. The color of the compound depends on the wavelength of the light that is absorbed.

Benzene, naphthalene, and anthracene are colorless because they absorb only ultraviolet light. The reflected or transmitted light is white light minus some of the ultraviolet. This still looks like white light to us because we cannot see ultraviolet light (even though it is capable of injuring our eyes!).

As the molecule and the cloud of π electrons is increased in size, the energy gaps between the π orbitals become smaller and light of lower energy—which means longer wavelength—is absorbed. The compound **naphthacene**, for example, absorbs blue light. Naphthacene is orange because orange is the color sensation produced by white light after the blue light has been removed.

naphthacene
orange crystals; mp 335°C

Many of the larger polynuclear aromatic hydrocarbons are carcinogenic.

This is by no means the only type of compound that can cause cancer. Because the incidence of cancer varies so much from one country to another and from one decade to another, it is estimated that 80% or more of the cases are caused by something in the environment and should therefore be preventable. The incidence of cancer also varies with occupation. Being a chimney sweep is hazardous because of the excessive contact with coal tar, which contains carcinogens such as benz[a]pyrene. The particular susceptibility of chimney sweeps to cancer was first pointed out in 1775.

benz[a]pyrene
yellow; mp 179°C
a carcinogenic hydrocarbon

Tobacco smoke is carcinogenic and has been shown to contain several four and five-ring aromatic hydrocarbons, including benz[a]pyrene.

One source of difficulty in recognizing causes of cancer is the long delay that may intervene between exposure to a carcinogen and the actual appearance of a tumor. Although it has been shown beyond any reasonable doubt that cigarette smoking does in fact cause cancer as well as a number of other pathological conditions, the tobacco industry still denies that there is any connection. They are like certain tribes in a remote part of New Guinea who are said not to have recognized the sequence of cause and effect that results in childbirth.

As the size of the polynuclear aromatic hydrocarbon is increased still further, the color deepens because red light begins to be absorbed as well as blue. By the time the molecule reaches the size of circumanthracene, it absorbs all of the light and is black.

circumanthracene
black; mp above 480°C; 97% C, 3% H

Another thing that happens as the size of the molecule increases is that the proportion of hydrogen, which is just a ring of atoms at the edge of the molecule, decreases. In the limit the molecule is pure carbon in the form of graphite, a black substance melting above 3500°C.

Graphite is a lubricant because the large, planar molecules are stacked on top of each other. The stack is easily deformed by a sliding motion, much like the spreading out of a deck of cards. The "lead" in a lead pencil is graphite mixed with clay to give it the desired degree of hardness.

SUMMARY

1. The structure of benzene: The benzene molecule is a planar, regular hexagon in which all of the C—C bonds have the same length.
    (a) The six π electrons of benzene occupy orbitals formed by the side-to-side overlap of a ring of six p_z orbitals.
    (b) Benzene is more stable than a typical alkene.
2. Reactions of benzene
    (a) They are like the reactions of alkenes except that substitution (which preserves the benzene π structure) is favored over addition.
    (b) The substitution reactions involve electrophilic reagents formed by the action of catalysts such as $FeCl_3$ or H_2SO_4.
    (c) Typical substitution reactions are chlorination, bromination, nitration, sulfonation, and alkylation.
3. Substituted benzenes
    (a) The first substituent, already on the ring, may be classified as either an ortho-para director or a meta director.
    (b) The second substituent enters the position indicated by the ortho-para or meta directing nature of the first substituent.
        (i) Ortho-para and meta directing effects can be explained in terms of electron-release and electron-withdrawal by the directing substituent.
4. Polynuclear hydrocarbons: Like benzene, these are planar. They make an hexagonal pattern of carbon atoms which, carried to the limit, is graphite, a form of pure carbon.
    (a) Important examples are naphthalene, anthracene, and phenanthrene.
    (b) Certain polynuclear hydrocarbons are carcinogenic.

EXERCISES

1. Draw scale diagrams of molecules of 1,3,5-cyclohexatriene and benzene showing the relative lengths of the C—C bonds.

2. Give structural formulas for toluene, naphthalene, and tetralin showing all hydrogen atoms explicitly.

3. Write structural formulas for (a) 2-phenylpentane; (b) *m*-chlorotoluene; (c) 1,2-dichloro-4-nitrobenzene; and (d) benzyl chloride.

4. Write equations for two addition reactions of benzene, including reaction conditions or catalysts.

5. Write equations (overall, not the mechanism of the reaction) for four substitution reactions of benzene.

6. One of the isomeric tribromobenzenes, $C_6H_3Br_3$, gives only one monochlorination product $C_6H_2Br_3Cl$. What is the structure of the original tribromobenzene?

7. How many isomers are there for (a) monomethylbenzene; (b) dimethylbenzene; (c) mononitronaphthalene; (d) dinitrobenzene; (e) dichloronaphthalene?

8. Name all of the compounds in question 7.

9. Show how to make each of the following from benzene, in two steps.

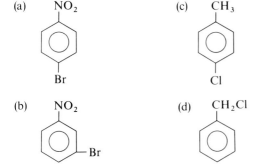

10. Give the structure, systematic name, and two civilized and nonbarbaric uses for TNT.

11. Explain briefly why TNT contains so much energy.

12. Name the following:

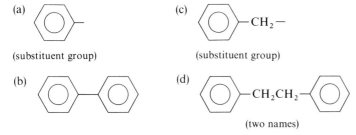

13. Explain why it is desirable that insecticides should be unstable enough to be degraded when they are ingested by living organisms.

14. Give the formal name and structure of DDT.

15. Explain why the colors of polynuclear aromatic hydrocarbons change as the molecule is made larger.

16. Explain why you should not allow coal tar to remain on your skin.

17. Give the names and structures of two three-ring aromatic hydrocarbons.

Chapter 7
Alcohols, Ethers, and Phenols

Alcohols

If one hydrogen atom of a water molecule is replaced by an alkyl group, the resulting compound is called an alcohol.

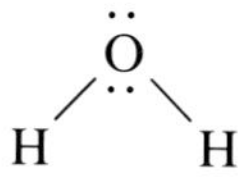

water   methyl alcohol or methanol CH_3OH
colorless liquid; bp 64.7°C

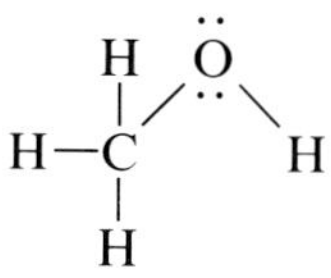

Alcohols are classified as primary, secondary, or tertiary according to the number of carbon atoms directly attached to the carbon bearing the OH group. Primary alcohols have one carbon atom directly attached to the carbon carrying the OH group. Note that ethyl alcohol and 1-propanol are both primary alcohols. The *indirectly* attached carbon atoms do not count.

Primary Alcohols, RCH_2OH

CH_3CH_2OH $CH_3CH_2CH_2OH$
ethyl alcohol or ethanol   1-propanol or n-propyl alcohol
bp 78.3°C   bp 97°C

Secondary Alcohols, $R-\underset{\underset{R'}{|}}{\overset{\overset{H}{|}}{C}}-OH$, RCHOHR'

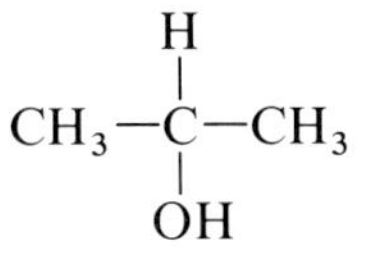

$CH_3-\underset{\underset{OH}{|}}{\overset{\overset{H}{|}}{C}}-CH_3$ $CH_3CHOHCH_3$ $CH_3-\underset{\underset{OH}{|}}{\overset{\overset{H}{|}}{C}}-CH_2CH_3$ $CH_3CHOHCH_2CH_3$

isopropyl alcohol or
2-propanol

bp 82.5°C

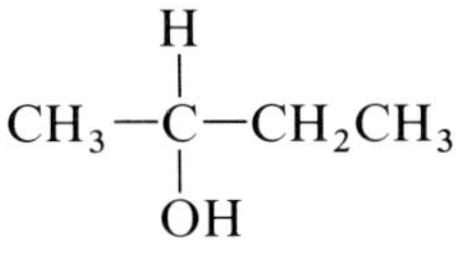

2-butanol or
sec-butyl alcohol

bp 99.5°C

Tertiary Alcohols, $R-\underset{\underset{R''}{|}}{\overset{\overset{R'}{|}}{C}}-OH$, RC(R')OHR''

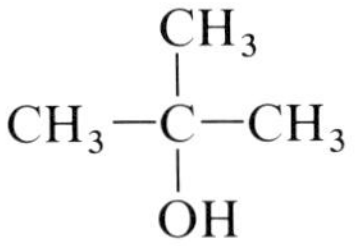

$CH_3-\underset{\underset{OH}{|}}{\overset{\overset{CH_3}{|}}{C}}-CH_3$ $CH_3C(CH_3)OHCH_3$ $CH_3-\underset{\underset{OH}{|}}{\overset{\overset{CH_3}{|}}{C}}-CH_2CH_3$ $CH_3C(CH_3)OHCH_2CH_3$

tertiary (*t*)-butyl alcohol or
2-methyl-2-propanol

bp 83°C; mp 25.5°C

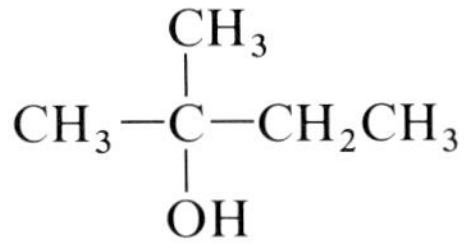

2-methyl-2-butanol or
tertiary (*t*)-pentyl alcohol

bp 102°C

Nomenclature

The systematic way of naming an alcohol is illustrated by the names **methanol** and **ethanol**. In these names the terminal **e** of the corresponding alkane has been replaced by **ol**. For larger molecules it is necessary to include a number to indicate the position of the OH group. This number is kept as small as possible.

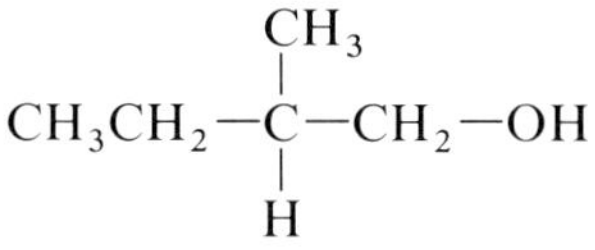

$CH_3CH_2-\underset{\underset{H}{|}}{\overset{\overset{CH_3}{|}}{C}}-CH_2-OH$

2-methyl-1-butanol

bp 128°C

The parent name is taken from the **longest chain that contains the hydroxyl group**. This is not always the longest chain. For example, in 3-propyl-2-hexanol the longest chain contains seven carbon atoms.

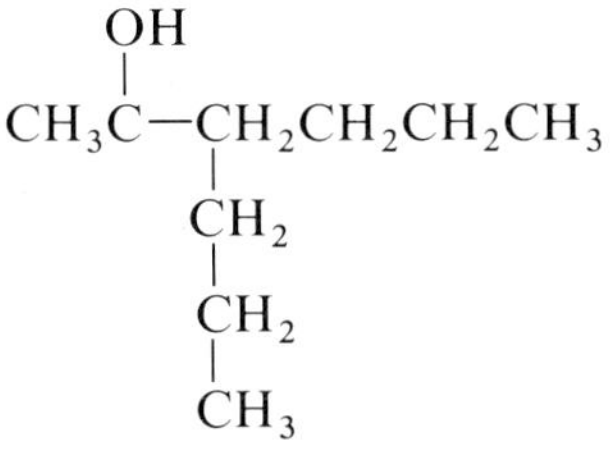

$CH_3\overset{\overset{OH}{|}}{C}-\underset{\underset{\underset{\underset{CH_3}{|}}{CH_2}}{\underset{CH_2}{|}}}{\overset{}{CH}}_2CH_2CH_2CH_3$

3-propyl-2-hexanol

Alcohols, Ethers, and Phenols

Naming more complicated molecules sometimes requires the use of the prefix **hydroxy**.

benzyl alcohol
bp 205°C

o-hydroxybenzyl alcohol
bp 86°C

Both benzyl alcohol and *o*-hydroxybenzyl alcohol are used as local anaesthetics. Benzyl alcohol is a constituent of jasmine and ylang-ylang oils. It has a very faint jasminelike odor.

In everyday use there are special terms for alcohols that reflect the substances from which they were made at one time. These names sometimes cause dangerous confusion between poisonous and relatively harmless alcohols and have been responsible for more than one death.

For example, methanol or methyl alcohol, is sold under the name "wood alcohol" because it used to be made by distilling wood chips. Methanol is quite poisonous. As little as 30 ml can cause death and lesser amounts can cause irreversible blindness.

Ethanol, the alcohol of alcoholic beverages, is sometimes called "grain alcohol" because one way of making it is the fermentation of grain. Ethanol boils at 78.3°C but forms a constant boiling **azeotropic** mixture with 5% water. The mixture, which boils at 78.13°C, cannot be separated by ordinary distillation but can be separated if benzene is added. The resulting 100% (or 200 proof) alcohol, called "absolute alcohol," is likely to be toxic because of remaining traces of benzene.

Finally, we have "denatured alcohol," which is ethyl alcohol to which has been added some substance such as benzene, aviation gasoline, or methyl alcohol to make it unpleasant as well as unsafe to drink. The U.S. Government permits denatured alcohol to be sold tax-free, thus making it cheap enough for use as a radiator antifreeze or as a solvent. It is possible, though illegal, to remove the denaturants from denatured alcohol. It is also too difficult to be worth the trouble. Even pure ethanol is moderately poisonous, to say nothing about ethanol from some dubious source. The oral LD_{50} for rats is 13.7 g/kg.

"Rubbing alcohol" is usually isopropyl alcohol.

Physical Properties

As the size of the alkyl group in an alcohol increases, the physical properties become less waterlike and more alkanelike. Thus methyl, ethyl, and *n*-propyl alcohols are miscible with water in any proportion; so is *t*-butyl alcohol. On the other hand *n*-butyl alcohol or 1-butanol is soluble only to the extent of 1 g/100 g of water. Water-soluble inorganic salts will usually dissolve in methanol to some extent but are less soluble or insoluble in ethanol.

The boiling points of the alcohols are higher than those of alkanes of about the same molecular weight because, like water, the alcohol molecules are held together by hydrogen bonds between the OH groups.

Many alcohols have distinctive odors or flavors and are found naturally in fruits and flowers. Synthetic alcohols are often added to foods to give them a particular flavor. Farnesol, whose systematic name shows how complex the system can be, has a lily-of-the-valley odor.

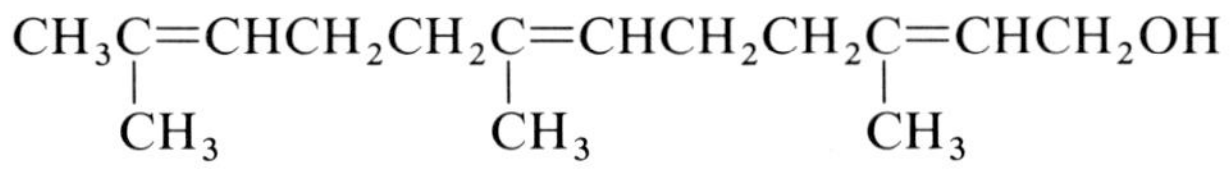

$$CH_3\underset{\underset{CH_3}{|}}{C}=CHCH_2CH_2\underset{\underset{CH_3}{|}}{C}=CHCH_2CH_2\underset{\underset{CH_3}{|}}{C}=CHCH_2OH$$

Farnesol or
3,7,11-trimethyl-2,6,10-dodecatriene-1-ol

Menthol produces a physiological reaction on the tongue that resembles a cool feeling and at the same time has a peppermintlike taste.

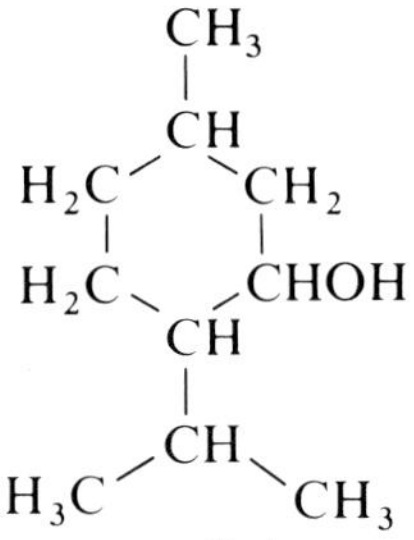

menthol
mp 41–43°C

Alcohols do not absorb light in the visible part of the spectrum and are therefore colorless. They do absorb infrared light; although this does not give rise to a visible color, the absorption can be measured with an infrared spectrophotometer. The presence of the OH functional group of an alcohol causes an **absorption band** somewhere in the 3610–3640 cm^{-1} range of infrared light frequencies.[1] The energy absorbed by the alcohol in this part of the infrared spectrum excites a stretching vibration of the O—H bond (Figure 7-1). If the pure alcohol is examined rather than a dilute

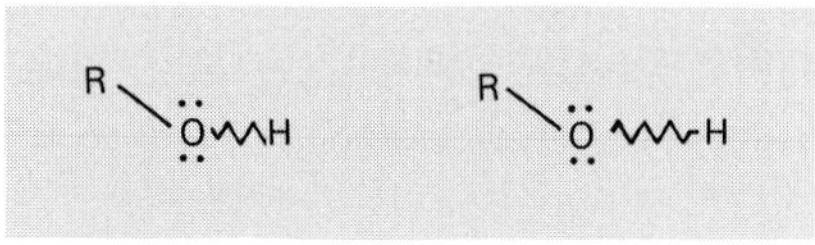

FIGURE 7-1. Stretching vibration of the O—H bond.

solution, a broad absorption band, extending from 3200 cm^{-1} to 3600 cm^{-1}, is obtained instead of the sharp band caused by the O—H stretching vibration of isolated OH molecules. This broad band is due to a similar motion of hydrogen atoms in hydroxyl groups that absorb at slightly different frequencies because they are involved in **hydrogen bonds** between two alcohol molecules (Figure 7-2).

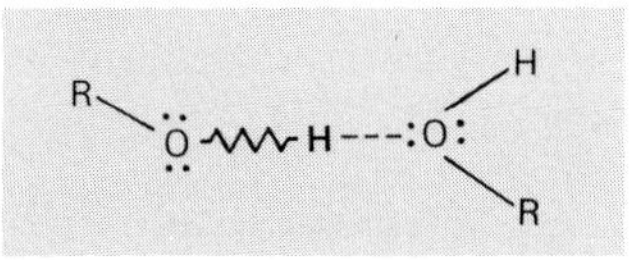

FIGURE 7-2. A pair of hydrogen-bonded alcohol molecules.

[1] 2760–2750 nm.

The attraction of the hydrogen nucleus (bold face H in Figure 7-2) for the oxygen electrons of the other alcohol molecule makes it easier to pull the hydrogen further away from its oxygen atom, which is just what is required to excite the O—H stretching vibration. This less energetic vibration can be caused by the absorption of a less energetic photon of light and accounts for the shifted position of the O—H band in the spectrum.

Water also has a broad O—H stretching absorption in the infrared. The reason that a sufficiently deep body of pure water looks blue is that an overtone of the O—H stretching vibration causes an absorption band that intrudes slightly into the visible part of the spectrum

Reactions of Alcohols

ALCOHOLS AS BASES OR ACIDS

The acid-catalyzed elimination of water from alcohols to give alkenes (Chapter 5) depends on the fact that a strong acid is able to donate a proton to the hydroxyl group. The hydroxyl group, in other words, is a base, if we define a base as a substance that can accept a proton. Like water, the alcohols are *very weak* bases and will accept a proton only from a comparatively strong acid (equations 7-1 and 7-2).

$$R-\ddot{O}-H + H^+ \rightleftharpoons R-\overset{+}{O}(H)(H) \quad (7\text{-}1)$$
(from a strong acid)

$$H-\ddot{O}-H + H^+ \rightleftharpoons H-\overset{+}{O}(H)(H) \quad (7\text{-}2)$$

Alcohols are also acids, though *very weak* ones, since they will give up their hydroxylic proton only to a strong base. Note that

$$CH_3CH_2-\ddot{O}-H + CH_3CH_2Na \rightarrow Na^{+\,-}O-CH_2CH_3 + CH_3CH_3 \quad (7\text{-}3)$$

very weak acid — ethylsodium, a very strong base — sodium ethoxide

ethane is too weak an acid to permit reaction 7-3 to go in the reverse direction.

The acidic hydrogen of an alcohol can be displaced by sodium, a reaction like that of water with sodium (equations 7-4 and 7-5).

$$CH_3OH + Na \rightarrow H_2 + Na^{+\,-}O-CH_3 \quad (7\text{-}4)$$
sodium methoxide

$$H_2O + Na \rightarrow H_2 + Na^{+\,-}OH \quad (7\text{-}5)$$

REPLACEMENT OF THE HYDROXYL GROUP BY CHLORINE

When *t*-butyl alcohol is shaken with concentrated HCl at room temperature, *t*-butyl chloride is formed as an oily layer floating on top of the acid (equation 7-6). The reason that the reaction goes so

$$\text{CH}_3-\underset{\underset{\text{CH}_3}{|}}{\overset{\overset{\text{CH}_3}{|}}{\text{C}}}-\text{OH} + \text{HCl} \rightarrow \text{CH}_3-\underset{\underset{\text{CH}_3}{|}}{\overset{\overset{\text{CH}_3}{|}}{\text{C}}}-\text{Cl} + \text{H}_2\text{O} \quad (7\text{-}6)$$

t-butyl alcohol conc. t-butyl chloride or 2-chloro-2-methylpropane bp 51°C

easily with t-butyl alcohol (and also with other tertiary alcohols) is that it involves the formation of a carbonium ion (equation 7-7).

$$\text{CH}_3-\underset{\underset{\text{CH}_3}{|}}{\overset{\overset{\text{CH}_3}{|}}{\text{C}}}-\text{OH} \xrightarrow{\text{H}^+} \text{CH}_3-\underset{\underset{\text{CH}_3}{|}}{\overset{\overset{\text{CH}_3}{|}}{\text{C}^+}} + \text{H}_2\text{O} \quad (7\text{-}7)$$

a carbonium ion

$$\text{CH}_3-\underset{\underset{\text{CH}_3}{|}}{\overset{\overset{\text{CH}_3}{|}}{\text{C}^+}} + \text{Cl}^- \text{ (from the HCl)} \rightarrow \text{CH}_3-\underset{\underset{\text{CH}_3}{|}}{\overset{\overset{\text{CH}_3}{|}}{\text{C}}}-\text{Cl} \quad (7\text{-}8)$$

In general, tertiary carbonium ions are more stable and more easily formed than secondary or primary carbonium ions, because of the alkyl substituents.[2]

Because primary and secondary carbonium ions are less easily formed, primary and secondary alcohols require vigorous reaction conditions to convert them to the corresponding chlorides.

$$\text{CH}_3\text{CH}_2\text{OH} + \text{HCl} \xrightarrow[\text{heat}]{\text{ZnCl}_2} \text{CH}_3\text{CH}_2\text{Cl} + \text{H}_2\text{O} \quad (7\text{-}9)$$

REPLACEMENT OF THE HYDROXYL GROUP BY SULFATE OR NITRATE

Sulfuric acid and nitric acid will also react with alcohols. These reactions are much like the reaction with HCl but give sulfates (equation 7-10) or nitrates instead of chlorides. Alkyl hydrogen

$$\text{CH}_3\text{CH}_2\text{OH} + \text{H}_2\text{SO}_4 \rightarrow \text{H}_2\text{O} + \text{CH}_3\text{CH}_2-\text{O}-\text{SO}_3\text{H} \quad (7\text{-}10)$$

ethyl hydrogen sulfate

sulfates are strong acids and form salts very easily. The sodium salt of lauryl hydrogen sulfate is a typical synthetic detergent.

$$\text{CH}_3\text{CH}_2\text{CH}_2\text{CH}_2\text{CH}_2\text{CH}_2\text{CH}_2\text{CH}_2\text{CH}_2\text{CH}_2\text{CH}_2\text{CH}_2-\text{O}-\overset{\overset{\text{O}^-}{|}}{\underset{\underset{\text{O}}{\|}}{\text{S}^+}}-\text{O}^-\text{Na}^+$$

sodium lauryl sulfate

Sodium lauryl sulfate has a long hydrophobic alkyl group which tends to make it dissolve in media like gasoline or greases. However, it also has a typical saltlike group at the other end of the molecule, which tends to make it dissolve in water. When both grease and water are present, sodium lauryl

[2] See Chapter 6 for other examples.

sulfate dissolves in both phases at the same time. As a result, a layer of the detergent molecules forms at the interface between the grease or hydrocarbon and the water. The hydrocarbon tails extend into the greasy phase and the salt extends into the water. Old-fashioned soap, made from fats (Chapter 10) acts in the same way (Figure 7-3) except that in soap the saltlike end of the molecule is COO^-Na^+ instead of $SO_3O^-Na^+$. The layer of detergent or soap reduces the amount of work needed to break up the grease into small droplets, which form an emulsion and are washed away.

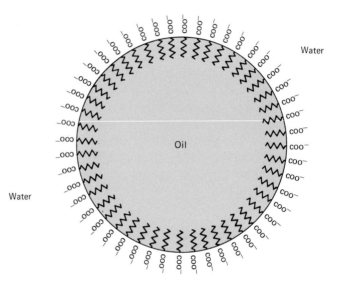

FIGURE 7-3. The hydrocarbon end (∿∿) of the soap molecule dissolves in the oil, the salt end (COO^-) dissolves in the water.

One important difference between soap and many synthetic detergents is that the soaps are quickly degraded by microorganisms. The presence of non-biodegradable detergents in streams is a serious part of the pollution problem. An advantage of the alkyl sulfate detergents is that they work in hard water. The Ca^{2+} ions of hard water form an insoluble precipitate with ordinary soap and make it necessary to use a great deal more than would otherwise be needed.

An example of the replacement of the hydroxyl group by nitrate (ONO_2) is the conversion of glycerol to glycerol trinitrate (equation 7-11).

$$\underset{\substack{\text{glycerol or} \\ \text{1,2,3-propanetriol} \\ \text{viscous liquid; mp 17.9°C}}}{\begin{array}{c} CH_2-CH-CH_2 \\ |\quad\;\; |\quad\;\; | \\ OH\;\; OH\;\; OH \end{array}} + 3\,HNO_3 \xrightarrow{H_2SO_4} \underset{\substack{\text{glycerol trinitrate or} \\ \text{nitroglycerine} \\ \text{heavy liquid (1.6 g/ml);} \\ \text{mp 13.2°C; explodes 260°C}}}{\begin{array}{c} CH_2-CH-CH_2 \\ |\quad\;\; |\quad\;\; | \\ O\quad\; O\quad\; O \\ |\quad\;\; |\quad\;\; | \\ NO_2\; NO_2\; NO_2 \end{array}} + 3\,H_2O \qquad (7\text{-}11)$$

Glycerol, sometimes also called glycerine, is obtained by the hydrolysis of fats (Chapter 10). It is used in the laboratory to lubricate rubber tubing to make it easier to slip on to a glass tube. It is also used in cosmetics to soothe and soften the skin.

Nitroglycerine in pills is used by heart patients as a vasodilator to make the blood circulate more easily. It is also a high explosive, sensitive both to shock and to high temperature. Although liquid nitroglycerine is used to put out oil well fires by the effect of the blast, most of it is used in the form of dynamite. Dynamite, invented by Alfred Nobel in 1866, consists of nitroglycerine soaked up in a porous material. The porous material may be either an inert substance such as diatomaceous earth, or another explosive such as cellulose nitrate (Chapter 13). Dynamite is much safer to handle than nitroglycerine, but it can easily be detonated by means of a small amount of primary explosive. The primary explosive in the "dynamite cap" is set off either by a fuse or electrically. The signs near construction projects urging you not to use your radio transmitter are there because some dynamite caps can be set off by a strong radio signal from a nearby transmitter.

At higher temperatures the reaction of alcohols with sulfuric acid leads to the formation of alkenes (Chapter 5) or ethers (see page 128).

Preparation of Alcohols

Methanol is prepared industrially by a special reaction starting from carbon monoxide.

$$2\,H_2 + CO \xrightarrow[400°C]{\text{metal oxides}} CH_3OH \qquad (7\text{-}12)$$

Ethanol is made by the fermentation of carbohydrates (Chapter 13) with yeast. Any carbohydrate will do, but sugar-cane molasses is a popular starting material (equation 7-13). The alcohol is separated from the water by distillation. If the sugar in fruits is used as the

$$\underset{\text{cane sugar}}{C_{12}H_{22}O_{11}} + H_2O \xrightarrow{\text{yeast}} 4\,CH_3CH_2OH + 4\,CO_2 \qquad (7\text{-}13)$$

raw material, the distillate usually contains traces of other compounds that give it a distinctive flavor; these include esters (Chapter 10) and aldehydes (Chapter 8).

Fermentation of starch gives not only ethyl alcohol but also a mixture of rather toxic alcohols known as fusel oil. These can be at least partly removed by distillation or by treatment with activated charcoal.

$$CH_3CH_2CH_2OH \qquad \underset{\underset{CH_3}{|}}{CH_3CHCH_2OH} \qquad \underset{\underset{CH_3}{|}}{CH_3CH_2CHCH_2OH}$$

Fusel oil components

The bacterium *Clostridium acetobutylicum* converts starch into 1-butanol (60%), ethanol (10%), and acetone (30%). The butyl alcohol is used to make a solvent for automobile lacquers.

$$\underset{\text{starch}}{(C_6H_{12}O_6)_n} \xrightarrow{\text{bacterium}} \begin{cases} CH_3CH_2CH_2CH_2OH \\ CH_3CH_2OH \\ \underset{}{CH_3-\overset{\overset{O}{\|}}{C}-CH_3} \end{cases} \qquad (7\text{-}14)$$

HYDROLYSIS OF ALKYL HALIDES

The hydrolysis of an alkyl halide is a general method of making alcohols. For tertiary alcohols it is sufficient just to treat the alkyl halide with water, but for primary and secondary alcohols the reaction is run with dilute sodium hydroxide (equation 7-15). This

$$\text{HO}^- + \text{CH}_3\text{CH}_2\text{CH}_2\text{Cl} \rightarrow \text{CH}_3\text{CH}_2\text{CH}_2\text{OH} + \text{Cl}^- \quad (7\text{-}15)$$
(NaOH) (NaCl)

reaction is a **nucleophilic displacement** in which the strong nucleophile HO⁻ displaces the weaker nucleophile Cl⁻ from the carbon nucleus.

$$\text{HO}^- \quad \overset{H}{\underset{\underset{\underset{CH_3}{|}}{CH_2}}{C}}\!\!-\!\!\overset{H}{}\text{Cl} \rightarrow \text{HO}\!-\!\overset{H}{\underset{\underset{\underset{CH_3}{|}}{CH_2}}{C}}\!\overset{H}{} \quad \text{Cl}^-$$

HYDRATION OF ALKENES

When an alkene is dissolved in cold sulfuric acid, the acid adds to the double bond to form an alkyl hydrogen sulfate (equation 7-16). The reaction mixture is then diluted with water and heated

$$\text{CH}_3\text{CH}=\text{CH}_2 + \text{H}_2\text{SO}_4 \rightarrow \underset{\underset{\underset{SO_3H}{|}}{O}}{\text{CH}_3\text{CHCH}_3} \quad (7\text{-}16)$$

without bothering to isolate the alkyl hydrogen sulfate (equation 7-17). The overall effect of these two operations, (a) dissolving the

$$\underset{\underset{\underset{SO_3H}{|}}{O}}{\text{CH}_3\text{CHCH}_3} + \text{H}_2\text{O} \xrightarrow{\text{heat}} \underset{\underset{OH}{|}}{\text{CH}_3\text{CHCH}_3} + \text{H}_2\text{SO}_4 \quad (7\text{-}17)$$

alkene in cold sulfuric acid and (b) diluting the mixture and heating it, is to add H₂O across the double bond as in equation 7-18.

$$\text{CH}_3\text{CH}=\text{CH}_2 + \text{H}_2\text{O} \xrightarrow{\text{H}_2\text{SO}_4} \underset{\underset{OH}{|}}{\text{CH}_3\text{CHCH}_3} \quad (7\text{-}18)$$

Note that Markovnikov's rule is obeyed.

HYDROBORATION

To convert an alkene to the alcohol *not* predicted by Markovnikov's rule a sequence of reactions starting with diborane (B₂H₆) is used. For example, propylene can be converted to 1-propanol in this way (equations 7-19 to 7-22).

$$3\,\text{CH}_3\text{CH}=\text{CH}_2 \xrightarrow{B_2H_6} (\text{CH}_3\text{CH}_2\text{CH}_2)_3\text{B} \quad \text{not} \quad \left(\underset{\underset{CH_3}{|}}{\overset{CH_3}{|}}{\text{H}\!-\!\text{C}\!-\!}\right)_{\!\!3}\!\text{B} \quad (7\text{-}19)^3$$

---

[3] The boron-hydrogen (B-H) compound adds across the double bond in such a way as to put the hydrogen rather than the bulkier boron atom in the position having the alkyl substituent. Thus the hydrogen is found in the **anti**-Markovnikov position.

$$(CH_3CH_2CH_2)_3B \xrightarrow{H_2O_2} (CH_3CH_2CH_2O)_3B + H_2O \quad (7\text{-}20)$$

$$(CH_3CH_2CH_2O)_3B \xrightarrow[H_2O]{NaOH} 3\ CH_3CH_2CH_2OH + B(OH)_3 \quad (7\text{-}21)$$

The overall result therefore is

$$CH_3CH=CH_2 \xrightarrow{B_2H_6} \xrightarrow{H_2O_2} \xrightarrow{H_2O} CH_3CH_2CH_2OH \quad (7\text{-}22)$$

Enols

The systematic name for a compound with both a double bond and an hydroxyl group in the molecule has both the **en** ending for the double bond and the **ol** ending for the hydroxyl group. An example is 2-propen-1-ol, or allyl alcohol.

$$CH_2=CHCH_2OH$$
2-propen-1-ol or allyl alcohol
colorless liquid; pungent odor;
irritates the eyes; bp 96–97°C

The chemical behavior of allyl alcohol can be sufficiently described merely by saying that it gives the typical reactions of *both* an ordinary alkene and an ordinary alcohol. However, when the hydroxyl group of an unsaturated alcohol is attached directly to one of the carbon atoms of the double bond, the properties of the compound are quite different. Most such compounds are unstable, and all attempts at making them lead instead to an isomer having a C=O double bond.

$$\underset{\substack{\text{the enol form}\\\text{of acetone}}}{CH_2=\underset{\underset{\text{OH}}{|}}{C}-CH_3} \rightleftarrows \underset{\substack{\text{acetone}\\\text{colorless liquid; bp 56.1°C}}}{CH_3-\underset{\underset{\text{O}}{\|}}{C}-CH_3} \quad (7\text{-}23)$$

Acetone, for example, contains only 0.00025 % of its enol isomer in equilibrium with the C=O or keto form. The importance of this minor isomer lies in the fact that many of the reactions of acetone and related compounds (Chapter 8) are actually reactions of the enol; the acetone is slowly converted to the enol by acidic or basic catalysts, and the enol then reacts rapidly with reagents such as bromine (equation 7-24). Thus a reaction that appears to be a

$$\underset{\text{acetone}}{CH_3-\underset{\underset{\text{O}}{\|}}{C}-CH_3} \xrightarrow[\text{(slow)}]{\text{acid cat.}} CH_2=\underset{\underset{\text{OH}}{|}}{C}-CH_3 \quad (7\text{-}24)^4$$

$$CH_2=\underset{\underset{\text{OH}}{|}}{C}-CH_3 + Br_2 \xrightarrow{\text{fast}} \left[\underset{\underset{\text{Br}}{|}}{CH_2}-\underset{\underset{\text{Br}}{|}}{\overset{\overset{\text{OH}}{|}}{C}}-CH_3\right] \xrightarrow{\text{fast}} \underset{\underset{\text{Br}}{|}}{CH_2}-\underset{\underset{\text{O}}{\|}}{C}-CH_3 + HBr$$

substitution reaction of acetone is in fact an addition reaction of the enol.

---

[4] In the acid-catalyzed enolization, a proton from the catalyst bonds to the oxygen and is replaced by one that departs from the CH_3 group.

$$CH_3-\overset{\overset{H^+}{\curvearrowright}}{\underset{\underset{}{\|}}{C}}-CH_2-H \rightarrow CH_3-\underset{\underset{}{|}}{\overset{\overset{O-H}{|}}{C}}=CH_2 \quad H^+$$

Alcohols, Ethers, and Phenols

A few enols are stable. An example is benzoylacetone, which exists 99 % in the enol form (equation 7-25). Benzoylacetone gives a

$$\underset{99\%}{\underset{\text{}}{\text{Ph}-\overset{OH}{\underset{|}{C}}=\overset{}{\underset{H}{C}}-\overset{O}{\underset{||}{C}}-CH_3}} \rightleftharpoons \underset{1\%}{Ph-\overset{O}{\underset{||}{C}}-CH_2-\overset{O}{\underset{||}{C}}-CH_3} \quad (7\text{-}25)$$

red color with ferric chloride ($FeCl_3$), a reaction characteristic of enols.

An example of a naturally occurring compound that is stable in the enol form is **ascorbic acid** or vitamin C. Ascorbic acid is widely distributed in

$$\begin{array}{c} HO \quad OH \\ | \quad \; | \\ C=C \\ / \quad \quad \backslash \\ H-C \quad \quad C=O \\ | \quad \; O \; \; / \\ CHCH_2OH \\ | \\ OH \end{array}$$

ascorbic acid

colorless solid; mp 190–192°C

plants and animals, but to be a useful source of the vitamin the food must be reasonably fresh because ascorbic acid is slowly oxidized by air. Scurvy, the dietary deficiency disease caused by lack of ascorbic acid is painful and debilitating and can cause death. Although scurvy is uncommon now, it almost always appeared during the long sailing voyages of the seventeenth and eighteenth centuries when ships were at sea for many months at a time. A particularly interesting description is given in *Voyage Around the World* (Anson's expedition), by Richard Walter, who was a member of the expedition. This book, though printed in 1753, is not particularly rare and may be in your local library. Although it is printed in the old style with a symbol that looks like f in place of the letter s at the beginning of words, it is easily readable and a first rate adventure story. At one point in the voyage, the ship was within sight of an island for several days but could not get there for lack of men with enough strength to trim the sails. In 1753 it was not generally known that fresh food would cure scurvy, but it had been noticed that the symptoms always seemed to disappear when the ship reached port. That the disease was caused by the lack of a specific vitamin was not recognized until 1917. Curiously, man is one of the few mammals that require ascorbic acid in the diet. The need for ascorbic acid is only one of many ways in which man is dependent on other species to sustain his own life.

Ethers

Compounds derived from water by replacing both hydrogen atoms with alkyl or aromatic groups are called ethers. Because they have no OH group to form hydrogen bonds, ethers boil at lower temperatures than alcohols of the same molecular weight. As we have already seen, dimethyl ether is a gas, whereas its isomer ethyl alcohol is a liquid at room temperature. The ethers R—O—R and

CH₃—O—CH₃
dimethyl ether
bp −24°C

CH₃CH₂—O—CH₂CH₃
diethyl ether
bp 34.6°C

the corresponding alkanes of formula R—CH$_2$—R rarely boil more than 10°C apart.

Low molecular weight ethers, and compounds with several ether groups in the same molecule, are soluble in water. The reason for their solubility is hydrogen bonding of the OH group of the water molecule to the ether oxygen (Figure 7-4). This is not sufficient to make compounds with large alkyl groups soluble in water, however. Diethyl ether, solubility 7.5 g/100 g is a borderline case; *t*-butyl ethyl ether is insoluble.

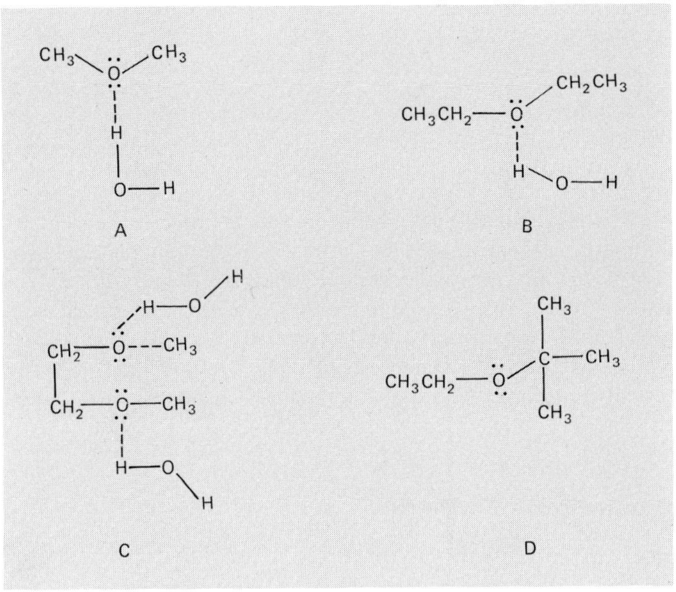

FIGURE 7-4. The larger the proportion of the alkyl part of the molecule to the oxygen part of the molecule, the less soluble the ether is in water. (**A**) Dimethyl ether; soluble, bp −24°C. (**B**) Diethyl ether; 7.5 g/100 g water, bp 34.6°C. (**C**) 1,2-Dimethoxyethane; soluble, bp 82°C. (**D**) *t*-Butyl ethyl ether; insoluble, bp 68°C.

Ethers are colorless unless other light-absorbing groups are also present in the molecule. The infrared spectrum has a broad absorption band at about 9000 nm,[5] corresponding to a C—O stretching vibration.

Diethyl ether is used in the laboratory as a solvent and in hospitals as an inhalation anaesthetic. It is the ether that is meant when the word ether is used alone, just as alcohol usually means ethyl alcohol.

Ether became generally used as an anaesthetic for surgical operations in about 1850, before which the patient was just held down by strong men. Because of the flammability of ether and the explosion hazard from ether-air mixtures, care must be taken to avoid sparks from static electricity.

[5] 1100 cm^{-1}.

Ether is also used for starting cold Diesel or gasoline engines. A piece of cloth wet with ether is simply held near the air intake of the engine and the vapor is sucked into the cylinders. Ether volatilizes and ignites so much more readily than gasoline or Diesel fuel that the engine usually starts instantly.

The ether linkage is part of the structure of many biologically important natural or synthetic compounds.

cineol
camphorlike odor;
bp 174–177°C

Cineol, for example, is the chief constituent of the oil from eucalyptus leaves. It is also found, usually with seven to ten other substances, in the volatile fragrances of many orchids. The fragrance has the important biological function of attracting bees to the flower to pollinate it. Because different species of orchids can be crosspollinated to produce hybrids, the continued existence of pure strains requires some mechanism that prevents crosspollination. Each orchid species has its own special fragrance, corresponding to a particular mixture of compounds. The special fragrance attracts only one kind of bee which visits, and pollinates, only that kind of orchid.

Preparation of Ethers

First, 1 mole of alcohol is converted to ethyl hydrogen sulfate (see equation 7-10), then a second mole of alcohol is added slowly at about 140°C.

$$CH_3CH_2O-\overset{\overset{O}{\|}}{\underset{\underset{O^-}{|}}{S}}-O-H + CH_3CH_2OH \xrightarrow{140°C} CH_3CH_2-O-CH_2CH_3 + H_2SO_4 \quad (7\text{-}26)$$

A more general procedure for making ethers is the Williamson synthesis, equation 7-27.

$$CH_3CH_2O^-Na^+ + CH_3Cl \rightarrow CH_3CH_2-O-CH_3 + Na^+Cl^- \quad (7\text{-}27)$$

sodium ethoxide   methyl ethyl ether
            bp 7.9°C

Notice that reaction 7-27 is essentially the same as the reaction between sodium hydroxide and methyl chloride to give methyl alcohol and sodium chloride (see equation 7-15). That is, Cl^- is displaced from carbon by a stronger nucleophile, ethoxide ion ($CH_3CH_2O^-$) in one reaction and OH^- in the other.

Methyl ethyl ether can also be called methoxyethane, using the substituent name **methoxy** for CH_3O-.

Vinyl ethyl ether can be made by adding an alcohol to acetylene.

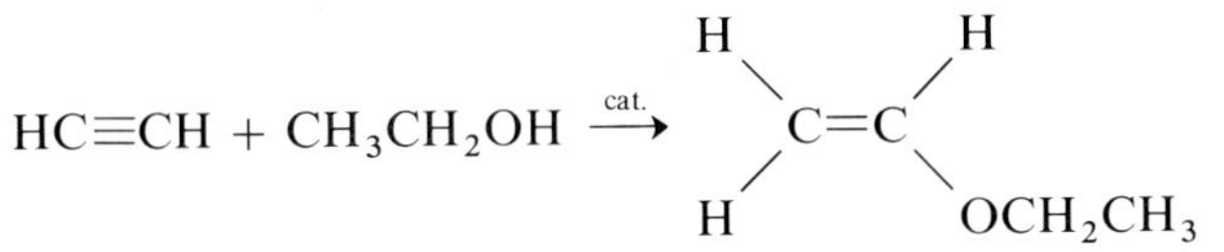

$$HC\equiv CH + CH_3CH_2OH \xrightarrow{cat.} \underset{\substack{\text{vinyl ethyl ether} \\ \text{bp } 35.5°C}}{\overset{H}{\underset{H}{>}}C=C\overset{H}{\underset{OCH_2CH_3}{<}}} \qquad (7\text{-}28)$$

Reactions of Ethers

The ether functional group is not very reactive, and the reactions of ethers are usually just the ordinary reactions of some other functional group that happens to be present in the molecule. Thus **anisole**, or methyl phenyl ether, can be brominated in the ring. The methoxy substituent (CH_3O) is a powerful ortho-para director, as can be seen from equation 7-29.

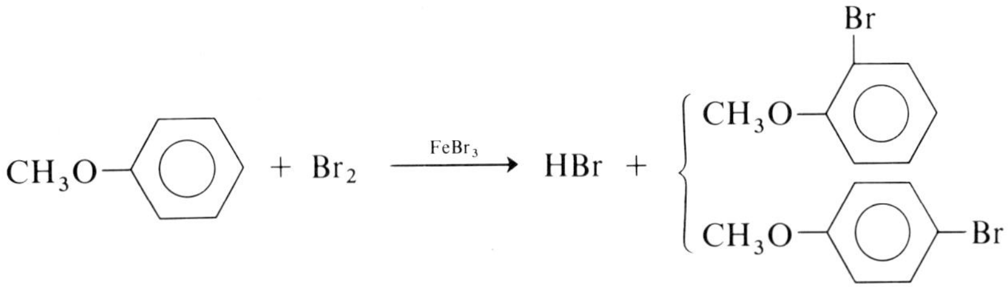

$$CH_3O-C_6H_5 + Br_2 \xrightarrow{FeBr_3} HBr + \begin{cases} CH_3O-C_6H_4-Br \text{ (ortho)} \\ CH_3O-C_6H_4-Br \text{ (para)} \end{cases} \qquad (7\text{-}29)$$

The most important reaction of ethers in which something actually happens to the ether functional group itself is the cleavage by HBr or HI. This reaction does not occur rapidly, and usually requires heating (equation 7-30). Excess HBr converts the alcohol to a second

$$CH_3CH_2-O-CH_2CH_3 + HBr \xrightarrow[\text{heat}]{\text{pressure}} CH_3CH_2Br + CH_3CH_2OH \qquad (7\text{-}30)$$

molecule of ethyl bromide.

Phenols

Phenol is both the name for the simplest member of a class of compounds and the name for the class. Any compound with a hydroxyl group directly attached to an aromatic ring is a phenol.

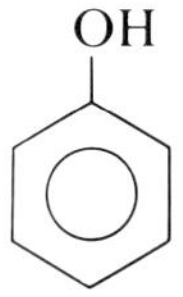

phenol
colorless; mp 40.85°C;
solubility, 1 g/15 ml of water

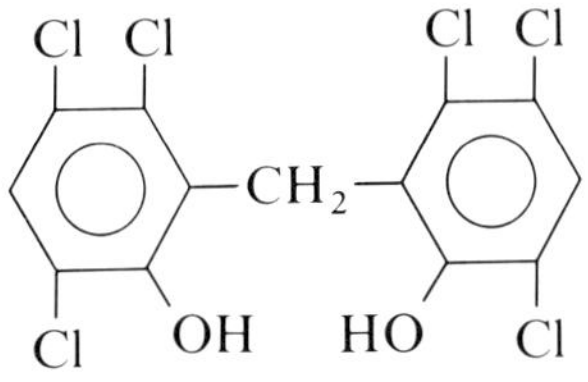

hexachlorophene
mp 164–165°C

Hexachlorophene has been used in germicidal soaps, such as pHisohex (trade name), and as a disinfectant.

Note that benzyl alcohol (page 118) is not a phenol, since it does not have an OH group *directly* attached to the benzene ring. On the other hand *o*-hydroxybenzyl alcohol is both a phenol and an alcohol.

Reactions of Phenols

There are several reasons for classifying phenols as a group of compounds distinct from the alcohols. The main reason is the fact that they are moderately acidic whereas the alcohols are essentially neutral. Thus both alcohols and phenols will react with sodium to give sodium salts and hydrogen (equations 7-31 and 7-32), but only the phenols are acidic enough to react with NaOH to form sodium salts. The formation of a sodium salt of a phenol by reaction with

$$CH_3CH_2CH_2CH_2CH_2OH + Na \rightarrow H_2 + CH_3CH_2CH_2CH_2CH_2O^-Na^+ \qquad (7\text{-}31)$$

$$\text{(ethylphenol)} + Na \rightarrow H_2 + \text{(sodium ethylphenoxide)} \qquad (7\text{-}32)$$

aqueous NaOH (equation 7-33) provides a very convenient way to separate phenols from less acidic and neutral compounds.

$$H_3C\text{-}C_6H_4\text{-}OH + NaOH \rightarrow H_2O + H_3C\text{-}C_6H_4\text{-}O^-Na^+ \qquad (7\text{-}33)$$

p-methylphenol or *p*-cresol
almost insoluble in water

colorless solid; mp 35.5°C

sodium salt of *p*-cresol
very soluble in water

For example, suppose we want to separate a mixture of benzyl alcohol and *p*-cresol. Neither substance is very soluble in water, so the mixture cannot just be washed with water in the hope that one of them will dissolve. However, these compounds can be separated by dissolving the mixture in ether and washing it with a water solution of NaOH. The phenol will react with the NaOH to form the water-soluble sodium salt; this will dissolve in the water layer of the separatory funnel while the alcohol will remain dissolved in the ether layer.

To recover a phenol from its sodium salt, use is made of the principle that a stronger acid will react with the salt of a weaker acid (Chapter 2).

SALT OF WEAKER ACID	STRONG ACID	WEAKER ACID	SALT OF STRONGER ACID

$$H_3C\text{-}C_6H_4\text{-}O^-Na^+ + HCl \rightarrow H_3C\text{-}C_6H_4\text{-}OH + NaCl \qquad (7\text{-}34)$$

sodium salt of *p*-cresol;
water soluble

After the strong acid has been added to the solution of the sodium salt of the phenol, the phenol either crystallizes out or can be extracted with ether.

Phenols are weaker acids than carboxylic acids. Thus a mixture of p-cresol and benzoic acid can be separated by dissolving the benzoic acid in the weak base, $NaHCO_3$; in this case p-cresol will not react and remains undissolved.

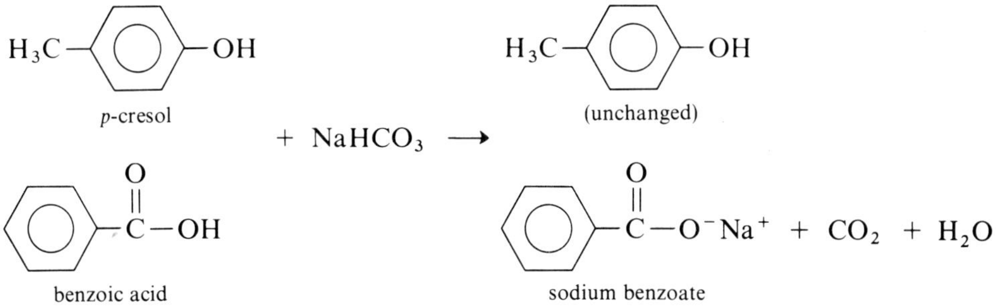

$$\text{p-cresol} + NaHCO_3 \longrightarrow \text{(unchanged p-cresol)}$$

$$\text{benzoic acid} \longrightarrow \text{sodium benzoate} + CO_2 + H_2O \qquad (7\text{-}35)$$

Another reason for considering phenols as a group distinct from that of the alcohols, besides their acidity, is the colored iron salt that they give with ferric chloride. This reaction is used as a color test for the presence of either a phenol or an enol. That both enols and phenols give such a test is not surprising when we compare their structures, using a double-bond structure for the benzene ring of the phenol. Enols are also more acidic than alcohols (see ascorbic acid, for example).

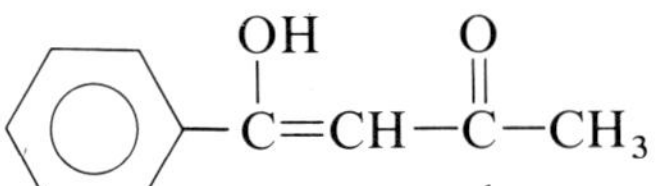

an enol

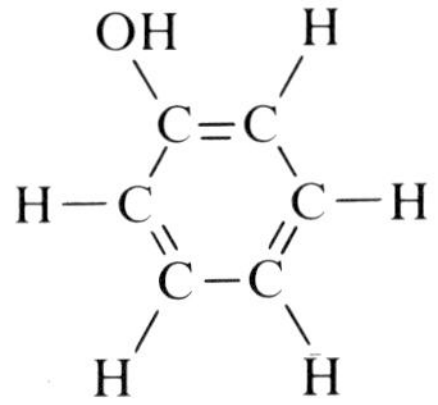

phenol

Unlike most enols, phenols exist almost exclusively in the hydroxy form. The isomer with a C=O instead of one of the C=C of the benzene ring would sacrifice much of the special stability of the benzene ring. In fact, when attempts are made to synthesize the keto form, a phenol is obtained instead (equation 7-36).

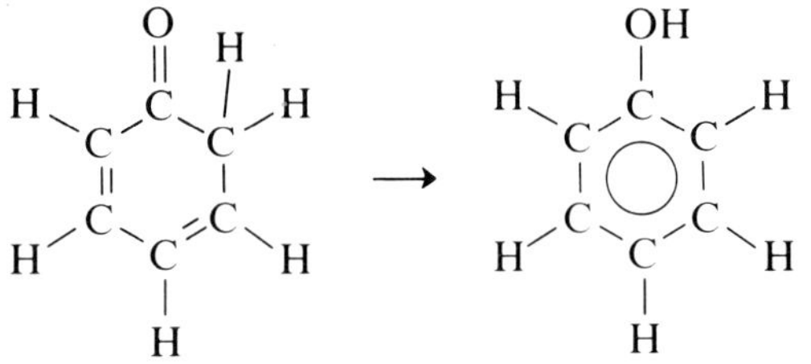

$$\text{unstable} \longrightarrow \text{stable} \qquad (7\text{-}36)$$

Phenol, which is also called **carbolic acid**, and the cresols are used widely as disinfectants and can be recognized by their characteristic sharp odors or "hospital smell." They should not be taken internally or applied to large areas of the skin.

Alcohols, Ethers, and Phenols

Phenols readily undergo electrophilic aromatic substitution reactions. The nitration of phenol with excess HNO_3 (H_2SO_4 catalyst) gives picric acid (equation 7-37).

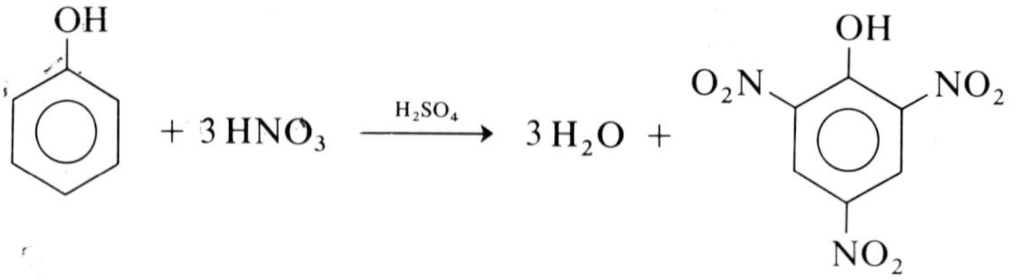

$$\text{C}_6\text{H}_5\text{OH} + 3\,HNO_3 \xrightarrow{H_2SO_4} 3\,H_2O + \text{2,4,6-trinitrophenol} \quad (7\text{-}37)$$

2,4,6-trinitrophenol
or **picric acid**
mp 122–123°C
(explodes if heated rapidly)

Like nitroglycerine, which also contains its own oxidizing agent in the form of NO_2 groups, picric acid is an explosive. It is also an antiseptic and has been used in burn ointments. Reaction of picric acid with alkali or other bases converts it into bright yellow salts. Proteins such as those in wool, silk fibers, or human skin, have basic amino groups ($-NH_2$, Chapter 11). Because of the presence of the NH_2 groups, proteins are dyed yellow by picric acid. If you get it on your hands in the laboratory, you must expect your hands to remain yellow for several days until the upper layer of skin wears off.

Mixed fibers can be detected by means of picric acid. A mixed cotton and wool thread, dyed with picric acid, and examined under a microscope will show both yellow fibers (the wool) and colorless fibers (the cotton). Picric acid has a unique collection of properties. One who makes a petard or bomb out of picric acid can be hoist, dyed yellow, and have his burns anointed, all in a single operation.

TABLE 7-1. *Other Reactions of Phenols*

Reaction	Equation
Bromination	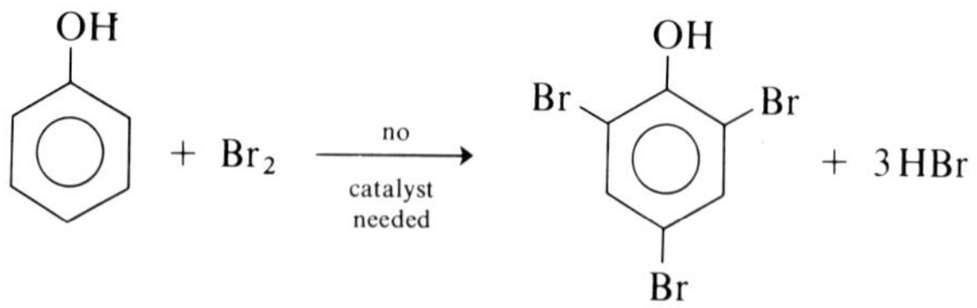 phenol + Br_2 $\xrightarrow{\text{no catalyst needed}}$ 2,4,6-tribromophenol + $3\,HBr$
Diazo coupling	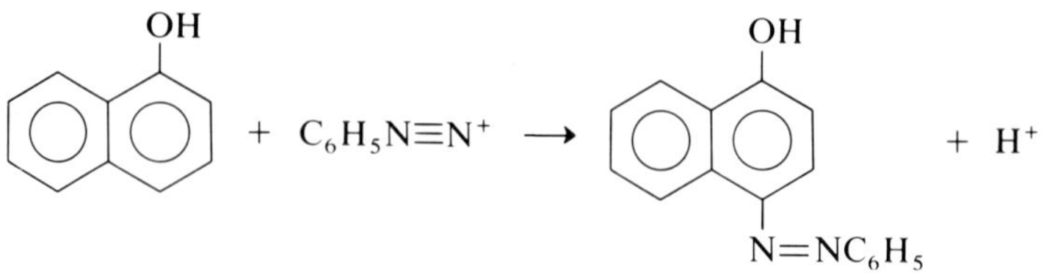 naphthol + $C_6H_5N{\equiv}N^+$ $\rightarrow$ 1-(phenylazo)-2-naphthol ($N=NC_6H_5$) + H^+
Williamson synthesis	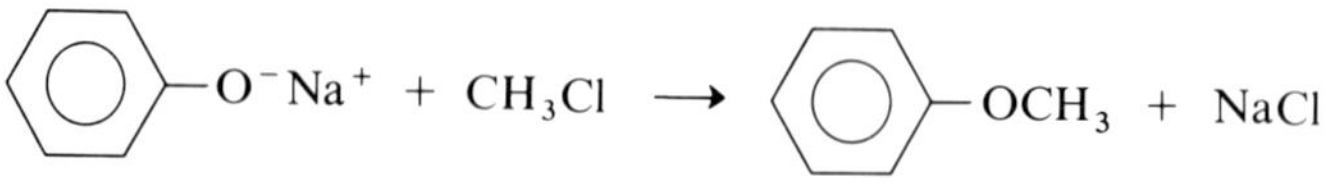 $C_6H_5{-}O^-Na^+ + CH_3Cl \rightarrow C_6H_5{-}OCH_3 + NaCl$

The reactions of α- and β-naphthol are similar to those of phenol.

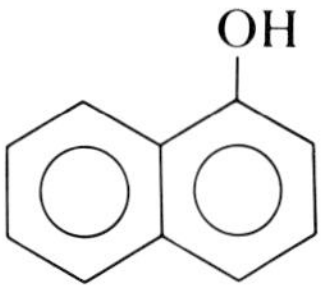

α-naphthol
colorless solid; mp 96°C

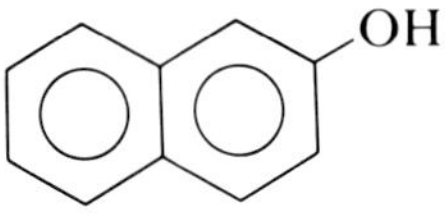

β-naphthol
colorless solid; mp 121–123°C

Preparation of Phenols

Phenols can be prepared by fusing the sodium salt of the corresponding sulfonic acid with NaOH.

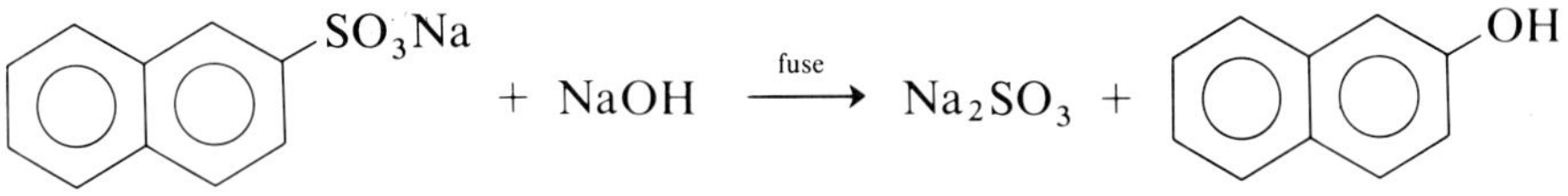

$$\text{C}_{10}\text{H}_7\text{SO}_3\text{Na} + \text{NaOH} \xrightarrow{\text{fuse}} \text{Na}_2\text{SO}_3 + \text{C}_{10}\text{H}_7\text{OH} \tag{7-38}$$

SOME PHENOLS

A number of simple substituted phenols are found in plants. **Eugenol**, which occurs in eucalyptus leaves, is used as an analgesic. Dentists sometimes put a small amount of eugenol under a filling as a sort of built-in pain reliever.

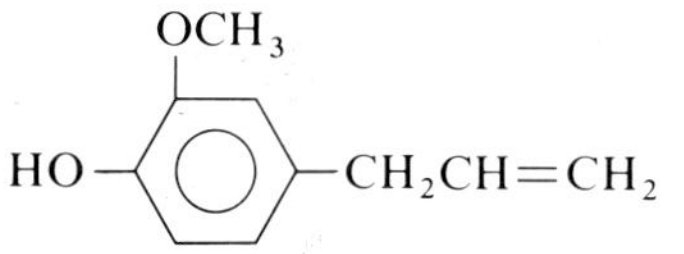

eugenol
bp 255°C; odor of cloves

The major euphoric ingredient of marijuana (Figure 7-5) or hashish from the plant *Cannabis sativa* is Δ^1-tetrahydrocannabinol. The Δ^1 in the name refers to the position of the double bond. In cannabinol itself both rings are aromatic.

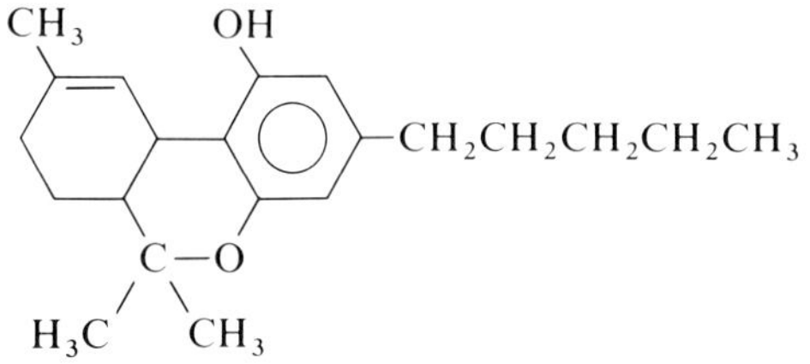

Δ^1-tetrahydrocannabinol
liquid; bp 155–157°C
(in vacuum at 0.05 mm Hg pressure)

Cholesterol, an unsaturated alcohol belonging to a group of natural products known as the **steroids**,[6] is found in fats and in the artery-narrowing deposits characteristic of the disease arteriosclerosis.

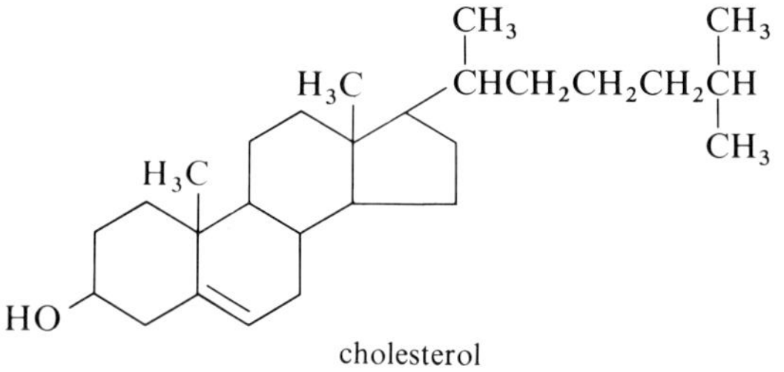

cholesterol

[6] For other examples of steroids, see page 112 (Figure 6-8) and page 143.

FIGURE 7-5. Marijuana leaves and berries. [Courtesy of the Florida Crime Laboratory]

Hydroquinone, or *p*-dihydroxybenzene, is used as a photographic developer. It reduces the exposed silver chloride to metallic silver and is itself oxidized to quinone. Photographers are not the only ones in the business of making

$$\underset{\substack{\text{hydroquinone}\\ \text{colorless; mp 170--171°C}}}{\text{HO–C}_6\text{H}_4\text{–OH}} \xrightarrow{[O]} \text{H}_2\text{O} + \underset{\substack{\text{quinone}\\ \text{yellow; mp 115.7°C}}}{\text{O=C}_6\text{H}_4\text{=O}} \tag{7-39}$$

quinone from hydroquinone. The bombadier beetle (*Brachinus*) uses an enzymatically catalyzed oxidation of hydroquinone by H_2O_2 as a defence mechanism (equation 7-40). The hydroquinone and hydrogen peroxide are stored in one part of the gland and the enzymes are stored in a sort of vestibular compartment. To start the reaction, the beetle injects the hydroquinone plus peroxide into the vestibule, where the ensuing violent reaction sprays the products out of the abdomen[7] (Figure 7-6). The quinone is irritating to the eyes even when cold.

$$\underset{\text{OH}}{\underset{|}{\bigcirc}}\text{OH} + H_2O_2 \xrightarrow{\text{enzymes}} \underset{\text{O}}{\underset{\|}{\bigcirc}}\text{O} + O_2 + H_2O \quad \text{(as steam, } 100°C\text{)} \qquad (7\text{-}40)$$

FIGURE 7-6. Bombardier beetle discharging. A thermocouple was used to prod the beetle and trigger the photographic flash unit to take the picture. A thermocouple can be used because the discharge comes out at the unusually high temperature (for living systems) of 100°C. [From D. J. Aneshansley, T. Eisner, J. Widom, and B. Widom, *Science*, **165**:61 (4 July 1969). Copyright 1969 by the American Association for the Advancement of Science]

[7] A catalyzed decomposition of H_2O_2 into steam and O_2 was used in German submarines and rockets to drive high-speed pumps during World War II.

SUMMARY

A. Alcohols
   1. Alcohols are named after the corresponding alkane with an -ol ending.
   2. They are classified as primary, secondary, or tertiary according to the number of carbon atoms directly attached to the ⩾C—OH group.
   3. The lower alcohols are soluble in water. Because of hydrogen bonding, they have higher boiling points than do the corresponding alkanes.
   4. Reactions
      (a) Alcohols are very weak acids, but will react with sodium.
      (b) The hydroxyl group may be replaced by chlorine by treatment with reagents such as $HCl/ZnCl_2$. It can also be replaced by nitrate or sulfate groups.
      (c) Treatment with acid and heat converts alcohols into alkenes and/or ethers.
   5. Preparation
      (a) Aqueous NaOH brings about the displacement of halide ion in an alkyl halide by hydroxide.
      (b) Cold dilute H_2SO_4 converts alkenes to alcohols by addition of H_2O across the double bond.
   6. Enols
      (a) Acetone exists mostly in the carbonyl or keto form.
      (b) The acid-catalyzed bromination of acetone is a reaction with the enol, formed by the catalyst.

B. Ethers
   1. The lower ethers are soluble or partly soluble in water. Ethers have low boiling points, about like those of alkanes of similar molecular weight.
   2. Ethers are made
      (a) from the alcohol and H_2SO_4.
      (b) from alkyl halides by the Williamson synthesis (a displacement reaction).
   3. Ethers are cleaved by HBr or HI. In aromatic substitution, the methoxy group is a powerful ortho-para director.

C. Phenols
   1. A phenol is a compound with an HO group directly attached to an aromatic ring.
   2. Phenols are weakly acidic, forming salts with NaOH but not with $NaHCO_3$.
   3. The HO group is a powerful ortho-para director in aromatic substitution reactions. Nitration of phenol gives picric acid.
   4. Phenols can be made by alkali fusion of sodium salts of sulfonic acids.

EXERCISES

1. Classify each of the following as a primary, secondary, or tertiary alcohol or as a phenol.

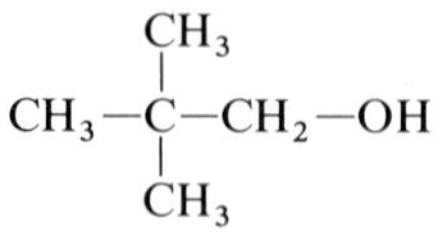

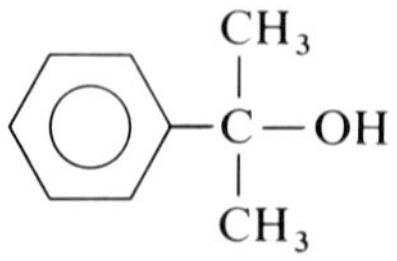

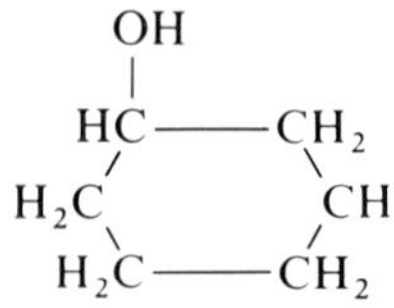

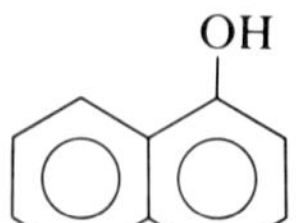

2. Define or explain (a) wood alcohol, (b) denatured alcohol, (c) absolute alcohol, (d) grain alcohol, (e) fusel oil.

3. At what point in the series $H(CH_2)_nOH$, with increasing n, does the compound become only partly soluble in water?

4. Write structures for (a) diphenyl ether, (b) 2,3-butanediol, (c) 2-methyl-2-propanol, (d) t-butyl alcohol, (e) 3-ethyl-2-pentanol.

5. Explain why alcohols have higher boiling points than alkanes of about the same molecular weight.

5. Explain why the $ZnCl_2$ catalyst is not needed to convert t-butyl alcohol to t-butyl chloride.

7. What are the essential features of a molecule for it to be an effective soap or detergent? Give one advantage and one disadvantage of sodium alkyl sulfate detergents.

8. Reactions 7-15 and 7-27 and the reaction shown below have a lot in common. All three are examples of what broad class of reactions?

$$I^- + CH_3CH_2Cl \rightarrow CH_3CH_2I + Cl^-$$

9. What reagents can be used to convert propene into (a) 2-propanol and (b) 1-propanol?

10. Give the structure of a naturally occurring enol. Circle the structural feature that makes it an enol and tell what the biological function of the compound is.

11. Refer to the boiling points of alkanes in Chapter 4. Predict the boiling point of dibutyl ether.

12. Predict the products of the following reactions:
    (a) Menthol (page 119) with HCl plus $ZnCl_2$.
    (b) Hexachlorophene (page 129) with aqueous NaOH at room temperature.
    (c) Δ^1-tetrahydrocannabinol (page 133) with aqueous NaOH at room temperature.

13. Predict the product of reaction of cineol (page 128) with excess HBr.

14. Write reaction sequences for the following conversions, putting the reagents or catalysts over the arrows.
    (a) 1-propanol $\rightarrow$ $\rightarrow$ 2-propanol
    (b) 2-pentanol $\rightarrow$ $\rightarrow$ pentane
    (c) 1-propanol $\rightarrow$ $\rightarrow$ propyne

15. Explain how a mixture of an alcohol and phenol can be separated. Would the method work for an enol and a phenol?

16. Give the structures and names of two high explosives.

Chapter 8
Aldehydes, Ketones, and Quinones

Compounds whose functional group is the carbonyl group ($>$C=O) occur widely in nature and play many important roles. When the carbonyl group is directly connected to at least one hydrogen atom, the compound is classified as an **aldehyde**. When the carbonyl group is directly connected to two carbon atoms, the compound is a **ketone**.

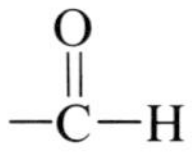

$$\begin{matrix} & O \\ & \| \\ -& C-H \end{matrix} \qquad \begin{matrix} & O & \\ & \| & \\ -C-& C-& C- \end{matrix}$$

aldehyde part structure 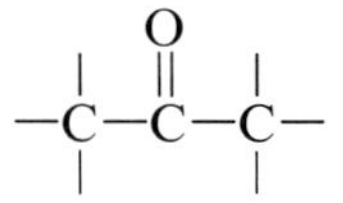   ketone part structure

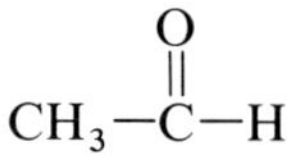

$$CH_3-\overset{O}{\underset{\|}{C}}-H \quad CH_3CHO \qquad H-\overset{O}{\underset{\|}{C}}-H \quad HCHO \qquad CH_3-\overset{O}{\underset{\|}{C}}-CH_3 \quad CH_3COCH_3$$

acetaldehyde or ethanal   formaldehyde   acetone, dimethyl ketone, or 2-propanone

colorless liquid; pungent odor; bp 21°C; soluble in water   colorless gas; bp −21°C; soluble in water   bp 56.5°C; soluble in water

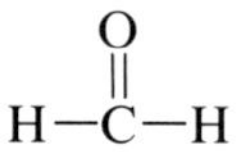

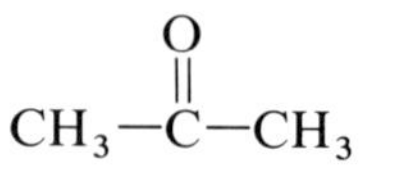

The Carbon-Oxygen Double Bond (C=O)

The C=O double bond is like the C=C double bond in some respects. A side-to-side overlap of a p_z orbital on carbon with a p_z orbital on oxygen forms a π bonding orbital (Figure 8-1). Because oxygen is more electronegative or electron-attracting than carbon, the pair of shared electrons in the π bonding orbital is closer to oxygen. This polarization makes the oxygen end of the carbonyl

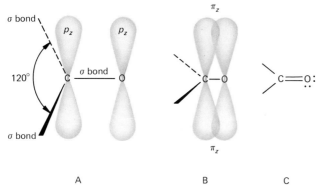

FIGURE 8-1. (A) Carbon and oxygen p_z orbitals. (B) The π bonding orbital. (C) The carbonyl group and unshared electron pairs.

more negative and the carbon end more positive.

$$\overset{\delta+}{\underset{}{>}}C=\overset{\delta-}{O}$$

In addition to the greater share of the π bonding electrons, the oxygen atom also has two pairs of unshared electrons as indicated by the dots in Figure 8-1C. The unshared electron pairs make the oxygen basic, although this basicity is noticeable only in rather strong acids.

$$H^+ + {>}C=\ddot{O}{:} \rightleftarrows {>}C=\overset{+}{\underset{}{O}}{:}{\diagup}^{H} \qquad (8\text{-}1)$$

Formaldehyde is used as an aqueous solution (called formalin) in embalming and for pickling noteworthy zoological specimens. Formaldehyde not only has a sharp, irritating odor of its own but also tends to mask the odors of other substances. This is not due primarily to reactions with the other substances but to its interference with the sense of smell. Commercial room deodorants consisting of a wick dipping into a solution of formaldehyde do not really clean the air—they hold your nose for you. Colds and tobacco smoking also interfere with the sense of smell.

As the sense of smell returns after a cold, substances may have odors differing from their normal ones, perhaps because some parts of the smelling organ recover before others. The nose is also subject to fatigue, and an odor will disappear more or less rapidly unless the nose is given a rest. Some presumably pure substances have more than one odor, an initial odor reported by the rested nose and a different one reported by the fatigued nose. This suggests that there may be more than one kind of odor-sensing receptor and that these differ not only in the kind of signal sent to the brain but also in the rate at which the receptors become fatigued.

Different substances also fatigue the nose at different rates, and an expert perfume-sniffer uses this phenomenon to detect one ingredient after another. The first sniff gives him the most powerful ingredient but, as the sense of smell becomes fatigued in so far as that ingredient is concerned, the odors of less powerful but less fatiguing ingredients can be detected.

A person unable to smell is said to be **anosmic**. Most anosmia is only partial; usually an anosmic person is unable to detect certain odors while still sensitive to others. Just as there are different kinds of color blindness so are there different kinds of anosmia, and the affliction is probably hereditary.

Acetone (Figure 8-2) is used industrially and in the laboratory as a solvent for organic compounds. It is also sometimes added to gasoline tanks to dissolve small amounts of water that may have condensed in the tank. The characteristic odor of acetone in the breath is one symptom of diabetes.

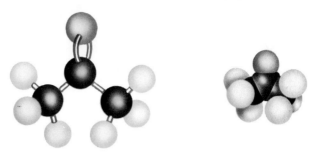

FIGURE 8-2. Models of an acetone molecule.

The carbonyl group is easily detected by its infrared absorption, associated with the stretching vibration of the bond, and by its absorption of ultraviolet light, associated with the "promotion" of electrons to orbitals of higher energy. The infrared absorption band is usually near 6000 nm,[1] aldehydes have another band near 3600 nm[2] corresponding to the C—H vibration. The ultraviolet absorption band of carbonyl compounds is near the visible range of wavelengths. In fact, when the carbonyl group is conjugated with double bonds as in *p*-benzoquinone (yellow) and 1,4-naphthoquinone (red), the absorption is in the visible range and the compound is colored.

Nomenclature of Aldehydes and Ketones

As usual, the simpler members of the series have very old and well-established trivial names such as formaldehyde, acetaldehyde, and acetone. The names **formaldehyde** and **acetaldehyde** refer to the fact that these aldehydes are readily oxidized to **formic** and **acetic** acids.

$$\underset{\text{formaldehyde}}{\text{H}-\overset{\overset{\text{O}}{\|}}{\text{C}}-\text{H}} + [\text{O}] \rightarrow \underset{\text{formic acid}}{\text{H}-\overset{\overset{\text{O}}{\|}}{\text{C}}-\text{OH}} \qquad (8\text{-}2)$$

$$\underset{\text{acetaldehyde}}{\text{CH}_3-\overset{\overset{\text{O}}{\|}}{\text{C}}-\text{H}} + [\text{O}] \rightarrow \underset{\text{acetic acid}}{\text{CH}_3-\overset{\overset{\text{O}}{\|}}{\text{C}}-\text{OH}} \qquad (8\text{-}3)$$

The systematic name for an aldehyde is derived from that of the corresponding alkane by changing the ane ending to **anal**. Thus the systematic name for formaldehyde, the one-carbon compound, would be methan**al**, although it is never used. Similarly, the systematic name for acetaldehyde (two carbons) is ethan**al**. The names for more complicated, substituted aldehydes have numbers indicating the position of the substituent; the aldehyde group is always numbered carbon 1.

[1] 1700 cm^{-1}
[2] 2800 cm^{-1}

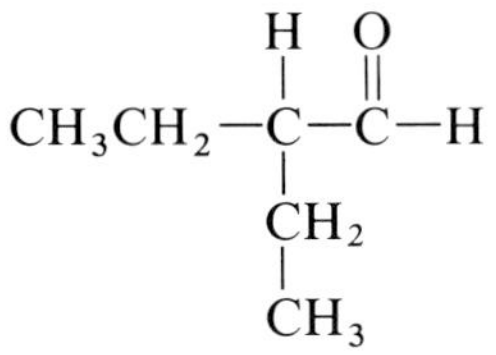

$$CH_3CH_2-\underset{\underset{CH_3}{\underset{|}{CH_2}}}{\overset{H}{\underset{|}{C}}}-\overset{O}{\underset{}{\overset{\|}{C}}}-H$$

2-ethylbutanal

colorless; bp 116°C;
only slightly soluble in H_2O

$$CH_3-\underset{\underset{CH_3}{|}}{\overset{H}{\underset{|}{C}}}-CH_2-\overset{O}{\overset{\|}{C}}-H$$

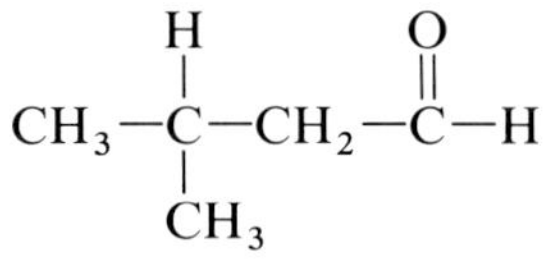

3-methylbutanal

colorless; bp 92.5°C;
only slightly soluble in H_2O

The position of a double bond in an unsaturated aldehyde is indicated by a number corresponding to the end of the C=C nearest the aldehyde group. Note the **enal** rather than **anal** ending.

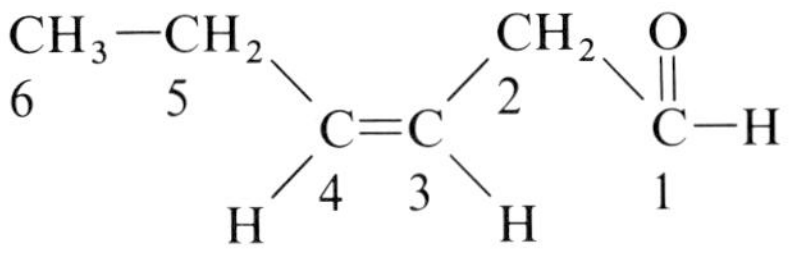

cis-3-hexenal

Cis-3-hexenal has a fresh "green" odor. It is found in many kinds of leaves and flowers. Its isomers, with the double bond trans or in a different position, have quite different odors.

The odor of fresh cucumbers is due to the **dienal**, *trans,cis*-2,6-nonadienal.

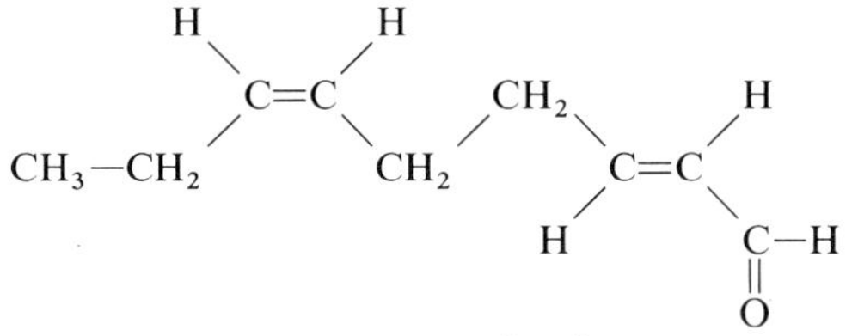

trans,cis-2,6-nonadienal

This substance is also partly responsible for the "freshness" of the odor of violets.

Simple ketones can be named adequately by using the names of the alkyl groups attached to the carbonyl, followed by the word "ketone." Acetone, for example, would be called dimethyl ketone. More complicated ketones are named systematically by a method like that used for aldehydes. Carbon 1 is the end carbon atom closer to the carbonyl group, and the ane ending of the alkane is replaced by the ending **anone**. Carbon 1 in a cyclic ketone is the carbonyl carbon.

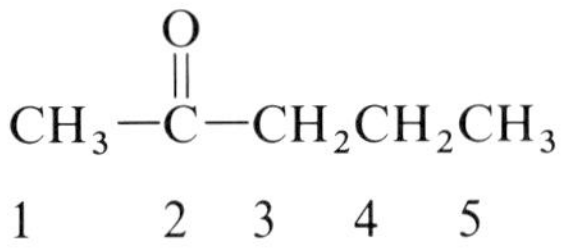

$$\underset{1 \quad 2 \quad 3 \quad 4 \quad 5}{CH_3-\overset{O}{\overset{\|}{C}}-CH_2CH_2CH_3}$$

methyl propyl ketone or
2-pentanone

bp 101.7°C; colorless, only
very slightly soluble in water

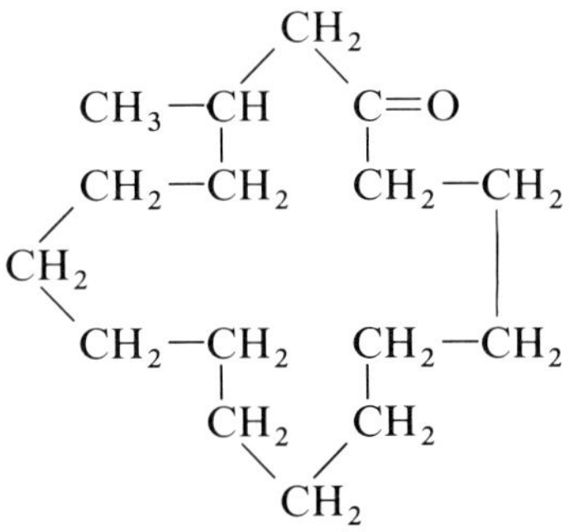

muscone or
3-methylcyclopentadecanone

bp 328°C

Aldehydes, Ketones, and Quinones

Muscone is found in the prenuptial or musk gland of the Himalayan musk deer, a small species about 20 in. high. It is also used in perfumes, where it allegedly serves a similar function. Most perfumes, and even the fragrances of flowers, contain small proportions of substances that by themselves are rather malodorous. Without such components the fragrance is sweet but insipid and uninteresting.

FIGURE 8-3. A connoisseur.

The gland of the musk deer contains only about 1% muscone, and about 60,000 musk deer were killed during 1 year. It is said that the deer are enticed into traps by the music of native flutes; however, since the perfume industry is somewhat given to exaggeration, it may be proper to treat this statement with reserve. Because of the understandable expense of any ingredient involving things like flute music, most perfumes now use cheaper synthetic musks rather than the naturally occurring muscone.

Another animal whose odor has been an asset to man but a liability to the animal is the Ethiopian civet-cat, the African equivalent of the skunk. Civet contains skatole, indole, and civetone. The last compound is a ketone with a large ring, structurally very much like muscone. Since a happy and contented civet-cat secretes less of the valuable civet, the caged animals lead an existence that no milk cow would envy.

A large number of unsaturated ketones are found in plant fragrances, where they are responsible for the sweet or "floral" component of the odor.[3]

ketone from the
Japanese chrysanthemum

phenolic ketone largely
responsible for the odor
of raspberries

The keto-alcohols estrone and testosterone are important female and male sex hormones. Both compounds belong to a class of

[3] In line formulas each junction of two lines and each line end stands for a carbon atom.

natural products known as **steroids**. The characteristic steroid pattern of three fused six-membered rings and one five-membered ring is also found in cholesterol (page 133).

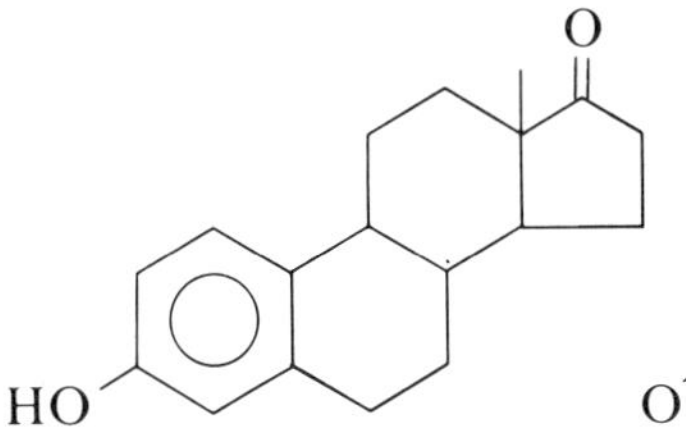

estrone, an estrogen or female sex hormone

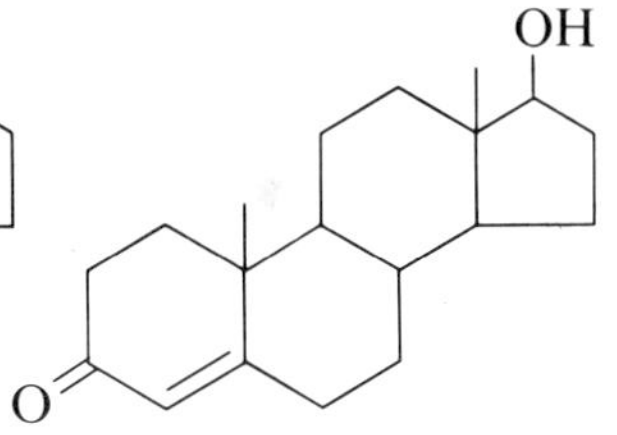

testosterone, an androgen or male sex hormone

Preparation of Aldehydes and Ketones

Aldehydes and ketones can be prepared by the removal of two atoms of hydrogen from an alcohol by oxidation. Thus primary and secondary alcohols react with oxidizing agents such as potassium dichromate ($K_2Cr_2O_7$) to give aldehydes (equation 8-4) and ketones (equation 8-5), respectively. They can also be made by the

$$R-CH_2-OH + [O] \rightarrow R-\overset{O}{\underset{\|}{C}}-H + H_2O \qquad (8\text{-}4)$$
(from an oxidizing agent)

$$R-\overset{OH}{\underset{|}{C}H}-R + [O] \rightarrow R-\overset{O}{\underset{\|}{C}}-R + H_2O \qquad (8\text{-}5)$$

hydrolysis of dihalides in which the two halogen atoms are on the same carbon. Although the product might be expected to be a diol, diols with the two hydroxyl groups on the same carbon atom are usually unstable and lose water to form a carbonyl group (equation 8-6).

$$R-\overset{Cl}{\underset{Cl}{\overset{|}{C}}}-R \xrightarrow[H_2O]{NaOH} \left[R-\overset{OH}{\underset{OH}{\overset{|}{C}}}-R \right] \rightarrow R-\overset{O}{\underset{\|}{C}}-R + H_2O \qquad (8\text{-}6)$$

Addition Reactions of the Carbonyl Group

ADDITION OF HYDROGEN

Like the C=C double bond, the C=O double bond can be converted to a single bond by addition reactions. The simplest of these is the addition of hydrogen, usually brought about by means of a platinum or nickel catalyst. The resulting compound is a primary alcohol if the carbonyl compound is an aldehyde (equation 8-7), a secondary alcohol if a ketone is used (equation 8-8).

$$\underset{CH_3}{\overset{H}{\diagdown}}C=O + H_2 \xrightarrow{catalyst} CH_3CH_2OH \qquad (8\text{-}7)$$

acetaldehyde   ethyl alcohol

$$CH_3-\overset{O}{\overset{\|}{C}}-CH_2CH_3 \xrightarrow[\text{catalyst}]{H_2} CH_3-\underset{H}{\overset{OH}{\underset{|}{\overset{|}{C}}}}-CH_2CH_3 \qquad (8\text{-}8)$$

2-butanone   2-butanol

ADDITION OF WATER OR ALCOHOLS

When unsymmetrical reagents are added across the C=O double bond the question arises of which part of the unsymmetrical reagent goes on oxygen and which on carbon. This is easy to predict: the more positively charged or electrophilic part of the reagent goes on oxygen and the more negatively charged or nucleophilic part on carbon (equation 8-9).

$$\overset{\delta+}{>}\overset{\delta-}{C=O} + \overset{\delta+}{X}-\overset{\delta-}{Y} \rightarrow \overset{Y\ X}{\underset{|\ |}{>C-O}} \qquad (8\text{-}9)$$

For example, when water adds to trichloroacetaldehyde (chloral) the product is chloral hydrate (equation 8-10).

$$\underset{\underset{Cl\ H}{|\ |}}{Cl-\overset{Cl}{\overset{|}{C}}-\overset{\delta-}{C}\overset{O}{\overset{\|}{\diagup}}} + :\overset{\overset{\delta+}{H}}{\underset{H}{\overset{\diagup}{O}}}: \rightleftharpoons Cl-\overset{Cl}{\underset{Cl}{\overset{|}{\underset{|}{C}}}}-\overset{O-H}{\underset{\underset{O-H}{|}}{\overset{\diagup}{C}}} \qquad (8\text{-}10)$$

chloral
mp −57.5°C

Chloral hydrate is used as an anesthetic for horses. It is also the active ingredient of a Mickey Finn.

Water will add across the C=O double bonds of other aldehydes and ketones also, but usually the equilibrium lies on the carbonyl side of the equation rather than on the hydrate side. The stability of the hydrate in the case of chloral is unusual.

The addition of alcohols across the C=O double bond gives compounds known as hemiacetals (equation 8-11). In a solution of

$$CH_3-\underset{H}{\overset{O^{\delta-}}{\overset{\diagdown}{\underset{\diagdown}{C^{\delta+}}}}} + :\underset{CH_2CH_3}{\overset{H^{\delta+}}{\overset{|}{O}:^{\delta-}}} \rightleftharpoons CH_3-\underset{\underset{O\diagdown CH_2CH_3}{H}}{\overset{OH}{\overset{\diagup}{C}}} \qquad (8\text{-}11)$$

acetaldehyde   ethanol   the hemiacetal

acetaldehyde in ethyl alcohol, for example, infrared and nuclear magnetic resonance spectroscopy show the presence of considerable amounts of the hemiacetal.

Heating an aldehyde with excess alcohol and a trace of acid forms a second ether linkage and converts the hemiacetal into a more stable and easily isolable acetal (equation 8-12).

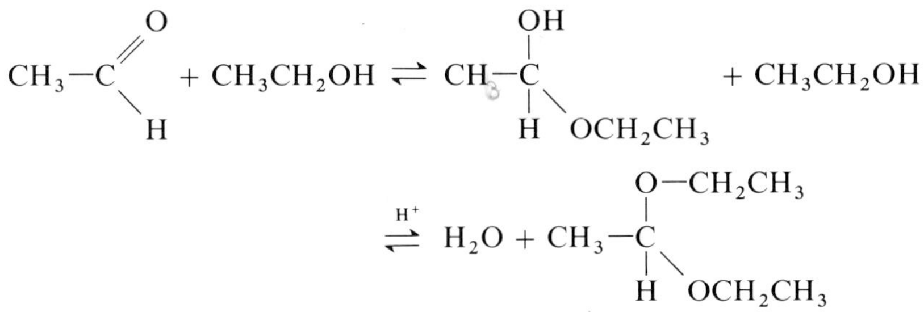

$$CH_3-\overset{O}{\underset{H}{C}} + CH_3CH_2OH \rightleftharpoons CH_3-\underset{H}{\overset{OH}{\underset{|}{C}}}\diagdown_{OCH_2CH_3} + CH_3CH_2OH$$

$$\overset{H^+}{\rightleftharpoons} H_2O + CH_3-\underset{H}{\overset{O-CH_2CH_3}{\underset{|}{C}}}\diagdown_{OCH_2CH_3}$$

(8-12)

The formation of hemiacetals and acetals is important in the chemistry of sugars and carbohydrates which we will study later. Glucose, a typical sugar, exists mostly in the form of a cyclic hemiacetal. The OH group on carbon 5 adds across the C=O double bond at carbon 1.

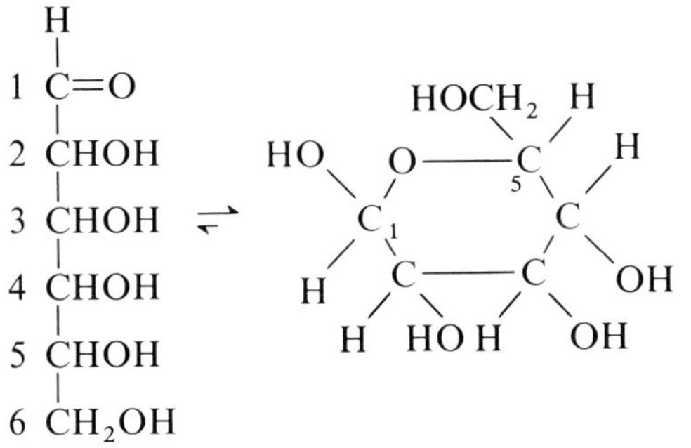

(8-13)

ADDITION OF GRIGNARD REAGENTS

When an alkyl or aryl halide is stirred with metallic magnesium in dry ether, the halide reacts with the metal to form a Grignard reagent, RMgX (equations 8-14 and 8-15). Grignard reagents react

$$CH_3Br \xrightarrow[\text{dry ether}]{Mg} CH_3MgBr \qquad (8\text{-}14)$$

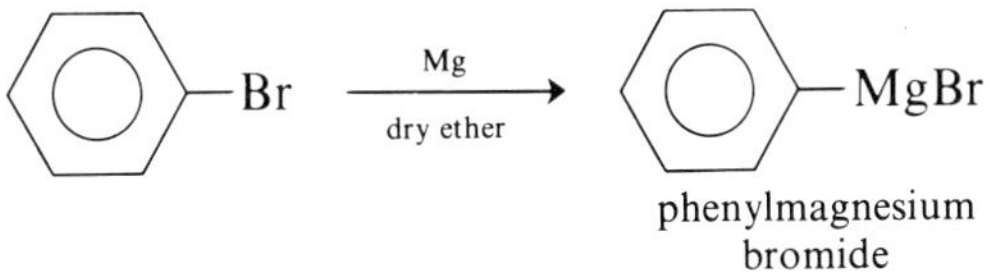

methylmagnesium bromide

$$\text{C}_6\text{H}_5\text{-Br} \xrightarrow[\text{dry ether}]{Mg} \text{C}_6\text{H}_5\text{-MgBr} \qquad (8\text{-}15)$$

phenylmagnesium bromide

with water as though they were carbon anions, R^-, which abstract a proton from the water (equation 8-16).

$$\overset{\delta-}{C}H_3-\overset{\delta+}{M}gBr + H_2O \rightarrow CH_4 + Mg^{2+}OH^-Br^- \qquad (8\text{-}16)$$

The direction of addition of the Grignard reagent across a C=O double bond is predictable. The Grignard reagent is polarized like $R^{\delta-}-^{\delta+}Mg$; therefore, the strongly nucleophilic alkyl group R should bond to the carbon of the C=O, and the magnesium should bond to the oxygen (equation 8-17). The addition product is the

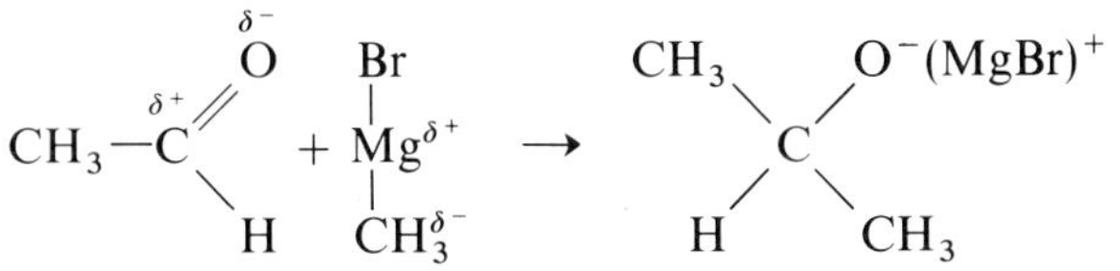

$$CH_3-\overset{\delta+}{\underset{H}{C}}\overset{\overset{\delta-}{O}}{\diagup} + \underset{CH_3^{\delta-}}{\overset{Br}{\underset{|}{Mg^{\delta+}}}} \rightarrow \underset{H}{\overset{CH_3}{\diagdown}}C\underset{CH_3}{\overset{O^-(MgBr)^+}{\diagup}} \qquad (8\text{-}17)$$

Aldehydes, Ketones, and Quinones

magnesium salt of an alcohol. It is easily decomposed to the alcohol by treatment with water (equation 8-18). The use of both reactions

$$\underset{\text{very strong base}}{\underset{H}{\overset{CH_3}{>}}\underset{CH_3}{\overset{O^-(MgBr)^+}{<}}C} + H_2O \rightarrow \underset{\text{weaker base}}{\underset{H}{\overset{CH_3}{>}}\underset{CH_3}{\overset{OH}{<}}C} + HO^-(MgBr)^+ \quad (8\text{-}18)$$

in sequence allows the conversion of acetaldehyde into 2-propanol, as in equation 8-19. Because any aldehyde can be employed instead

$$CH_3-\overset{O}{\underset{\|}{C}}-H \xrightarrow{CH_3MgBr} CH_3-\underset{CH_3}{\overset{OMgBr}{\underset{|}{C}}}-H \xrightarrow{H_2O} CH_3-\underset{CH_3}{\overset{OH}{\underset{|}{C}}}-H \quad (8\text{-}19)$$

of acetaldehyde and almost any Grignard reagent instead of methylmagnesium bromide, reactions like equation 8-19 can be used to make a large assortment of alcohols. Notice that the alcohols are always *secondary alcohols* unless the aldehyde is formaldehyde; in that case *primary alcohols* are formed.

$$R-\overset{O}{\underset{\|}{C}}-H + R'MgBr \rightarrow R-\underset{R'}{\overset{OMgBr}{\underset{|}{C}}}-H \xrightarrow{H_2O} R-\underset{R'}{\overset{OH}{\underset{|}{C}}}-H \quad (8\text{-}20)$$

To make a tertiary alcohol, the synthesis begins with a ketone (equation 8-21).

$$CH_3-\overset{O}{\underset{\|}{C}}-CH_2CH_3 + CH_3CH_2CH_2CH_2MgBr \rightarrow$$

$$CH_3-\underset{CH_2CH_2CH_2CH_3}{\overset{OMgBr}{\underset{|}{C}}}-CH_2CH_3 \xrightarrow{H_2O} CH_3-\underset{CH_2CH_2CH_2CH_3}{\overset{OH}{\underset{|}{C}}}-CH_2CH_3 \quad (8\text{-}21)$$

Synthesis of a tertiary alcohol

Synthesis by known reaction paths is useful in proving the structures of natural products and other molecules. For example, suppose that a tertiary alcohol with eight carbon atoms, is suspected to be either

$$CH_3-\underset{CH_2CH_2CH_2CH_3}{\overset{OH}{\underset{|}{C}}}-CH_2CH_3 \quad \text{or} \quad CH_3CH_2-\underset{CH_2CH_2CH_3}{\overset{OH}{\underset{|}{C}}}-CH_2CH_3$$

3-methyl-3-heptanol   3-ethyl-3-hexanol

The former is synthesized using the reaction of equation 8-21, which we know will give this product. Similarly, 3-ethyl-3-hexanol is synthesized by the sequence in equation 8-22. The two synthetic alcohols will each have a unique

$$CH_3CH_2-\overset{O}{\underset{\|}{C}}-CH_2CH_3 \xrightarrow{CH_3CH_2CH_2MgBr} \xrightarrow{H_2O} CH_3CH_2-\underset{CH_2CH_2CH_3}{\overset{OH}{\underset{|}{C}}}-CH_2CH_3 \quad (8\text{-}22)$$

set of physical properties (such as boiling point, density, light absorption, and so on). These properties can then be compared with those of the naturally occurring alcohol being investigated. The same sequence of reactions can be extended to obtain alkanes of known structure (equations 8-23 and 8-24).

$$CH_3-\underset{CH_2CH_2CH_2CH_3}{\underset{|}{\overset{OH}{\overset{|}{C}}}}-CH_2CH_3 \xrightarrow{-H_2O} \begin{bmatrix} CH_3\underset{CH_2CH_2CH_2CH_3}{\overset{|}{C}}=CHCH_3 \\ + \\ CH_3\underset{\overset{\|}{CHCH_2CH_2CH_3}}{\overset{}{C}}-CH_2CH_3 \end{bmatrix} \xrightarrow[\text{cat.}]{H_2} CH_3\underset{CH_2CH_2CH_2CH_3}{\overset{|}{C}H}CH_2CH_3$$

(8-23)

$$CH_3CH_2-\underset{CH_2CH_2CH_3}{\underset{|}{\overset{OH}{\overset{|}{C}}}}-CH_2CH_3 \xrightarrow{-H_2O} \begin{bmatrix} CH_3CH_2\underset{CH_2CH_2CH_3}{\overset{|}{C}}=CHCH_3 \\ + \\ CH_3CH_2\underset{\overset{\|}{CHCH_2CH_3}}{\overset{}{C}}CH_2CH_3 \end{bmatrix} \xrightarrow[\text{cat.}]{H_2} CH_3CH_2\underset{CH_2CH_2CH_3}{\overset{|}{C}H}CH_2CH_3$$

(8-24)

The general idea is to start with a simple molecule of known structure, apply a series of known reactions, and end up with a more complicated molecule. The structure of the more complicated molecule is then known as well. It was largely by such methods that the science of organic chemistry was built up.

ADDITION OF AMINES

Like water and alcohol, amines (RNH_2) are also nucleophilic and will also add to carbonyl compounds. However, because the nitrogen has a second proton attached to it, the addition of the amine to the carbonyl group is followed by a second reaction in which a molecule of water is split out (equation 8-25). The overall reaction is therefore

$$\underset{\text{acetone}}{\overset{H_3C}{\underset{H_3C}{>}}C=O} + \underset{\text{phenylhydrazine}}{H_2\ddot{N}NHC_6H_5} \rightarrow \begin{bmatrix} \overset{H_3C}{\underset{H_3C}{>}}\overset{OH}{\underset{}{\overset{|}{C}}}-NHNHC_6H_5 \end{bmatrix}$$

$$\xrightarrow{H^+} H_2O + \underset{\text{acetone phenylhydrazone}\atop\text{mp 42°C}}{\overset{H_3C}{\underset{H_3C}{>}}C=NNHC_6H_5}$$

(8-25)

that shown in equation 8-26.

$$\overset{H_3C}{\underset{H_3C}{>}}C=O + H_2NNHC_6H_5 \xrightarrow{H^+} H_2O + \overset{H_3C}{\underset{H_3C}{>}}C=NNHC_6H_5$$

(8-26)

Reactions like that of equation 8-26 convert C=O double bonds into C=N double bonds. The formation of phenylhydrazones, in particular, is used as an aid in identifying ketones and aldehydes. The phenylhydrazone is usually a solid and its melting point can be compared with that of a phenylhydrazone of known structure.

ADDITION OF HCN

Hydrogen cyanide reacts with aldehydes and ketones to give adducts called **cyanohydrins**. The proton, being an electrophile, adds to the electron-rich nucleophilic oxygen end of the carbonyl group. The cyanide ion, which is quite nucleophilic, adds to the carbon or electrophilic end of the carbonyl (equation 8-27).

$$\text{Ph-CHO} + \text{HCN} \rightleftharpoons \text{Ph-CH(OH)(CN)} \qquad (8\text{-}27)$$

benzaldehyde
colorless oil; bp 179°C;
odor of almonds

benzaldehyde cyanohydrin
or mandelonitrile
mp −10°C

Further reaction of a cyanohydrin with an alcohol gives a cyano-ether (equation 8-28).

$$\text{Ph-CH(OH)(CN)} + \text{ROH} \xrightarrow{H^+} \text{Ph-CH(OR)(CN)} + \text{H}_2\text{O} \qquad (8\text{-}28)$$

The benzaldehyde found in bitter almonds and in the roots of peach trees is actually present as **amygdalin**, a compound like the product of reaction 8-28 in which the alcohol is a large sugar molecule.[4]

Amygdalin (two glucose units linked to benzaldehyde cyanohydrin)

$$\text{Amygdalin} \xrightarrow[\text{enzymes}]{\text{H}_2\text{O}} 2\ \beta\text{-D-glucose} + \text{benzaldehyde cyanohydrin}$$

$$\text{benzaldehyde cyanohydrin} \rightleftharpoons \text{benzaldehyde} + \text{HCN}$$

Amygdalin is poisonous because it hydrolyzes in the digestive tract to give HCN. Cyanogenetic or cyanide-generating glycosides are found in many plants, including some used for food. An example is manioc or cassava, from which tapioca is made. The tubers are specially treated to remove the HCN before they are safe to eat.

The evolutionary value of a cyanogenetic glycoside to the plant may be to inhibit competing species from growing too close to it.

Benzaldehyde cyanohydrin is found in the reactor glands of millipedes, who store HCN in this form. When attacked, the millipede ejects its benzaldehyde cyanohydrin past an enzyme which hydrolyzes it to benzaldehyde plus HCN on the way out. The millipede *Apheloria corrugata* produces enough HCN to kill a mouse. Man did not invent chemical warfare after all.

[4] The heavy lines indicate the edge of the molecule closest to the reader.

Enolization

In Chapter 7 it was pointed out that certain enols or unsaturated alcohols can be formed by changing the positions of a proton and the double bond in carbonyl compounds (equation 8-29). Many of

$$CH_3-\underset{\text{acetone}}{\overset{O}{\overset{\|}{C}}}-CH_3 \rightleftharpoons \underset{\text{acetone enol}}{CH_2=\overset{OH}{\overset{|}{C}}-CH_3} \quad (8\text{-}29)$$

the reactions of carbonyl compounds are really reactions of the enol isomer. Even though the amount of enol present at equilibrium is very small in most cases, more is formed from the carbonyl compound as it is used up by a reagent.

The iodination of acetone (equation 8-30) is an example. Careful

$$CH_3-\overset{O}{\overset{\|}{C}}-CH_3 + I_2 \xrightarrow{H^+} CH_3-\overset{O}{\overset{\|}{C}}-CH_2I + HI \quad (8\text{-}30)$$

studies of the reaction have shown that the first step does not involve the iodine at all, but merely the formation of enol (equation 8-31).

$$CH_3-\overset{O}{\overset{\|}{C}}-CH_3 \xrightarrow{H^+} CH_2=\overset{OH}{\overset{|}{C}}-CH_3 \quad (8\text{-}31)$$

The enol reacts with iodine as fast as it is formed (equation 8-32).

$$CH_2=\overset{OH}{\overset{|}{C}}-CH_3 + I_2 \rightarrow \left[\overset{OH}{\underset{I}{\overset{|}{C}H_2-\overset{|}{\underset{I}{C}}-CH_3}} \right] \rightarrow CH_2-\overset{O}{\overset{\|}{C}}-CH_3 + HI$$

$$\underset{I}{} \quad (8\text{-}32)$$

Bases also catalyze the formation of enol, or rather of enolate ion, since enols are appreciably acidic. Although the structure of enolate

$$CH_3-\overset{O}{\overset{\|}{C}}-CH_3 + HO^- \rightleftharpoons H_2O + CH_3-\overset{O^-}{\overset{|}{C}}=CH_2 \quad (8\text{-}33)$$

ion is usually written as in equation 8-33, the negative charge is not completely localized on the oxygen atom. Acetone enolate ion undergoes many reactions in which it acts as though it is a carbanion with a carbonyl group next to the charged carbon atom.

$$\underset{\text{enolate structure}}{CH_3-\overset{O^-}{\overset{|}{C}}=CH_2} \leftrightarrow \underset{\text{carbanion structure}}{CH_3-\overset{O}{\overset{\|}{C}}-CH_2^-}$$

An important reaction of the carbanion, a powerful nucleophile, is with other carbonyl compounds, as in equation 8-34, for example.

$$C_6H_5-\overset{\delta-}{\overset{O}{\overset{\|}{\underset{\delta+}{C}}}}\underset{H}{} + \overline{C}H_2-\overset{O}{\overset{\|}{C}}-CH_3 \rightleftharpoons C_6H_5-\overset{O^-}{\overset{|}{\underset{H}{C}}}-CH_2-\overset{O}{\overset{\|}{C}}-CH_3$$

$$(8\text{-}34)$$

Reaction 8-34 is completed by neutralizing the new —O⁻ group with a proton from the solvent (equation 8-35). Thus we have as an

$$C_6H_5-\underset{H}{\underset{|}{\overset{O^-}{\overset{|}{C}}}}-CH_2-\overset{O}{\overset{\|}{C}}-CH_3 + H^+ \rightarrow C_6H_5-\underset{H}{\underset{|}{\overset{OH}{\overset{|}{C}}}}-CH_2-\overset{O}{\overset{\|}{C}}-CH_3 \quad (8\text{-}35)$$

overall result the **condensation** of two carbonyl compounds to form a more complex molecule (equation 8-36). Although equation 8-36

$$C_6H_5-\overset{O}{\overset{\|}{C}}-H + CH_3-\overset{O}{\overset{\|}{C}}-CH_3 \xrightarrow{\text{base}} C_6H_5-\underset{H}{\underset{|}{\overset{OH}{\overset{|}{C}}}}-CH_2-\overset{O}{\overset{\|}{C}}-CH_3 \quad (8\text{-}36)$$

may seem surprising at first, on closer examination it is seen to be just like the addition of HCN to benzaldehyde, and also like the addition of a Grignard reagent.

$$C_6H_5-\overset{O}{\overset{\|}{C}}-H + \underset{\text{(a carbanion)}}{R^-} \rightarrow C_6H_5-\underset{R}{\underset{|}{\overset{O^-}{\overset{|}{C}}}}-H$$

$$+ CN^- \rightarrow C_6H_5-\underset{CN}{\underset{|}{\overset{O^-}{\overset{|}{C}}}}-H$$

$$+ {}^-CH_2-\overset{O}{\overset{\|}{C}}-CH_3 \rightarrow C_6H_5-\underset{H}{\underset{|}{\overset{O^-}{\overset{|}{C}}}}-CH_2-\overset{O}{\overset{\|}{C}}-CH_3 \quad (8\text{-}37)$$

Condensation reactions are frequently used in the synthesis of drugs or in the syntheses of natural products for the purpose of proving their structures. Because the product of the condensation reaction is an alcohol as well as a ketone, it can be expected to undergo typical alcohol reactions such as dehydration. Thus cinnamaldehyde, the flavoring material in the cinnamon plant, can be made in the laboratory by the sequence of reactions shown in equation 8-38. Cinnamaldehyde is used in flavors and perfumes.

$$\underset{\text{benzaldehyde}}{C_6H_5-\overset{O}{\overset{\|}{C}}-H} + CH_3-\overset{O}{\overset{\|}{C}}-H \xrightarrow{\text{NaOH}} \left[C_6H_5-\underset{H}{\underset{|}{\overset{OH}{\overset{|}{C}}}}-CH_2-\overset{O}{\overset{\|}{C}}-H \right]$$

$$\xrightarrow{\text{heat}} \underset{\text{cinnamaldehyde}}{C_6H_5-CH=CH-\overset{O}{\overset{\|}{C}}-H} + H_2O \quad (8\text{-}38)$$

mp −7.5°C; bp 252°C; odor of cinnamon

FIGURE 8-4. In the sixteenth century, perfume ingredients were obtained by distillation of essential oils from plants. [From *Liber de Arte Destillandi*, Strasbourg, 1516]

Other aldehydes used in flavoring and in perfumes are vanillin (from vanilla) and piperonal (heliotrope).

vanillin
bp 285°C

piperonal
bp 263°C

Vanillin is found in the pods of the vanilla plant, but it can also be made from the waste product lignin that is discarded in the manufacture of paper pulp. Although the prevailing odor downwind from a paper mill is primarily that of sulfur dioxide and hydrogen sulfide, there is also a sickly sweet component.

The aldehyde

is a sort of perfume for bollweevils, for whom it acts as a sex attractant.

Aldehydes, Ketones, and Quinones

Oxidation of Aldehydes and Ketones

In the oxidation of a primary alcohol to an aldehyde it is sometimes difficult to prevent the reaction from going further and oxidizing the aldehyde to a carboxylic acid, RCOOH.

$$CH_3CH_2OH \xrightarrow[H^+]{[O] \; K_2Cr_2O_7} H_2O + CH_3-\underset{\underset{}{}}{\overset{\overset{O}{\|}}{C}}-H \quad (8\text{-}39)$$

(RCH$_2$OH)   (RCHO)
primary alcohols   aldehydes

$$CH_3-\overset{\overset{O}{\|}}{C}-H \xrightarrow[H^+]{[O] \; K_2Cr_2O_7} CH_3-\overset{\overset{O}{\|}}{C}-OH \quad (8\text{-}40)$$

(RCHO)   (RCOOH)
aldehydes   carboxylic acids

The bonds to the carbonyl groups of ketones are much less easily cleaved by oxidizing agents, and terminating the oxidation of a secondary alcohol at the ketone stage is no problem.

$$Ph-\underset{CH_2CH_3}{\overset{OH}{\underset{|}{\overset{|}{C}}-H}} \xrightarrow{[O]} Ph-\overset{\overset{O}{\|}}{C}-CH_2CH_3 \quad (8\text{-}41)$$

propiophenone
mp 18.6°C; flowery odor

$$Ph-\overset{\overset{O}{\|}}{C}-CH_2CH_3 \xrightarrow[\substack{H^+ \\ \text{slow, requires a} \\ \text{high temperature}}]{K_2Cr_2O_7} Ph-\overset{\overset{O}{\|}}{C}-OH + CO_2 + H_2O \quad (8\text{-}42)$$

benzoic acid

Their ease of oxidation is the basis of tests for aldehydes. One reagent used for this purpose is ammoniacal silver nitrate. Aldehydes, but not ketones, reduce Ag^+ to metallic silver which deposits as a silver mirror on the wall of the test tube.

A similar reaction is used to silver glass for use in mirrors.

$$CH_3-\overset{\overset{O}{\|}}{C}-H + Ag(NH_3)_2^{+\,-}OH \xrightarrow{H_2O} Ag\downarrow + CH_3-\overset{\overset{O}{\|}}{C}-OH + NH_4^+ \quad (8\text{-}43)$$

acetaldehyde   mirror

Methyl ketones are easily oxidized by sodium hypochlorite (NaOCl) to a carboxylic acid and chloroform.

$$Ph-\overset{\overset{O}{\|}}{C}-CH_3 \xrightarrow{NaOCl} Ph-\overset{\overset{O}{\|}}{C}-OH + CHCl_3 \quad (8\text{-}44)$$

chloroform

The ease with which reaction 8-44 occurs, and the high yield of products, is in contrast with the difficult oxidation of acetophenone with reagents such as $K_2Cr_2O_7$ (equation 8-42). Sodium hypoiodite (NaOI) gives a similar reaction leading to iodoform (CHI_3), a yellow solid, mp 120°C. At one time iodoform was widely used as an antiseptic. Ketones other than methyl ketones do not give acids on treatment with hypohalites.

Quinones

A quinone is a compound with two carbonyl groups and two double bonds in a six-membered ring. One example is *p*-benzoquinone, part of the defensive armament of the bombardier beetle (Chapter 7). The other possible way of arranging two carbonyl groups and two double bonds in a six-membered ring corresponds to *o*-benzoquinone.

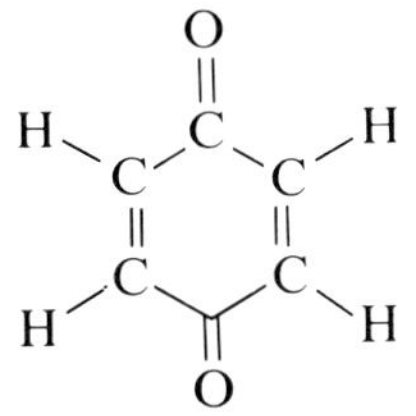

p-benzoquinone
yellow; mp 115.7°C

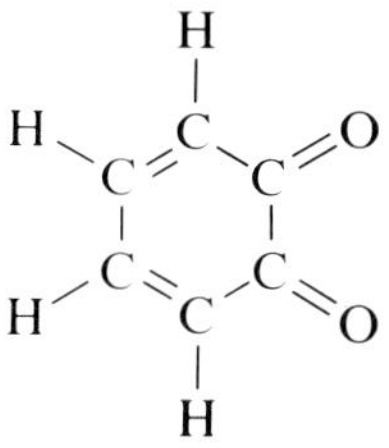

o-benzoquinone
red, unstable

Reduction of a quinone with hydrogen and a catalyst or with sodium hyposulfite ($Na_2S_2O_4$) gives the corresponding **hydroquinone**.

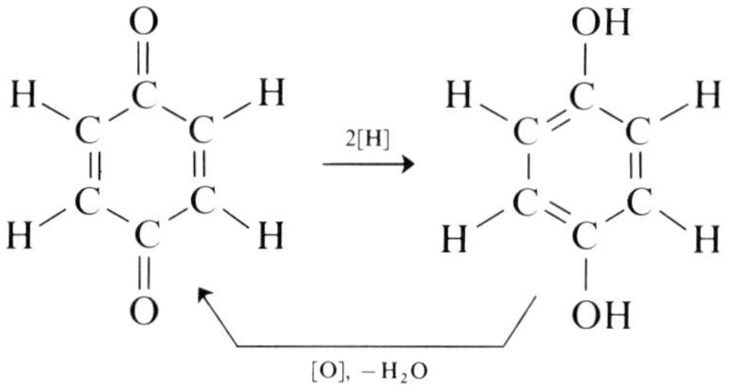

(8-45)

quinone → hydroquinone

[O], –H_2O

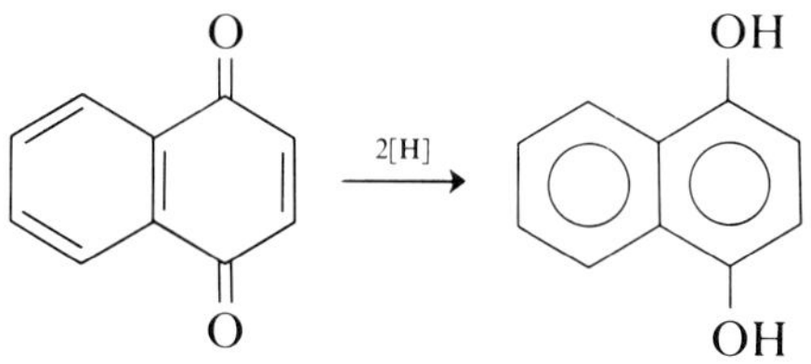

(8-46)

1,4-naphthoquinone
yellow; mp 194°C

1,4-dihydroxynaphthalene
or naphthohydroquinone
colorless, mp 176°C

Like phenols, hydroquinones are soluble in alkali. Thus a typical quinone can be made soluble by reducing it to the hydroquinone in alkaline solution (equation 8-47). The quinone can be recovered from the alkaline solution of the hydroquinone simply by adding an oxidizing agent. Exposure to air is usually sufficient to effect the oxidation.

$$\underset{\substack{\text{anthraquinone}\\\text{yellow; mp 286°C; insoluble in water}}}{\text{[anthraquinone structure]}} \xrightarrow[\substack{\text{NaOH}\\\text{adds two electrons}}]{\text{Na}_2\text{S}_2\text{O}_4} \underset{\text{soluble in water}}{\text{[disodium salt structure]}} \quad (8\text{-}47)$$

[O] removes two electrons

A few of the many quinones found in nature are listed in Table 8-1. Vitamin K, or phylloquinone, is an antihemorrhagic factor that helps blood to clot. Echinochrome A is a substance released by sea urchin eggs, apparently as a chemical trail to guide the sperm to the eggs.

All quinones are colored, and several naturally occurring quinones or quinonoid compounds used as dyes are also shown in Table 8-1. Juglone, a

FIGURE 8-5. Cortes.

TABLE 8-1. *Some Naturally Occurring Quinones*

Compound	Formula
Vitamin K or phylloquinone	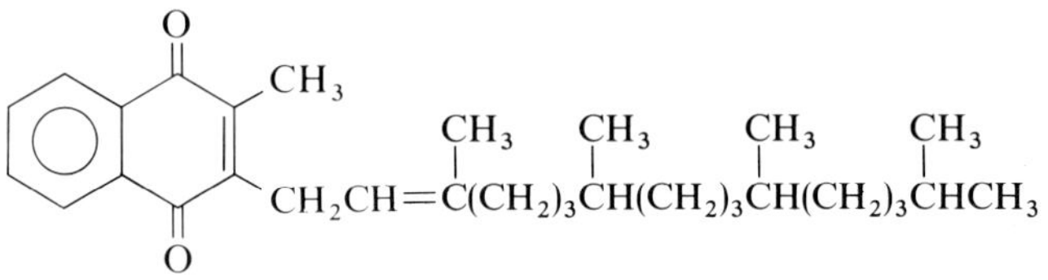 [2-methyl-3-phytyl-1,4-naphthoquinone structure with side chain $CH_2CH=C(CH_3)(CH_2)_3CH(CH_3)(CH_2)_3CH(CH_3)(CH_2)_3CHCH_3$]
Echinochrome A; red; mp 220°C	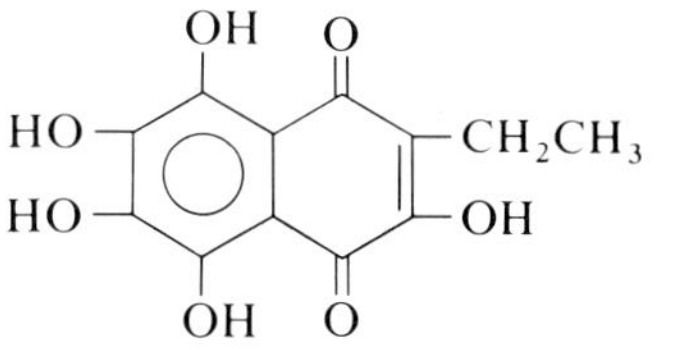 [naphthoquinone with OH groups and CH_2CH_3 substituent]
Alizarin	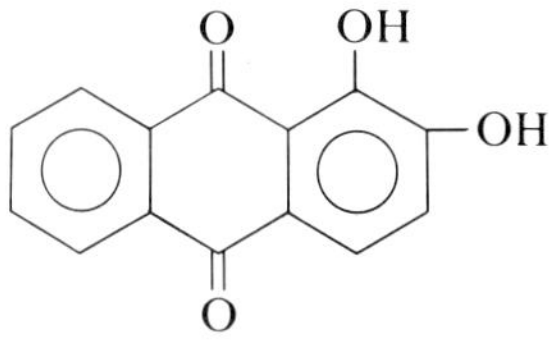 [1,2-dihydroxyanthraquinone]
Juglone; brown-red; mp 151–154°C	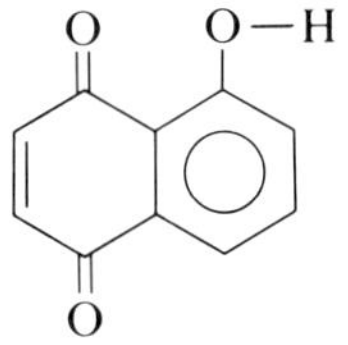 [5-hydroxy-1,4-naphthoquinone]
Carminic acid	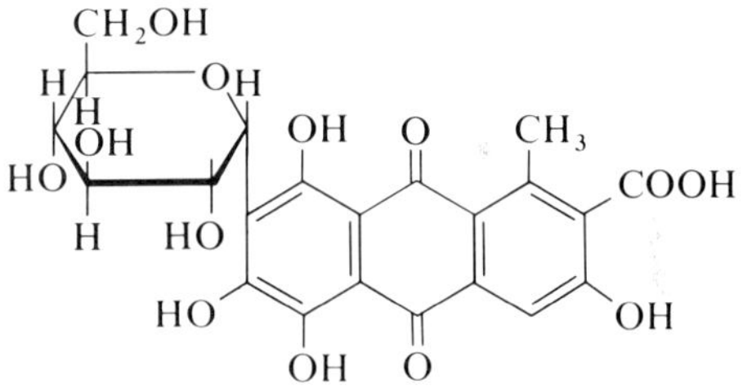 [anthraquinone with glucose, CH_3, COOH, and OH substituents]
Tyrian purple	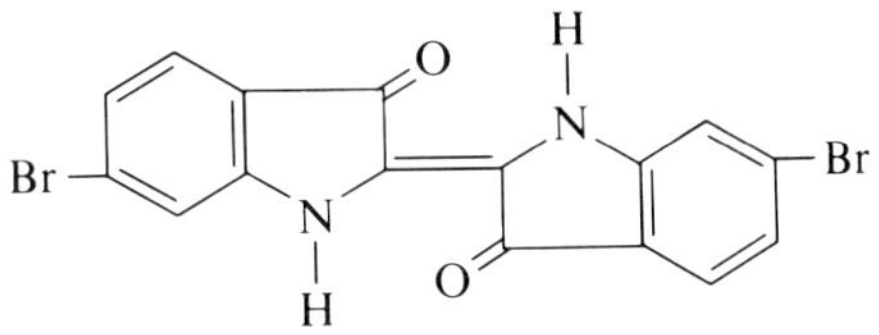 [6,6'-dibromoindigo structure]

dye obtained from walnut shells, was used in colonial times. Alizarin, from the madder plant, has been found in Egyptian mummy wrappings and was also used in the red coats of the British army.

The dyestuff, or crude dye, **cochineal** consists of the dried bodies of *Coccus cacti*, an insect found on a Mexican cactus. The pure dye is called carminic acid. It is worth noting that good dyes were once so rare and expensive that Cortes (Figure 8-5) was almost as much interested in extorting cochineal from the Aztecs as in extorting gold.

Indigo, found in indigo plant and in the British plant, woad, is not strictly a quinone but, like quinones, can be reduced to a colorless alkali-soluble form.

In the **vat dyeing** process, the cloth is dipped in the reduced (colorless or **leuco**) alkaline solution and becomes colored on being exposed to the oxidizing action of the air. Woad (Figure 8-6) was used for tattooing by the Picts at the time of the Roman invasion of Britain.

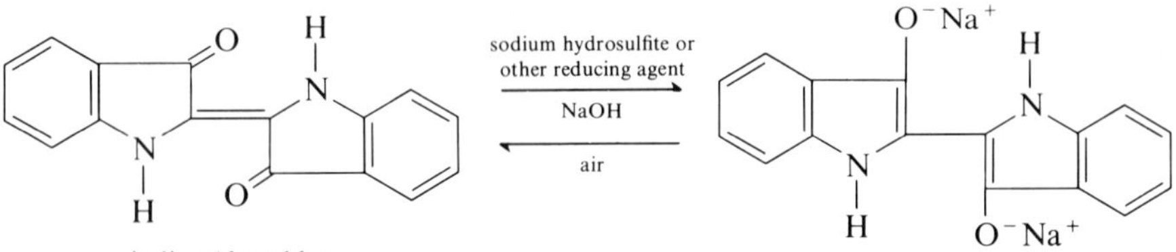

indigo (deep blue and insoluble) ⇌ (sodium hydrosulfite or other reducing agent, NaOH / air) the reduced (leuco) form of indigo (colorless but soluble)

FIGURE 8-6. Woad. [From J. W. Krutch, *Herbal*, Putnam, N.Y., 1965, and Pierandrea Mattioli, *Commentaries on the Six Books of Dioscorides*, Prague, 1563]

Tyrian purple, or dibromoindigo (Table 8-1) is found in the purple snails *Murex brandaris* and *Murex trunculus*; they contain a colorless material that is oxidized to the dye on exposure to air. The chemical study of the structure of Tyrian purple used only 1.4 g of the dye, but this required no less than 12,000 snails. The name royal purple was applied to the dye when its use was restricted to members of the imperial household by decree of the Roman emperors.

OTHER QUINONOID COMPOUNDS

The coloring materials of flowers such as roses, poppies, and cornflowers, is cyanin. The same pigment can have different colors in different flowers or in the same flower grown in different soils because the color depends not only on the H^+ concentration but also on the presence of metals such as iron with which it can form salts.

cyanin cation
red; the form present in acid solution

cyanin color base
violet; the form present in neutral solution

The compound fluorescein has that name because of its intense green fluorescence. Fluorescent molecules absorb light of one color (including colorless ultraviolet light) and **emit** or give off light of another color. Although solutions of fluorescein look bright green, the true orange color can be seen by looking *through* a layer of the solution at a light source (Figure 8-7). Dyes

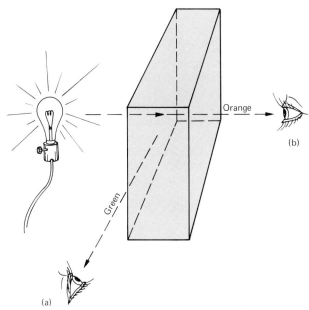

FIGURE 8-7. The solution appears bright green to the observer at (a) and orange to the observer at (b). The observer at (a) sees fluorescence radiation; the observer at (b) sees the actual color which is the original white light minus the light absorbed by the fluorescein.

that fluoresce with an orange color are used on diving equipment. The blue wavelengths that make these dyes fluoresce penetrate deeper into the water than does red or orange light. Ordinary red or orange objects look black at depths greater than a few feet.

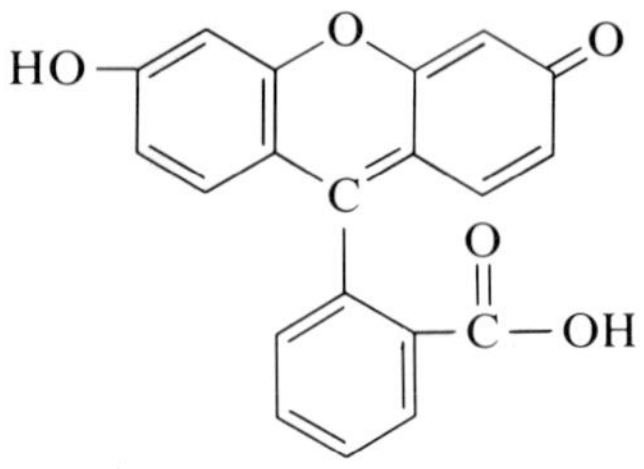

fluorescein

The sodium salt is an orange-red powder whose solutions appear bright green because of fluorescence

SUMMARY

A. Aldehydes and ketones
   1. Aldehydes have the structure $R-\overset{O}{\underset{\|}{C}}-H$; ketones have the structure $R-\overset{O}{\underset{\|}{C}}-R$.
   2. The systematic names of aldehydes end in -al, of ketones in -one, for example, ethanal and propanone.
   3. The C=O is polarized in the sense C^+-O^-. It reacts with nucleophiles at the carbon end and with electrophiles at the oxygen end.
   4. The addition reactions include
      (a) Hydrogenation to give alcohols.
      (b) Reversible addition of water to give hydrates and of 1 mole of alcohol to give hemiacetals.
      (c) Addition of two moles of alcohol to give acetals.
      (d) Addition of Grignard reagents to give Mg salts of alcohols (secondary from aldehydes, tertiary from ketones).
      (e) Addition of amines to give products $R_2C=NR$ plus H_2O.
      (f) Addition of HCN to give cyanohydrins.
   5. Enolization
      (a) Removal of a proton gives enolate ion.
      (b) Condensation reactions involve the addition of enolate ion to carbonyl groups.
   6. Oxidation
      (a) Aldehydes are readily oxidized to carboxylic acids; ketones resist oxidation.
      (b) Sodium hypochlorite, a special reagent for methyl ketones, oxidizes them to a carboxylic acid and chloroform.

B. Quinones
   1. Quinones are six-membered ring compounds containing two carbonyl groups and two double bonds.
   2. They are reduced to hydroquinones, phenols with two hydroxyl groups.
   3. The reduction of quinones and related compounds to the alkali-soluble hydroquinone form is used in vat dyeing.

4. The quinone structure and similar structures are found in many important naturally occurring compounds, including dyes.
    (a) Some compounds emit light (fluoresce) as well as absorb light.

EXERCISES

1. Give the common names, structures, and a short description of the physical properties of two simple aldehydes and the simplest ketone.

2. Write structures for
    (a) formaldehyde
    (b) acetaldehyde
    (c) ethanal
    (d) 2-ethylbutanal
    (e) 3-methylpentanal
    (f) *trans*-3-hexenal
    (g) 2-hexanone
    (h) 3-pentanone
    (i) 3-methylpentadecanone

3. Name

(a) $H-\overset{\overset{O}{\|}}{C}-H$

(b) $CH_3-\overset{\overset{O}{\|}}{C}-H$

(c) $CH_3-\overset{\overset{O}{\|}}{C}-CH_3$

(d) $CH_3CH_2-\overset{\overset{O}{\|}}{C}-CH_3$

(e) $CH_3-\overset{\overset{O}{\|}}{C}-CH_2CH_2CH_3$

(f) $CH_3CH_2-\overset{\overset{O}{\|}}{C}-CH_2CH_2-\underset{\underset{CH_3}{|}}{C}HCH_3$

4. Write the equation for the reaction of *trans,cis*-2,6-nonadienal (page 141) with ammoniacal silver nitrate, $Ag(NH_3)_2^+ \; {}^-OH$.

5. Write equations illustrating five different addition reactions to the C=O double bond.

6. Define and give structural formulas as examples for each. (a) Hemiacetals; (b) acetals; (c) cyanohydrins; (d) cyanogenetic glycosides; (e) enols.

7. Complete the following equations

(a) $Cl-\underset{\underset{Cl}{|}}{\overset{\overset{Cl}{|}}{C}}-\overset{\overset{O}{\|}}{C}-H + H_2O \rightleftarrows$

(b) 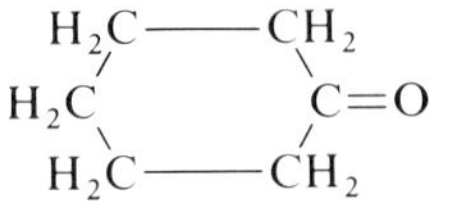
$$\begin{matrix} H_2C\!\!-\!\!\!-\!\!\!-\!\!CH_2 \\ \diagup \qquad \diagdown \\ H_2C \qquad\qquad C=O \\ \diagdown \qquad \diagup \\ H_2C\!\!-\!\!\!-\!\!\!-\!\!CH_2 \end{matrix} + H_2 \xrightarrow{catalyst}$$

(c) $CH_3CH_2-\overset{\overset{O}{\|}}{C}-H + 2CH_3OH \xrightarrow{H^+}$

(d) $CH_3MgBr + CH_3CH_2\overset{\overset{O}{\|}}{C}-H \rightarrow$

(e) $CH_3CH_2MgBr + CH_3-\overset{\overset{O}{\|}}{C}-CH_3 \rightarrow$

(f) $CH_3CH_2MgBr + H-\overset{\overset{O}{\|}}{C}-H \rightarrow$

(g) $CH_3-\overset{\overset{O}{\|}}{C}-H + H_2NNHC_6H_5 \rightarrow$

(h) 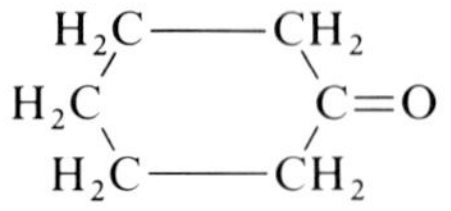 $\begin{array}{c} H_2C\text{———}CH_2 \\ H_2C \qquad\qquad C=O \\ H_2C\text{———}CH_2 \end{array} + HCN \rightleftarrows$

(i) $C_6H_5-\overset{\overset{O}{\|}}{C}-H + CH_3-\overset{\overset{O}{\|}}{C}-CH_3 \xrightarrow{base}$

(j) $C_6H_5-\overset{\overset{O}{\|}}{C}-H + CH_3CH_2-\overset{\overset{O}{\|}}{C}-CH_2CH_3 \xrightarrow{base}$

8. Describe a test to distinguish between

   (a) an aldehyde $R-\overset{\overset{O}{\|}}{C}-H$ and a ketone $R-\overset{\overset{O}{\|}}{C}-R'$

   (b) a ketone $R-\overset{\overset{O}{\|}}{C}-CH_3$ and a ketone $R-\overset{\overset{O}{\|}}{C}-CH_2CH_3$

9. Describe the process of vat dyeing with a quinone or quinonelike dye.

10. Write the equation for a simple reaction that could be used to distinguish muscone (page 141) from methyl cyclohexyl ketone.

11. Write equations showing how to convert muscone (page 141) into 1,3-dimethylcyclopentadecanol.

12. Piperonal (page 15) will react with phenylmagnesium bromide, followed by water, to give a secondary alcohol.
    (a) Write the equation for these reactions.
    (b) Suggest a reason why the same reaction will not work with vanillin (page 15).

13. Write an equation for a reaction that would convert Tyrian purple (page 155) into a colorless, water-soluble compound.

14. Juglone (Table 8-1) is insoluble in water. If you wished to extract some of this compound from walnut shells, what could be added to the water to make the juglone soluble?

15. (a) A substance $C_5H_{10}O$ reacts with NaOI (sodium hypoiodite, from NaOH and I_2) to give a yellow solid precipitate, mp 119°C.
    (b) The substance $C_5H_{10}O$ also reacts with H_2 in the presence of Ni to give a substance $C_5H_{12}O$.
    (c) The substance $C_5H_{12}O$ reacts with metallic sodium to give a substance $C_5H_{11}ONa$.
    (d) The substance $C_5H_{12}O$ reacts with concentrated H_2SO_4 to give a substance C_5H_{10}.

(e) The substance C_5H_{10} reacts with H_2 in the presence of platinum to give pentane.

What is the structure of the substance $C_5H_{10}O$?

Write equations for all the reactions mentioned.

15. What alcohols could be oxidized to the following aldehydes or ketones?
    (a) cyclohexanone
    (b) propanal
    (c) 2-propanone
    (d) benzophenone (diphenyl ketone)

Chapter 9
Optical Isomerism

Introduction

The isomers ethyl alcohol and dimethyl ether (Chapter 3) differ a great deal from each other both in chemical and physical properties. The isomers *cis*- and *trans*-2-butene differ less in chemical and physical properties, although they are still easy to tell apart. In this chapter we will take up a more subtle kind of isomerism in which the isomers (enantiomers) are mirror images of each other. As one might expect, most physical and chemical properties of mirror image molecules are identical. The exceptions, however, are extremely important, especially to living organisms. For example, one molecule might be a highly effective drug and its mirror image entirely ineffective. To give another example: if we were transported to a world in which all the molecules are mirror images of the molecules in this world, we would probably starve to death because our bodies would be unable to make any use of the mirror-image molecules in that world's potatoes and beef steaks.

Enantiomers

Two molecules that are different, but so much alike that one is the mirror image of the other, are called enantiomers. An example is the pair of glyceraldehyde enantiomers shown in Figure 9-1.[1] The

[1] The trivial name glyceraldehyde comes from the relationship of this aldehyde to the alcohol glycerol.

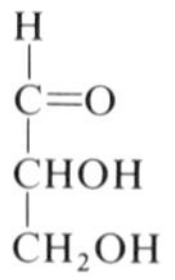

$$\begin{array}{c} H \\ | \\ C=O \\ | \\ CHOH \\ | \\ CH_2OH \end{array} \qquad \begin{array}{c} CH_2OH \\ | \\ CHOH \\ | \\ CH_2OH \end{array}$$

glyceraldehyde 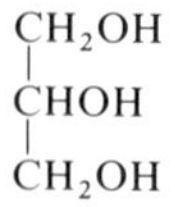   glycerol difference between the glyceraldehyde enantiomers is like the mirror image difference between a right hand and a left hand. The small capital D in the name D-glyceraldehyde comes from the Latin *dextro* meaning right and the small capital L in L-glyceraldehyde comes from the Latin *levo* meaning left.

Of course, the choice of which isomer to call D and which to call L is arbitrary and could have been done the other way round. However, chemists have agreed to call the molecule in Figure 9-1A

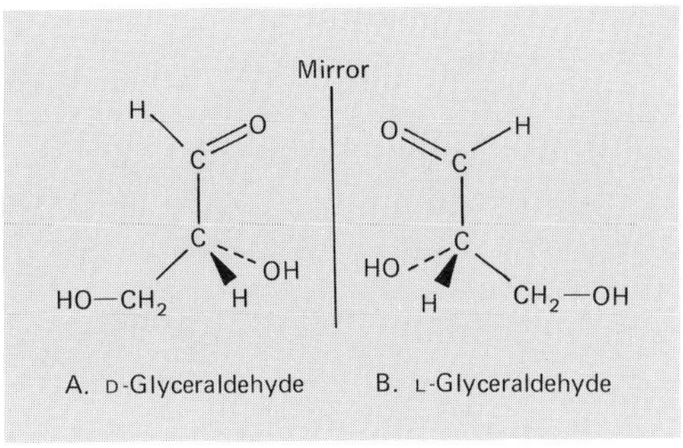

[FIGURE 9-1: Mirror-image structures of D-Glyceraldehyde (A) and L-Glyceraldehyde (B)]

[2]FIGURE 9-1. The mirror-image isomers of glyceraldehyde.

the D enantiomer and the one on the other side of the mirror the L enantiomer. Furthermore, any closely related molecule that looks like the D enantiomer of glyceraldehyde is also called D and its enantiomer is called L. An example is the pair of lactic acid enantiomers in Figure 9-2.

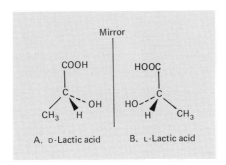

FIGURE 9-2. The lactic acid enantiomers.

THE FISCHER PROJECTION

In order to be able to represent particular D and L enantiomers on paper without using perspective drawings, chemists have adopted a convention suggested by Emil Fischer (Figure 9-3). In the Fischer projection for D-glyceraldehyde, for example (Figure 9-4), it is understood that the central carbon atom is in the plane of the paper,

---

[2] In perspective drawings a dotted bond indicates a group behind the plane of the paper and a heavy bond indicates a group in front of the plane of the paper. A wedge-shaped bond has its thick end closer to the reader and its thin end farther away; it may be either in front of the plane of the page or behind it.

Optical Isomerism

163

FIGURE 9-3. Emil Fischer (1852–1919) was awarded the Nobel prize in 1902. He is especially honored for his development of the organic chemistry of carbohydrates, proteins, and dyes. [Bettmann Archives]

the top and bottom carbon atoms behind the plane, and the H and HO groups projecting in front of the plane. The rule is to arrange the atoms of the carbon skeleton in a vertical line with the carbon of the most important functional group at the top. The D enantiomer then has the functional group of the middle carbon on the right; the L enantiomer has it on the left.

$$\begin{array}{c} \text{COOH} \\ | \\ \text{HO}-\text{C}-\text{H} \\ | \\ \text{CH}_2\text{OH} \end{array} \qquad \begin{array}{c} \text{COOH} \\ | \\ \text{H}-\text{C}-\text{NH}_2 \\ | \\ \text{CH}_3 \end{array}$$

L-lactic acid   D-alanine, an amino acid

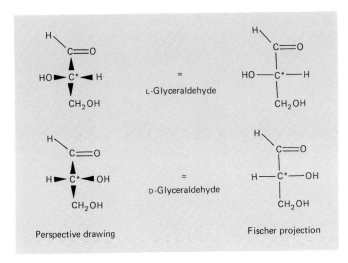

FIGURE 9-4. The Fischer projection.

Plane Polarized Light

In Table 9-1, which lists some properties of D and L lactic acids, only the last entry shows any difference between them. This physical property is the rotation of the plane of polarized light as it passes through a solution of lactic acid. The rotation, in degrees, depends on the wavelength of the light used and on the temperature. These are mentioned in the symbol for the specific rotation. Thus $[\alpha]_{546}^{21.5}$ refers to the **specific rotation** using light of wavelength 546 Å and at a temperature of 21.5°C.[3] Tables for other enantiomeric pairs would look much the same: of a long list of physical properties, only the specific rotation would show any difference.

TABLE 9-1. *The Lactic Acid Enantiomers*

Property	D-Lactic acid	L-Lactic acid
Melting point, °C	26.0	26.0
Boiling point (at 14 mm Hg), °C	122	122
Solubility (water)	very soluble	very soluble
Solubility (CHCl$_3$)	almost insoluble	almost insoluble
Acid dissociation constant (pK in water)	3.79	3.79
Specific rotation, $[\alpha]_{546}^{21.5}$	−2.6°	+2.6°

L-Lactic acid accumulates in muscles during exercise and is also found on the skin. It is said to attract mosquitoes.

Because the rotation of plane polarized light is such an indispensable property in working with enantiomers, a more detailed discussion of polarized light will be useful. In Chapter 3 it was pointed out that light is a wavelike electrical phenomenon and, in the

---

[3] Specific rotation is so named because the rotation also depends on the concentration of the substance in solution and the thickness of the sample through which the light passed. These factors are corrected for in the specific rotation.

discussion of wavelength and frequency, light waves were compared with water waves. Just as a cork bobs up and down in a water wave, an electron tends to move back and forth when an electrical wave passes it. The water wave is said to be "polarized in a vertical plane" because the cork (Figure 9-5) moves only in a vertical plane; that is, it moves only up and down. A polarized light wave tends to move a charged particle only in one direction, but that direction can be at any angle to the vertical and not just up and down (Figure 9-6).

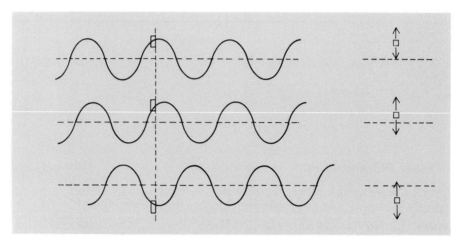

FIGURE 9-5. Motion of a cork in a water wave. As the wave passes from left to right, the cork moves only straight up and down. It does not move along with the wave, nor does it move from side to side.

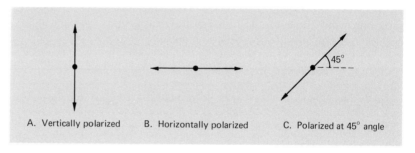

A. Vertically polarized   B. Horizontally polarized   C. Polarized at 45° angle

FIGURE 9-6. A few of the many possible directions of motion of an electron in a polarized light wave. The light wave is progressing out of the paper and towards the reader.

The light from an ordinary lamp is unpolarized. It is a mixture of light waves polarized in all possible directions and with the crests and troughs passing at all possible times (Figure 9-7). A beam of unpolarized light can be converted into a beam of polarized light by a device called a polarizer. The polarizer blocks all the components of the beam except those waves that are polarized in one particular direction (Figure 9-8).

A beam of light will ordinarily pass through two polarizers in succession only if they have their "polarizing axes" at the same angle. Thus we can see through both of the sheets of Polaroid in Figure 9-9A because they are aligned at the same angle, but the arrangement in Figure 9-9B causes the light to be blocked off.

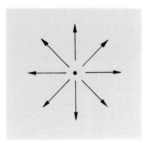

FIGURE 9-7. A beam of unpolarized light is a mixture of waves polarized at all possible angles.

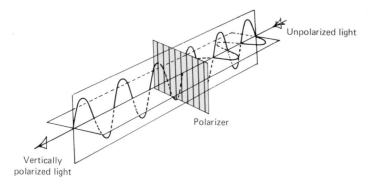

FIGURE 9-8. The beam of light entering the polarizer from the right is unpolarized, although only two of the many different components are shown. The polarizer, which might be a tourmaline crystal or a sheet of Polaroid, permits only the vertically polarized component to get through.

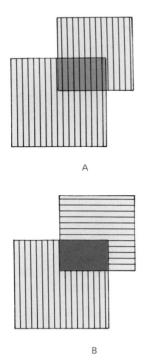

FIGURE 9-9. The Polaroid sheets in (**A**), with their polarizing axes parallel, transmit the light; when the axes are not parallel, as in (**B**), they block it off.

Optical Isomerism

The lenses in Polaroid sunglasses consist of sheets of polyvinyl alcohol (Chapter 12) which have been stretched to line the molecules up as long parallel chains. These oriented polymer molecules, like the regular array of ions in a tourmaline crystal, tend to transmit light polarized in one direction and absorb components of the light that are polarized at right angles to that direction. There are several different kinds of Polaroid. In some, the polyvinyl alcohol has been treated with iodine to give a complex that absorbs more of the light. In others, the stretched polyvinyl alcohol has been dehydrated.

The glare reflected from the surface of a body of water is partly polarized in a horizontal plane, whereas the sunglasses are oriented so as to pass light polarized only in a vertical plane. Hence the glare is blocked by the sunglasses. One can tell whether a pair of sunglasses really do act as polarizers by looking at the same sheet of water with the head first in a vertical position and then in a horizontal position. With the head in a vertical position, the water should look darker and reefs and fish should be more visible than they are when the head is held in the horizontal position.

THE POLARIMETER

If the experiment of Figure 9-9 were tried with a solution of D-lactic acid placed between the two polarizers, **it would be found that the front polarizer would have to be rotated through some angle to the left in order to transmit the maximum amount of light**. The reason for this is that D-lactic acid is **levorotatory**. On the way through the solution the light from the first polarizer, originally vibrating in a vertical plane, has had its plane of polarization rotated to the left. The front polarizer has to be rotated by the same amount to restore the original brightness.

A polarimeter (Figure 9-10) is simply a device consisting of two polarizers, a transparent container for the solution being investigated, and a protractor or circular scale for measuring the number of degrees through which the front polarizer had to be rotated.

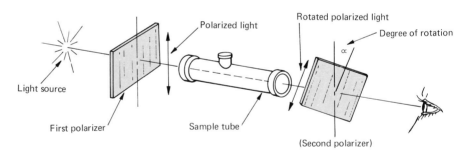

FIGURE 9-10. A polarimeter.

Optical Activity

A substance is said to be optically active if it rotates the plane of plane polarized light as the light passes through it. D-Lactic acid rotates the plane of polarized light to the left; therefore it is an example of a levorotatory substance. We should expect the mirror image L-lactic acid molecules to do the mirror image of anything that the D-lactic acid molecules do. Since the D-lactic acid molecules rotate the plane of polarized light to the left ($[\alpha] = -2.6°$), L-lactic acid rotates the plane of polarized light to the right, and by the same number of degrees ($[\alpha] = +2.6°$).

Most molecules are identical with their mirror images and are optically inactive; that is, they have a specific rotation of 0.000°. A mixture of *equal numbers* of D and L molecules of the same compound is also optically inactive. The reason these ($\pm$) mixtures are optically inactive is that the D and L components act independently: for every D molecule rotating the plane of polarized light in one direction, there is an L molecule that rotates it back again. The net result is no rotation at all.

A ($\pm$) mixture is also called a **racemic** mixture and the process of converting an optically active sample of a pure enantiomer into the ($\pm$) mixture is called **racemization**.

HOW TO PREDICT OPTICAL ACTIVITY

Superposition. A substance will exist as optically active D and L enantiomers if the molecules are different from their mirror images. The most direct way to test this is to make a model of a molecule and its mirror image and see if one model can be superposed on the other, matching each part with the corresponding part of the other molecule. If this can be done, there is only one form and the compound will be optically inactive. If it cannot be done, there are two forms and the substance will be optically active.[4]

Figure 9-11 shows the superposition test as applied to D- and L-lactic acids. In Figure 9-11A we have matched up the central

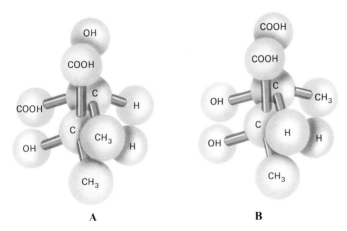

FIGURE 9-11. Futile attempts at superposing a D-lactic acid model on an L-lactic acid model.

carbon atoms, the hydrogen atoms, and the methyl groups, but the OH turns out to be next to the COOH of the other molecule instead of being next to the OH. In Figure 9-11B we have matched the HO with HO and the COOH with COOH but now the H is next to CH_3 of the other molecule instead of being next to H. The lactic acid mirror images simply cannot be superposed no matter what we do, and lactic acid exists in two distinct optically active D and L forms.

---

[4] Unless, of course, one happens to have a 50:50 mixture ($\pm$ mixture) of the two enantiomers.

Plane of Symmetry. Another way of deciding whether a molecule will be optically active is to see if it has a plane of symmetry. If the molecule has a plane of symmetry, the mirror images will be identical and the solutions will be optically inactive.

A plane of symmetry is a plane that divides a molecule into two halves such that one half is the mirror image of the other. If one half of the molecule can be replaced with a mirror without changing its appearance, the molecule has a plane of symmetry. An example is propanoic acid, Figure 9-12. Note that if one of the hydrogen atoms in Figure 9-12 were an OH group (as it is in lactic acid) the symmetry would be destroyed.

Figure 9-13 shows how the superposition test confirms the prediction from the plane of symmetry for propanoic acid.

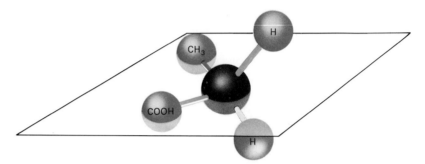

FIGURE 9-12. The propanoic acid molecule, CH_3CH_2COOH, has a plane of symmetry passing through the CH_3, the central C, the COOH, and half way between the two H atoms. If the model were cut in half along this plane and placed on a mirror, the half-model plus its image would look just like the original whole model.

Asymmetric Carbon Atoms. An asymmetric carbon atom is one that has **four different groups attached to it.** The starred carbon atoms in the structures shown below are examples.

All of these compounds are optically active, their mirror images are nonsuperposable, and they lack planes of symmetry. Looking for asymmetric carbon atoms is the easiest way to spot an optically active compound, but one must beware of compounds with *two*

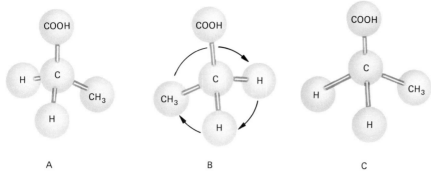

| A | B | C |

FIGURE 9-13. Models (**A**) and (**B**) of propanoic acid are mirror images of each other. If (**B**) is merely rotated 120° about the C—COOH bond as an axis, it will look like (**C**). It can be seen by inspection that (**C**) can be superposed on (**A**).

asymmetric carbon atoms. Compounds with two *mirror-image* asymmetric carbon atoms in the same molecule are called **meso** compounds (Figure 9-14). They have a plane of symmetry and are optically inactive.

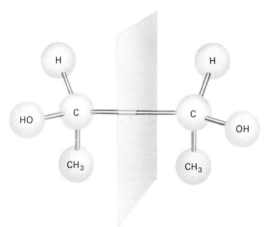

FIGURE 9-14. *meso*-2,3-Butandiol and its plane of symmetry.

Synthesis of Asymmetric Molecules

Many of the reactions already studied can be used to create asymmetric carbon atoms. Equations 9-1 through 9-4 are examples, and others can easily be invented. Although these and similar

$$CH_2=C\begin{matrix}CH_2CH_3\\|\\H\end{matrix} + HCl \rightarrow CH_3-\overset{*}{\underset{H}{\overset{Cl}{C}}}-CH_2CH_3 \quad (9\text{-}1)$$

$$CH_3-\overset{O}{\overset{\|}{C}}-CH_2CH_2CH_3 + CH_3CH_2MgBr \rightarrow CH_3-\overset{*}{\underset{CH_2CH_3}{\overset{O^-(MgBr)^+}{C}}}-CH_2CH_2CH_3$$

(9-2) Optical Isomerism

$$CH_3-\overset{O}{\overset{\|}{C}}-CH_2CH_2CH_3 + H_2 \xrightarrow{cat.} CH_3-\overset{OH}{\underset{H}{\overset{|}{\overset{*}{C}}}}-CH_2CH_2CH_3 \qquad (9\text{-}3)$$

$$CH_2=CHCH_3 + Br_2 \rightarrow BrCH_2-\overset{Br}{\underset{H}{\overset{|}{\overset{*}{C}}}}-CH_3 \qquad (9\text{-}4)$$

reactions give compounds with asymmetric carbon atoms, *the product is never optically active but always turns out to be a ($\pm$) mixture*. That is, for every molecule of HCl that reacts with 1-butene to give one of the enantiomeric 2-chlorobutanes, there is another molecule that reacts to give the other enantiomer. If a billion molecules of HCl react with a billion molecules of 1-butene to give 2-chlorobutane, the two enantiomers will be found in equal numbers and the product will be a ($\pm$) mixture.

This is just what one would expect. The 1-butene molecule has a plane of symmetry that passes through the plane of the double bond as shown in Figure 9-15. Both sides of the molecule are exactly alike,

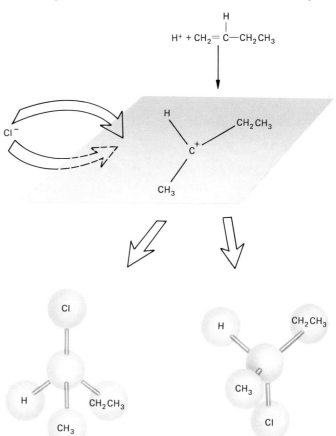

FIGURE 9-15. Because approach is equally probable from either side, the two enantiomers are formed in equal amounts.

so there is no reason for the Cl⁻ to approach one side in preference to the other. Reaction from one side gives one enantiomer and reaction from the other side gives the other enantiomer, and the result is a (±) mixture. The same conclusion is reached when other reactions are considered in the same detail.

The general conclusion is that ordinary chemical reactions cannot give optically active products, unless the reagents were optically active to begin with.

Even though ordinary reactions cannot give optically active products, most of the organic molecules found in living organisms are present only as one of the two possible enantiomers. In fact, construction from optically active materials is probably a requirement for life. One of the great unanswered questions is how this came about.

According to one speculation, a few molecules reacted to give an optically active product. This is possible with a reaction involving just a few molecules. For example, with one molecule of reagent, only one molecule of product is obtained and it has to be either D or L; with three molecules of reagent the worst that can be achieved is to get two molecules of one enantiomer and one of the other. In living systems, optically active catalysts preferentially form optically active products. If the primitive reaction happened to give an optically active product *that was also a catalyst for the formation of the same substance*, a reaction that once got started giving the D (or L) enantiomer would keep on doing so and the competing reaction giving the other enantiomer would not have much chance.

THE R AND S SYSTEM

Earlier in this chapter we used the prefixes D and L to describe the configurations of compounds that resembled D(+)-glyceraldehyde and L(−)-glyceraldehyde, respectively. Thus, D-lactic acid is the enantiomer that resembles D-glyceraldehyde (Figure 9-16).

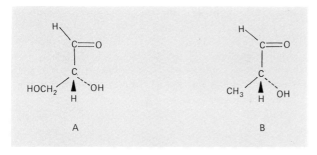

FIGURE 9-16. (**A**) D-Glyceraldehyde. (**B**) D-Lactic acid.

There is no difficulty in applying the D and L system of classifying enantiomers to lactic acid, because it is quite clear that the carboxyl group (COOH) corresponds to the aldehyde group (CHO) and the methyl group (CH₃) corresponds to the hydroxymethyl group (HOCH₂). Classification becomes less obvious when we try to name something like the 2-chlorobutane enantiomers (Figure 9-17) or the carvone enantiomers (Figure 9-18).

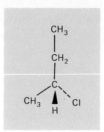

FIGURE 9-17. One of the 2-chlorobutane enantiomers.

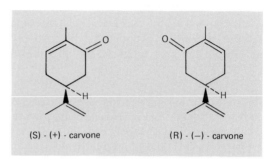

FIGURE 9-18. The carvone enantiomers.

In order to name configurations without reference to D-glyceraldehyde, a more general system has been developed. In this system, one of the enantiomers is considered to have the right-handed or R (Latin: *rectus*, right) chirality and the other the S (Latin: *sinister*, left) chirality.[5] For many purposes this is all that one needs to know about the system, and there are straight-forward though complicated rules for deciding which isomer is which.

Step 1. Assign a sequence of priority to the four groups or atoms attached to the asymmetric carbon atom. We will elaborate the rule for assigning the priority later, but for the present note that atoms of higher atomic number in the periodic table have higher priority. Hence in fluorochloroiodomethane, the sequence of priorities is I > Cl > F > H.

Step 2. Position the molecule so that the atom of lowest priority (hydrogen in this case) is pointing away from you.

Step 3. Move your eye or a pencil from the atom of highest priority, to the atom of second highest priority, to the atom of third highest priority. The movement will either be counterclockwise as in Figure 9-19A or clockwise as in 9-19B: Counterclockwise means that the configuration is to be called S; clockwise means that it is to be called R.

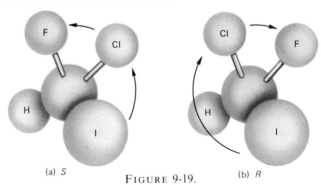

FIGURE 9-19.

[5] From the Greek *cheir*, hand.

THE PRIORITY RULES

1. For four different elements directly attached to the asymmetric carbon atom, the priority depends on the atomic number (Figure 9-20). For isotopes such as hydrogen and deuterium, the heavier isotope has the higher priority.

$$^3CH_3 - \overset{\overset{\displaystyle ^2O-CH_3}{|}}{\underset{\underset{\displaystyle ^1Cl}{|}}{C}} - {^4H}$$

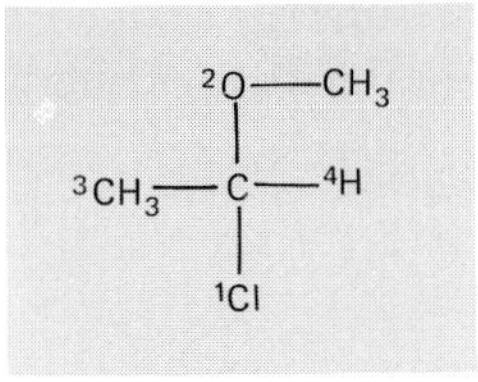

FIGURE 9-20.

2. If two groups start with the same element next to the asymmetric carbon atom, work outward along the chain to determine the priority (Figure 9-21).

   Which has the higher priority, CH_3 or CH_3CH_2?

   Answer: in CH_3 the second atoms are H, H, H; in CH_3CH_2 they are C, H, H. CH_3CH_2 has a higher priority than CH_3 because carbon has a higher atomic number than hydrogen.

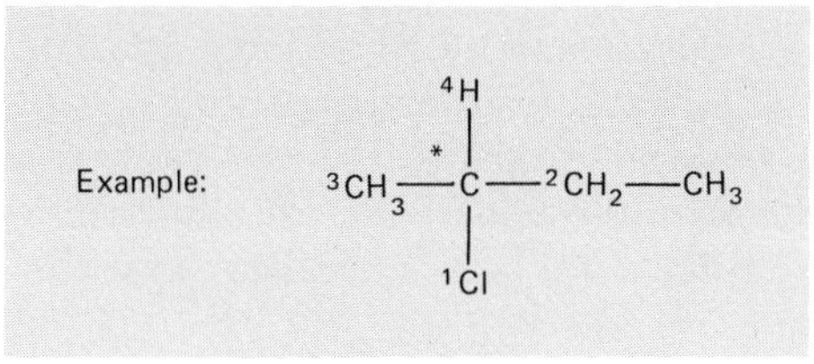

Example:
$$^3CH_3 - \overset{\overset{\displaystyle ^4H}{|}}{\underset{\underset{\displaystyle ^1Cl}{|}}{\overset{*}{C}}} - {^2CH_2 - CH_3}$$

FIGURE 9-21.

3. A double bond to an atom is the equivalent of *two* such atoms, a triple bond is the equivalent of *three* such atoms (Figure 9-22).

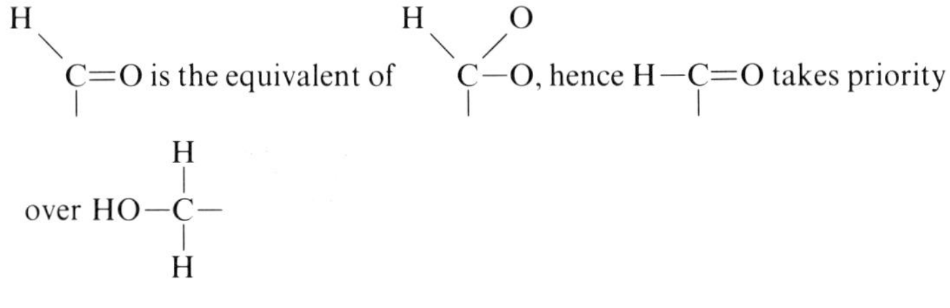

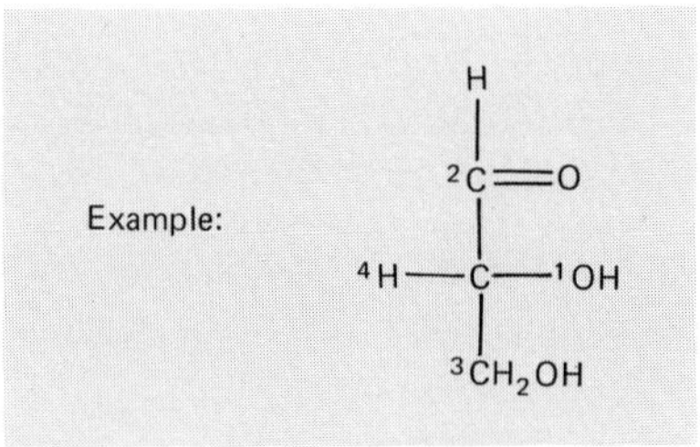

$$\underset{|}{\overset{H}{\diagdown}}C=O \text{ is the equivalent of } \underset{|}{\overset{H\diagdown \quad /O}{C-O}}, \text{ hence } H-\underset{|}{C}=O \text{ takes priority}$$

over $HO-\underset{\underset{\displaystyle H}{|}}{\overset{\overset{\displaystyle H}{|}}{C}}-$

Example:
$$^4H - \overset{\overset{\displaystyle H}{|} \quad \overset{\displaystyle ^2C=O}{}}{\underset{\underset{\displaystyle ^3CH_2OH}{|}}{C}} - {^1OH}$$

FIGURE 9-22.

Optical Isomerism

Diastereomers

In Table 9-1 identical physical and chemical properties were listed for D-lactic acid and its enantiomer L-lactic acid, with the single exception of the signs of the rotation of polarized light. When we consider compounds with two dissimilar asymmetric carbon atoms, however, the results are somewhat different. Take the compound 2-(acetylamino)-3-methylpentanoic acid, for example.

$$CH_3CH_2-\underset{\underset{CH_3}{|}}{\overset{\overset{H}{|}}{C^3}}-\underset{\underset{\underset{\underset{CH_3}{|}}{C=O}}{\underset{|}{NH}}}{\overset{\overset{H}{|}}{C^2}}-\overset{\overset{O}{\|}}{C}-O-H$$

This compound has two asymmetric carbon atoms, one at position 2 and the other at position 3. Since either of these can have the R or S configuration, there are not just two but four stereoisomers. The R and S designations and Fischer projection formulas are shown in Table 9-2, together with the physical properties of each isomer.

The isomers designated as 2(R),3(S) and 2(S),3(R) are obviously mirror images of each other and hence are called enantiomers. Of course, they also have identical physical properties except for the sign of $[\alpha]_D^{25}$.[6] The isomers 2(R),3(R) and 2(S),3(S) are also clearly mirror images of each other and constitute a pair of enantiomers. Again, the physical properties of the pair of enantiomers are identical except for the sign of $[\alpha]_D^{25}$. *But note that they are different from those of the other pair of enantiomers.* For example, both 2(R),3(R) and 2(S),3(S) melt at 156°C, but 2(R),3(S) and 2(S),3(R) melt at 150°C. The two pairs also have different solubilities in water, 4.15 for the 2(R),3(S) and 2(S),3(R) as compared to 4.37 for the 2(R),3(R) and the 2(S),3(S) pairs.

Isomers that are not enantiomers of each other, for example the 2(R),3(R) and 2(S),3(R) compounds in Table 9-2, are called diastereomers. All of the relationships between four such compounds are indicated in Figure 9-23.

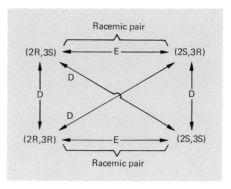

FIGURE 9-23. The relationships E (for enantiomer) and D (for diastereomer) among a set of four stereoisomers.

---

[6] The D subscript on the specific rotation indicates that the D line of a sodium flame was used as the light source.

TABLE 9-2. *The 2-(Acetylamino)-3-Methylpentanoic Acid Stereoisomers*

Structures 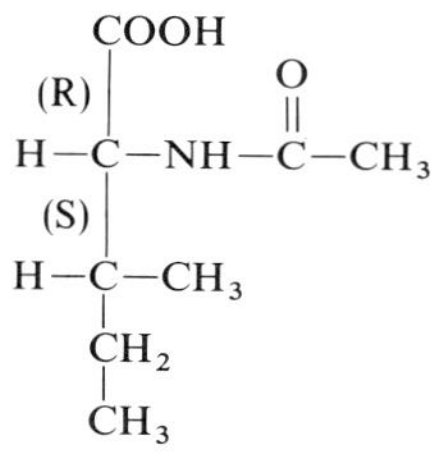 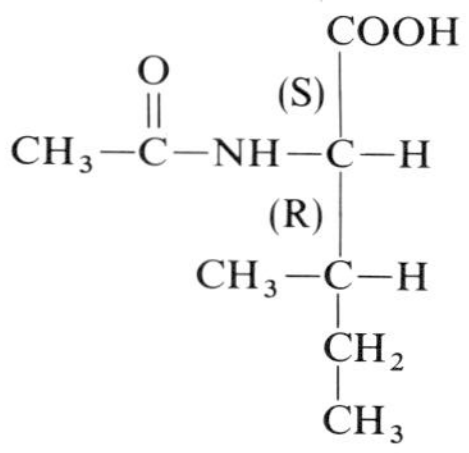 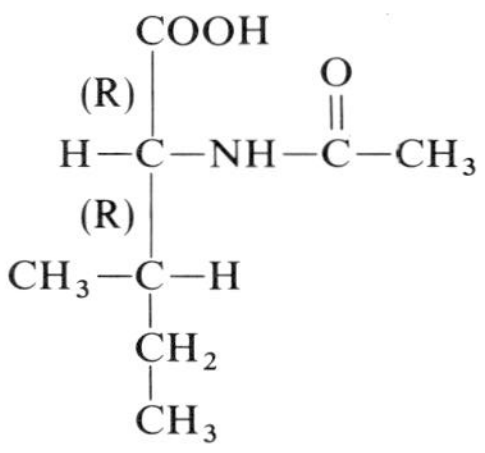 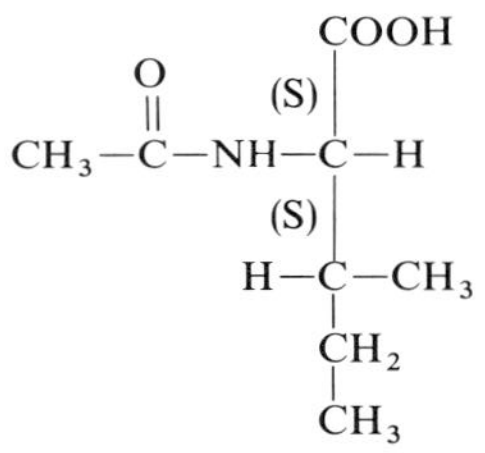			
COOH | H—C—NH—C—CH₃ (R) ‖ O | H—C—CH₃ (S) | CH₂ | CH₃	COOH | CH₃—C—NH—C—H (S) ‖ O | CH₃—C—H (R) | CH₂ | CH₃	COOH | H—C—NH—C—CH₃ (R) ‖ O | CH₃—C—H (R) | CH₂ | CH₃	COOH | CH₃—C—NH—C—H (S) ‖ O | H—C—CH₃ (S) | CH₂ | CH₃
(R,S) designations	2(R),3(S)	2(S),3(R)	2(R),3(R)	2(S),3(S)
Enantiomeric pairs	⎫———⎫		⎫———⎫	
$[\alpha]_D^{25}$	−15.5°	+15.5°	−21.5°	+21.5°
Melting point, °C	150	150	156	156
Solubility in H₂O, per 100 g	4.15	4.15	4.37	4.37

MESO COMPOUNDS—A SPECIAL CASE

In most cases, if a compound has two asymmetric carbon atoms it will occur as 2^2 or four optically isomeric forms. These can give rise to two optically inactive racemic or ($\pm$) mixtures. Thus the top pair of compounds in Figure 9-23 could be mixed to give one ($\pm$) mixture and the bottom pair of compounds could be mixed to give a second ($\pm$) mixture.

When the two ends of the molecule are alike, things become somewhat simpler. An example is the isomerism in the tartaric acids shown in Figure 9-24. There are still the same number of optically inactive forms, but one of the ($\pm$) mixtures has been replaced by the internally compensated meso compound. The meso compound has a plane of symmetry and it can be seen to be identical to its mirror image by turning one of the structures end-for-end.

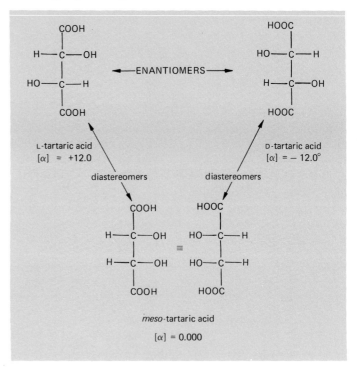

FIGURE 9-24. The tartaric acid optical isomers.

Resolution of Racemic Mixtures

Because enantiomers have the same solubility, they cannot be separated by crystallization. However, this difficulty can be circumvented by *temporarily converting the pair of enantiomers into a pair of diastereomers*, a tactic that we will illustrate by the resolution of a ($\pm$) mixture of glyceric acid enantiomers.

$$\begin{array}{cc}
\text{COOH} & \text{COOH} \\
| & | \\
\text{H—C—OH} & \text{HO—C—H} \\
| & | \\
\text{CH}_2\text{OH} & \text{CH}_2\text{OH} \\
\text{(R)-glyceric acid} & \text{(S)-glyceric acid}
\end{array}$$

Let us also suppose that we have available an asymmetric base, one of many that are found in various plants. To keep the formulas simple, we will not specify what the base is except to note that it has an asymmetric carbon atom and that our sample is one of the pure enantiomorphs. Let it be called (R)B: where the pair of electrons represents an unshared pair on a basic nitrogen atom.

The base (R)B: will form compounds with (R)- and (S)-glyceric acids as shown in equations 9-5a and 9-5b. The resulting salts

$$\underset{\text{(R)-glyceric acid}}{\text{HOCH}_2-\overset{\overset{\text{H}}{|}}{\underset{\underset{\text{OH}}{|}}{\text{C}}}-\overset{\overset{\text{O}}{\|}}{\text{C}}-\text{OH}} + :\text{B(R)} \rightarrow \underset{\text{(R)G(R)B salt}}{\text{HOCH}_2-\overset{\overset{\text{H}}{|}}{\underset{\underset{\underset{\text{H}}{|}}{\text{O}}}{\text{C}}}-\overset{\overset{\text{O}}{\|}}{\text{C}}-\text{O}^-\ldots\text{H}-\overset{+}{\text{B}}(\text{R})} \quad (9\text{-}5a)$$

$$\underset{\text{(S)-glyceric acid}}{\text{HOCH}_2-\overset{\overset{\text{OH}}{|}}{\underset{\underset{\text{H}}{|}}{\text{C}}}-\text{COOH}} + :\text{B(R)} \rightarrow \underset{\text{(S)G(R)B salt}}{\text{HOCH}_2-\overset{\overset{\overset{\overset{\text{H}}{|}}{\text{O}}}{|}}{\underset{\underset{\text{H}}{|}}{\text{C}}}-\overset{\overset{\text{O}}{\|}}{\text{C}}-\text{O}^-\ldots\text{H}-\overset{+}{\text{B}}(\text{R})} \quad (9\text{-}5b)$$

(R)G(R)B and (S)G(R)B are diastereomers of each other rather than enantiomers.[7] Since they are diastereomers, they will have different solubilities, and one can expect to be able to separate them by crystallization. The procedure, then, is to treat the ($\pm$) mixture of (R)- and (S)-glyceric acid with the single enantiomer (R)B: and separate the resulting mixture of diastereomeric salts. Pure (R)- and pure (S)-glyceric acid can then be obtained by treating the separated salts with a strong acid such as HCl, as in equations 9-6a and 9-6b.

$$(\text{R})\text{G}(\text{R})\text{B} + \text{HCl} \rightarrow (\text{R})\text{G} + (\text{R})\text{BHCl} \quad (9\text{-}6a)$$

$$(\text{S})\text{G}(\text{R})\text{B} + \text{HCl} \rightarrow (\text{S})\text{G} + (\text{R})\text{BHCl} \quad (9\text{-}6b)$$

A typical optically active base that is often used to serve the function of the compound (R)B: in the resolution of ($\pm$) mixtures of acids is the naturally occurring alkaloid quinine.

quinine
mp 177°C

---

[7] The *enantiomer* of (R)G(R)B would be (S)G(S)B, not (S)G(R)B.

This compound is found in the bark of the cinchona tree, which contains about 8% quinine. It is used in treatment of malaria. The specific rotation of natural quinine, using the D line of a sodium flame as the light source, is $[\alpha]_D^{15} = -169°$ in alcohol. It is very slightly soluble in water and has a bitter taste. Quinine fluoresces blue in ultraviolet light and so do drinks that have been mixed with "tonic water" containing quinine. Crystals of quinine are **triboluminescent**, which means that they give off flashes of light when they are ground or broken in the dark. Quinine is a base and its pK at 18°C is 5.07.

There is an historical connection between the names quinine and quinone. Quinone was given that name because it was first made by a vigorous oxidation of quinine.

Another method of separating ($\pm$) mixtures is a special one that will work only with certain compounds under certain conditions that might be hard to discover. An example is Pasteur's resolution of the tartaric acids. L-Tartaric acid is found in fruits and its acid potassium salt usually precipitates out in the dregs during wine making. The ($\pm$) tartaric acid mixture is known as racemic acid (Latin: *racemus*, bunch of grapes). Louis Pasteur discovered that the sodium ammonium salt of racemic acid does something very unusual. If this salt is crystallized at a temperature below 27°C, the crystals are found to consist of two visibly different kinds, shown in Figure 9-25.

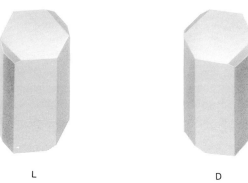

L   D

FIGURE 9-25. Crystals of D- and L-ammonium hydrogen tartrate formed from a solution of the ($\pm$) salt.

The crystals are asymmetric and the two kinds of crystal are nonsuperposable mirror images. When they are separated into two piles with tweezers and the two batches are dissolved separately, one batch is found to give a dextrorotatory solution and the other batch a levorotatory solution. Most ($\pm$) mixtures will not crystallize into separate ($+$) and ($-$) crystals, so Pasteur was lucky as well as observant.

Pasteur also discovered the chemical and biochemical methods of resolution. In the biochemical method a mold such as *Penicillium glaucum* destroys the L-enantiomer but leaves the D-enantiomer untouched. If Pasteur had been still luckier he might have discovered the antibiotic drug penicillin almost a century ahead of time.

Compounds with Many Asymmetric Carbon Atoms

A compound with n asymmetric carbon atoms can have $2^{(n-1)}$ optically inactive forms. If these are all ($\pm$) mixtures, as is usually the case, the total number of optical isomers is 2^n. Thus the aldohexose

FIGURE 9-26. Louis Pasteur, discoverer of methods of resolving ($\pm$) mixtures.

shown below (a sugar, see Chapter 13) has four asymmetric carbon atoms and exists as $2^3 = 8\,(\pm)$ pairs, or 16 isomers all together. One of them is D-glucose.

$$
\begin{array}{c}
\mathrm{H}\mathrm{O} \\
\diagdown\diagup \\
\mathrm{C} \\
| \\
\mathrm{*CHOH} \\
| \\
\mathrm{*CHOH} \\
| \\
\mathrm{*CHOH} \\
| \\
\mathrm{*CHOH} \\
| \\
\mathrm{CH_2OH} \\
\text{an aldohexose}
\end{array}
\qquad
\begin{array}{c}
\mathrm{H}\mathrm{O} \\
\diagdown\diagup \\
\mathrm{C} \\
| \\
\mathrm{H\!-\!C\!-\!OH} \\
| \\
\mathrm{HO\!-\!C\!-\!H} \\
| \\
\mathrm{H\!-\!C\!-\!OH} \\
| \\
\mathrm{H\!-\!C\!-\!OH} \\
| \\
\mathrm{CH_2OH} \\
\text{D-glucose}
\end{array}
$$

To reduce the number of names required, we assign only one name for each ($\pm$) pair. Thus one of the 16 aldohexose isomers is D-

FIGURE 9-27. Jean Baptiste Biot, discoverer of optical activity.

glucose, another is D-mannose, and so on. The mirror images are just called L-glucose and L-mannose instead of being given entirely new names. After all, the enantiomers are identical in most of their chemical and physical properties, even though they may be dramatically different in their biological properties.

Asymmetric Carbon Atoms in Living Systems

Pairs of enantiomers almost always differ from each other in their interactions with biological systems. For example, enantiomers may have different odors, or one may be a powerful drug and the other

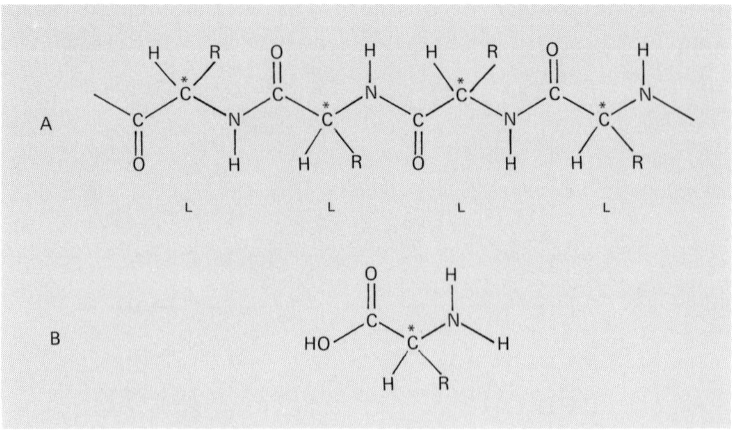

FIGURE 9-28. (**A**) All of the asymmetric carbon atoms in a naturally occurring polypeptide molecule have the L configuration. Only a small segment of the chain is shown here. (**B**) A molecule of the amino acid monomer.

without effect. The reason for this is that the materials of plants and animals are *themselves* optically active. The system (D enantiomer plus an organism) is therefore *not* the mirror image of the system (L enantiomer plus the same organism). The two systems are analogous to diastereomers rather than to enantiomers, and can be expected to behave differently from each other.

Many biologically important molecules (Chapters 13 through 15) consist of long chains in which there may be thousands of asymmetric carbon atoms. The chirality or **tacticity** of these asymmetric carbon atoms is never random. A typical polypeptide, for example, is made **entirely from L-amino acids**, as illustrated in Figure 9-28. In Chapter 12 we will refer to such compounds as **isotactic** polymers.

FIGURE 9-29. Bindweed. This vine always twines counterclockwise, in contrast to honeysuckle vines which always turn clockwise. The transmission of the preferred twining direction from one generation of plants to the next probably involves asymmetry at the molecular level. [From J. W. Krutch, *Herbal*, Putnam, N.Y., 1965, and Pierandrea Mattioli, *Commentaries on the Six Books of Dioscorides*, Prague, 1563]

It turns out that isotactic polymers have higher melting points and form stronger fibers than do polymers in which the sequence of D and L configurations is random.

Since all the organisms on our planet share this preference for L-amino acids as building blocks for polypeptides, the entire planet forms an ecological system in which each component species can provide food for the others. A system based exclusively on D-amino acids would probably work just as well.

Living systems are able to synthesize complicated molecules in which each asymmetric carbon atom has to have a particular configuration only because the reactions are catalyzed by **enzymes** (Chapter 15). Enzymes are highly specific catalysts that will catalyze the reaction of a molecule only if it has exactly the right shape, even to the point of rejecting the D enantiomer of a reagent when an L enantiomer is what is needed. Of course the enzymes themselves are optically active or they would not be able to show this kind of specificity.

Another example of enzyme specificity is Pasteur's use of the mold *Penicillium glaucum* to destroy L-tartaric acid, leaving the D-tartaric acid component of the ± mixture untouched. The mold produces enzymes for catalyzing reactions of the L enantiomer but not the D.

SUMMARY

1. Enantiomers are nonidentical mirror-image molecules.
    (a) They have different Fischer projection formulas (the D and L convention).
    (b) A pair of enantiomers have the same physical properties except that they rotate the plane of polarized light in opposite directions.
2. Polarized light is light whose electrical field component is oscillating only in a single plane.
    (a) A polarizer converts unpolarized light (oscillating in all possible planes) to polarized light.
    (b) A polarimeter measures the rotation of the plane of polarized light by a substance.
    (c) An optically active compound is one whose solution rotates the plane of polarized light passed through the solutions. It is either levorotatory (rotates to the left) or dextrorotatory (to the right).
3. Optical activity
    (a) Mixtures of equal numbers of D and L mirror image molecules do not rotate the plane of polarized light and are optically inactive. They are called (±) mixtures or racemic mixtures.
    (b) A substance will have nonidentical D and L enantiomeric forms if the molecule does not have a plane of symmetry.
4. Asymmetric carbon atoms
    (a) An asymmetric carbon atom is one with four different groups attached to it.
    (b) A compound with one asymmetric carbon atom has a pair of optically active enantiomeric forms.
5. Creation of asymmetric carbon atoms
    (a) Ordinary reactions always produce (±) mixtures.
    (b) Nevertheless, most substances in living organisms are optically active rather than (±) mixtures.

6. R and S chirality nomenclature
    (a) Can be used in cases where the D and L system breaks down.
7. Compounds with more than one asymmetric carbon atom
    (a) Compounds with several asymmetric carbon atoms can occur either as optically inactive meso compounds or as ($\pm$) mixtures.
    (b) The number of inactive forms [meso or ($\pm$)] is $2^{(n-1)}$, where n is the number of asymmetric carbon atoms.
    (c) A pair of optical isomers that are not enantiomorphs of each other are diastereomers.
    (d) Diastereomers have different physical properties.
8. Racemic mixtures can be resolved by
    (a) Temporarily converting them to a mixture of diastereomeric salts, which are then separated.
    (b) Separating mirror-image crystals by hand (in certain cases only).
    (c) Allowing a microorganism to destroy one of the enantiomers selectively.
9. The tacticity of large biological molecules with many asymmetric carbon atoms
    (a) Polypeptides are isotactic, made entirely from L-amino acids.
    (b) Isotactic polymers have higher melting points and make stronger fibers.
    (c) It is advantageous for an ecological system to use isotactic polymers, but the mirror-image isotactic system would probably work just as well.

EXERCISES

1. Define enantiomer.

2. Make a three dimensional drawing and also show the Fischer projection structure for
    (a) D- and L-glyceraldehydes
    (b) D- and L-lactic acids
    (c) D- and L-tartaric acids
    (d) *meso*-tartaric acid
    (e) D- and L-alanine

3. A fictitious compound, "D-novic acid," has the properties listed below. Predict the properties of its as yet unknown enantiomer, "L-novic acid."

Property	"D-novic acid"	"L-novic acid"
Melting point, °C	110	
Solubility in water, g/liter	7.0	
Density, g/ml	0.986	
pK	5.74	
$[\alpha]_{546}^{30}$	$-298.6°$	

4. The solubility of "D-novic acid" in optically active dextrorotatory 2-butanol is 2.07 g/liter. Think about this and make an intelligent guess:
    (a) Is it possible to predict the solubility of "L-novic acid" in dextrorotatory 2-butanol?
    (b) Is it possible to predict the solubility of "L-novic acid" in levorotatory 2-butanol?

5. Explain what is meant by plane polarized light and by unpolarized light.

6. Explain how two Polaroid sunglass lenses can be placed on top of each other and the combination made to look (a) clear; (b) black.

7. Draw a diagram of a polarimeter.

8. Define optical activity.

9. Do molecules related to glyceraldehyde and with the D configuration (a) always rotate the plane of polarized light to the right; (b) never; (c) sometimes?

10. What is the specific rotation of (a) a ($\pm$) mixture; (b) a meso compound; (c) 1-butanol?

11. Define (a) ($\pm$) mixture; (b) racemic mixture; (c) plane of symmetry; (d) asymmetric carbon atom.

12. Describe two ways of predicting whether a given molecule can be optically active.

13. Put stars by any asymmetric carbon atoms in the following molecules.

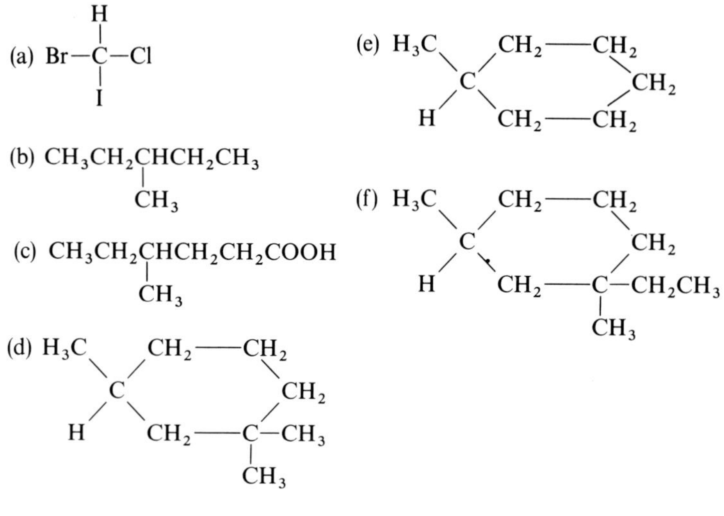

(a) $Br-\overset{\overset{H}{|}}{\underset{\underset{I}{|}}{C}}-Cl$

(b) $CH_3CH_2\overset{}{\underset{\underset{CH_3}{|}}{C}H}CH_2CH_3$

(c) $CH_3CH_2\overset{}{\underset{\underset{CH_3}{|}}{C}H}CH_2CH_2COOH$

(d) H₃C, CH₂—CH₂, C, CH₂, H, CH₂—C—CH₃, CH₃, CH₃

(e) H₃C, CH₂—CH₂, C, CH₂, H, CH₂—CH₂

(f) H₃C, CH₂—CH₂, C, CH₂, H, CH₂—C—CH₂CH₃, CH₃

14. Give an example of a meso compound other than those in this chapter.

15. When two of the four optically active optical isomers of ethyl hydrogen tartrate were hydrolyzed to give ethyl alcohol and tartaric acid, the optical activity disappeared.

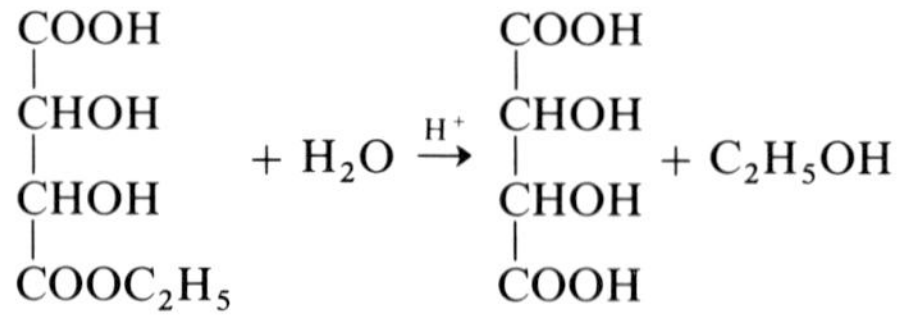

$$\begin{array}{c} COOH \\ | \\ CHOH \\ | \\ CHOH \\ | \\ COOC_2H_5 \end{array} + H_2O \xrightarrow{H^+} \begin{array}{c} COOH \\ | \\ CHOH \\ | \\ CHOH \\ | \\ COOH \end{array} + C_2H_5OH$$

The formulas above are noncommittal as to the configurations of the asymmetric carbon atoms.
 (a) Draw two structures, showing the configurations, that would have behaved in the way described here.
 (b) Explain why your two compounds behave this way.

16. A synthesis was devised that should give 2-chloro-3-bromobutane, and the product was distilled through a fractionating column. More than one fraction (sample of a particular boiling point) was obtained.
    (a) How many fractions (one, two, three, or four) would you predict?
    (b) Explain your choice.

17. (a) Write six reactions that would give rise to asymmetric carbon atoms, starting with nonasymmetric molecules.
    (b) Will the products be optically active?
    (c) Suppose one of the reactions can be run at a much lower temperature by feeding the reagents to one end of a rabbit and collecting the product from the other end (followed by appropriate purification of the product, of course). Will the product be optically active?

18. Explain why the reaction of
    (a) one molecule of HCl with one molecule of 1-butene gives optically active product.
    (b) a bottle of concentrated HCl and a bottle full of 1-butene gives optically inactive product.

19. Describe with diagrams, how you could use $(-)$-quinine to separate or resolve a $(\pm)$ mixture of lysergic acids. (You don't need to know what lysergic acid is in order to answer this question.)

20. Describe, with diagrams, how you might use D-lactic acid to separate a mixture of $(+)$- and $(-)$-quinine.

21. Define diastereomer.

22. Given the following properties for molecule A, predict as many properties as you can for B, C, and D. Explain your answer, including why you were not able to predict some of the properties. The group R does not contain any asymmetric carbon atoms.

```
      R                          R
      |                          |
   H—C—OH                     H—O—C—H
      |                          |
   Cl—C—R                     R—C—Cl
      |                          |
      H                          H

      A                          B
```

mp 62°C
$[\alpha]_{456}^{26} = -21.2°$

```
      R                          R
      |                          |
   H—C—OH                     HO—C—H
      |                          |
   R—C—Cl                     Cl—C—R
      |                          |
      H                          H

      C                          D
```

23. A certain alcohol has the empirical formula $C_4H_{10}O$. Draw all the possible structures for it. Which structure or structures are possible if the alcohol is optically active?

24. Define isotactic polymer.

25. Levorotatory or (R)-carvone has a strong spearmint odor whereas its enantiomer, dextrorotatory or (S)-carvone, has a caraway odor.

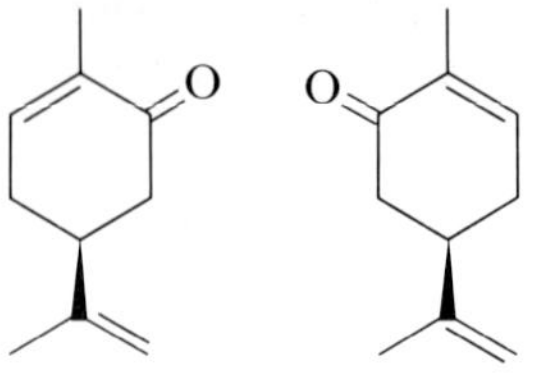

(S)(+)-carvone   (R)(−)-carvone

What does this tell us about the odor receptors in the nose?

26. (Open book) What is D-glyceraldehyde, R or S?

27. How could one tell the difference between a solution that is optically inactive and one that just happens to have a rotation of 360°?
    Hint: The rotation of an optically active compound is proportional to the concentration of the solution.

28. Draw the structure of the simplest monochloroalkane that can be optically active.

29. Explain what is meant by the (−) in a name such as D(−)-fructose.

Chapter 10
Carboxylic Acids, Esters, and Related Compounds

The **carboxylic acid** functional group (COOH) is already familiar from encounters in earlier chapters. In this chapter we will examine it in more detail and take up some additional functional groups that are closely related.

Carboxylic Acids

Nomenclature of Carboxylic Acids

As usual, the simpler members of the series have been known for a very long time and have common names indicating where they occur in nature, while the more complicated members have systematic names. The simplest carboxylic acid is formic acid (Latin: *formica*, ant). In medieval times it was actually obtained by the destructive distillation of ants. Formic acid is responsible for the sting of red ant bites and of nettles. Acetic acid, the next member of the series, is the acid responsible for the sour taste and sharp odor of vinegar (Latin: *acetum*, vinegar). Table 10-1 lists both the systematic and the common

$$H-\overset{\overset{O}{\|}}{C}-OH$$
formic acid
colorless liquid; mp 8.4°C; bp 100.7°C; soluble in water

$$CH_3-\overset{\overset{O}{\|}}{C}-OH$$
acetic acid
colorless liquid; mp 16.6°C; bp 118.1°C; soluble in water

names of some of the more important straight-chain carboxylic acids.

The common names of aromatic acids resemble the names of the related aromatic hydrocarbon. Thus we have **benzene** and **benzoic**

acid, naphthalene and α and β-**naphthoic acids**. In the case of benzene

benzoic acid
colorless solid; mp 122°C;
solubility in water,
0.27 g/100 ml at 18°C

α-naphthoic acid or
1-naphthoic acid
colorless solid; mp 160°C;
very slightly soluble
in hot water

β-naphthoic acid or
2-naphthoic acid
colorless solid; mp 185°C;
very slightly soluble
in hot water

and benzoic acid, the hydrocarbon took its name from that of the acid rather than the other way around. Benzoic acid is a constituent (20%) of a naturally occurring material called gum benzoin, a sticky gum or resin from various balsams of the *Styrax* genus which grow in southeast Asia. Gum benzoin has been used as an ointment for centuries.

As can be seen from Table 10-1, the systematic names for simple unbranched acids come from the name of the corresponding alkane. For example, the name hexadecanoic acid is constructed by replacing the terminal **e** in hexadecane with **-oic acid**.

TABLE 10-1. *Names of Some Common Carboxylic Acids*

Structure	Systematic Name	Common Name	Derivation
HCOOH	methanoic	formic	L. *formica*, ant
CH_3COOH	ethanoic	acetic	L. *acetum*, vinegar
CH_3CH_2COOH	propanoic	propionic	Gr. *protos*, first, *pion*, fat
$CH_3CH_2CH_2COOH$	butanoic	butyric	L. *butyrum*, butter
$CH_3(CH_2)_3COOH$	pentanoic	valeric	L. *valere*, powerful
$CH_3(CH_2)_4COOH$	hexanoic	caproic	
$CH_3(CH_2)_6COOH$	octanoic	caprylic	L. *caper*, goat
$CH_3(CH_2)_8COOH$	decanoic	capric	
$CH_3(CH_2)_{10}COOH$	dodecanoic	lauric	laurel
$CH_3(CH_2)_{14}COOH$	hexadecanoic	palmitic	palm oil
$CH_3(CH_2)_{16}COOH$	octadecanoic	stearic	Gr. *stear*, tallow

NAMING SUBSTITUTED BRANCHED-CHAIN ACIDS

Some branched-chain acids have the same common name as the straight-chain or normal isomer, except for a prefix **iso** indicating the presence of a branch at the carbon atom next to the end. The prefixes normal and iso have only a limited usefulness, however; hence they have been largely supplanted by more systematic names.

isobutyric acid, α-methylpropionic acid,
or 2-methylpropanoic acid
colorless liquid; bp 154°C;
partly soluble in water (20g/100 ml)

isovaleric acid, β-methylbutyric acid,
or 3-methylbutanoic acid
bp 178°C; slightly soluble in water
(4.2 g/100 ml)

Unfortunately there are *two* such systems still in use, although only one of them is approved by IUPAC (International Union of Pure and Applied Chemistry).

The older of the two systems uses Greek letters to designate the carbon atoms, starting with the one *next* to the COOH group. Thus we have names such as α-methylpropionic acid or β-methylbutyric acid. Note that these names are derived from the *common* names of the parent acids, hence α-methylpropionic acid and never α-methylpropanoic acid.

The newer (IUPAC) system uses the systematic name for the parent unsubstituted acid and indicates the substituents by means of prefixes and numbers. **Note that the numbers start with the carbon of the COOH, whereas the letters (of the old system) start at the next carbon.**

$$\overset{\varepsilon}{C}-\overset{\delta}{C}-\overset{\gamma}{C}-\overset{\beta}{C}-\overset{\alpha}{C}-COOH$$
$$(6)\ (5)\ (4)\ (3)\ (2)\ (1)$$

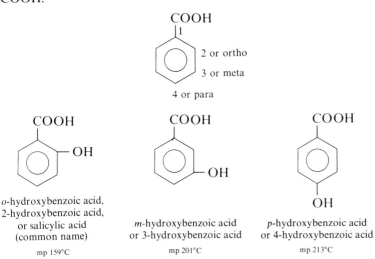

α-chloropropionic acid or
2-chloropropanoic acid

γ-hydroxybutyric acid or
4-hydroxybutanoic acid

Substituted benzoic acids are named either with the prefixes ortho, meta, and para or with numbers in which carbon 1 bears the COOH.

o-hydroxybenzoic acid,
2-hydroxybenzoic acid,
or salicylic acid
(common name)
mp 159°C

m-hydroxybenzoic acid
or 3-hydroxybenzoic acid
mp 201°C

p-hydroxybenzoic acid
or 4-hydroxybenzoic acid
mp 213°C

Dissociation and Physical Properties of Carboxylic Acids

ACIDITY

Carboxylic acids of low molecular weight such as formic acid (HCOOH) and acetic acid (CH_3COOH) are soluble in water. Their aqueous solutions are moderately acidic because at any given time a certain proportion of the molecules are **dissociated** into H_3O^+ ions and carboxylate ions (equation 10-1). As we have already noted

$$CH_3-\overset{O}{\underset{\|}{C}}-OH + H_2O \rightleftarrows H_3O^+ + CH_3-\overset{O}{\underset{\|}{C}}-O^- \quad (10\text{-}1)$$

in Chapter 2, carboxylic acids are stronger acids than alcohols. They are also stronger acids than phenols. A convenient way to describe the strength, or tendency to dissociate, of an acid is to use its dissociation constant.

$$K = \frac{[RCOO^-][H_3O^+]}{[RCOOH]} \quad (10\text{-}2)[1]$$

In equation 10-2 K is the dissociation constant, $[RCOO^-]$ is the concentration of the ion formed by a reaction like that of equation 10-1, $[RCOOH]$ is the concentration of acid that remains undissociated, and $[H_3O^+]$ is the concentration of hydronium ions. A big dissociation constant means a strong acid and a high concentration of hydronium ions (H_3O^+) in solution. Typical K values for carboxylic acids range from 1.8×10^{-5} for acetic acid to 3×10^{-1} for trichloroacetic acid. Notice that these K values are less than those for typical strong inorganic acids (all greater than 1) but are greater than that for phenol ($K = 1.1 \times 10^{-10}$) and still greater than those for alcohols.[2]

In Chapter 2 we noted that the carboxyl functional group, COOH, is much more acidic than the hydroxyl group, OH. We will now consider two reasons for the greater acidity of carboxylic acids.

Recall (from Chapter 2) that a major factor in determining the acidity of a hydrogen atom bonded to an atom of another element is the electronegativity of that atom. Water is more acidic than methane, for example. One reason that carboxylic acids are more acidic than alcohols is that the carbonyl group, C=O, resembles $^+C-O^-$ and pulls electrons away from the OH group. This, in turn, increases the electronegativity of the OH oxygen. Looking at it in another way, the C=O group demands a greater share of the electrons, and the oxygen atom passes on this demand to the O—H bond (Figure 10-1). This polarizes the O-H bond ($\overset{\delta-}{O}-\overset{\delta+}{H}$) and makes it easier to remove the hydrogen as H^+, leaving the electrons on oxygen.

$$R-\overset{O}{\underset{\|}{C}}-\overset{\delta-}{O}-\overset{\delta+}{H} \rightarrow R-\overset{O}{\underset{\|}{C}}-O^- + H^+ \quad (10\text{-}3)$$

A second reason for the greater acidity of the carboxyl group has to do with reluctance of the carboxylate **ion** (COO^-) to take the proton back once it has been removed. The reason for this lack of

---

[1] $[H^+]$ is often written instead of $[H_3O^+]$, the fact that the proton is actually bonded to a water molecule being understood.

[2] Biochemists like to describe the acidity of a substance by its pK, which is simply $-\log K$. Thus the pK of a typical phenol is about 10, that of acetic acid about 5, and that of trichloroacetic acid about 1. (The smaller the pK the stronger the acid.) This usage is an extension of using pH = $-\log[H_3O^+]$ in place of $[H_3O^+]$.

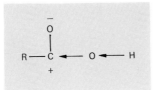

FIGURE 10-1. The arrows indicate a polarization, or shift of electrons, in the bonds. The electronegative carbon causes electrons to be shared unequally with the oxygen, as indicated by the first arrow. This makes the oxygen more electronegative and causes the electrons of the O—H bond to be shared unequally as indicated by the second arrow.

reactivity of COO⁻ is very much like the reason for the unusual stability of benzene: COO⁻ has a new kind of bond, intermediate between a single and a double bond (Figure 10-2). Just as all the C—C bonds in benzene have the same length, whereas C=C double bonds are shorter than C—C single bonds, the two C⋯O bonds of COO⁻ also have the same length. Furthermore, each oxygen atom has half of the negative charge of the ion.

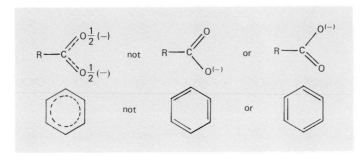

FIGURE 10-2. Carboxylate ion bonds and benzene bonds compared.

SOLUBILITY AND BOILING POINT

Beginning at butanoic acid, the straight-chain alkanoic acids become progressively less soluble in water. They continue to be "soluble" in aqueous sodium bicarbonate, however, because they are converted by that base into the water-soluble sodium salts.

$$\underset{\text{insoluble}}{R-\overset{O}{\underset{\|}{C}}-OH} + NaHCO_3 \rightleftharpoons \underset{\text{soluble}}{R-\overset{O}{\underset{\|}{C}}-O^-Na^+} + H_2O + CO_2$$

In Chapter 2 (equation 2-23 and Figure 2-2) we saw how the salt-forming reaction of cyclohexanecarboxylic acid could be used to separate it from a mixture with a nonacidic compound, the alcohol cyclohexanol. We can also use differences in acidity to separate carboxylic acids (more acidic) from phenols (less acidic) and from alcohols (least acidic). An example of this is shown in the flow sheet of Figure 10-3.

Paper chromatography is a particularly good technique for separating carboxylic acids from each other and is also useful for separating mixtures of other kinds. In this technique (Figure 10-4) a

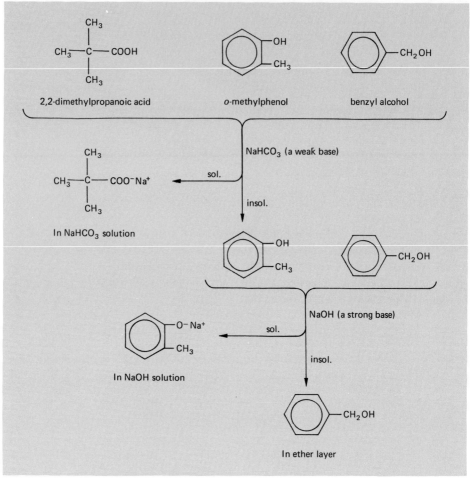

FIGURE 10-3. Separation of a carboxylic acid, a phenol, and an alcohol by means of the differences in their acidities.

drop of the mixture is placed on one end of a paper strip. The spot containing the mixture of acids is then **developed** by dipping the end of the strip into a suitable solvent. The solvent creeps along the paper and, although it washes the carboxylic acids along with it, they move at different rates. Hence the original spot, consisting of a mixture of all the acids, is separated into several spots, each containing just one acid.

A convenient way of describing the relative ease with which various substances move along the paper is to cite their R_f values. The quantity R_f is simply the ratio of the distance that the spot has moved to the distance that the solvent has moved. The R_f values depend both on the nature of the material being chromatographed and on the solvent but, in general, large molecules and highly polar molecules are more strongly adsorbed on the paper and have smaller R_f values.

Just as the solubility of the carboxylic acids in water decreases with increasing length of the carbon chain, the boiling points increase. The boiling point goes up about 20°C for each methylene

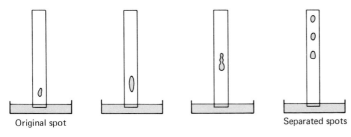

FIGURE 10-4. As the solvent moves up the paper strip (because of capillary attraction), the acids also move, but at different rates. After the acids have moved to different spots on the paper, the paper is dried, the spots are cut out with a pair of scissors, and the acids are extracted with a good solvent such as ether.

(CH_2) group. Thus the smaller molecules are volatile enough to smell; formic and acetic acids have a sharp odor, and the acids with from four to ten carbon atoms have an obnoxious, rancid odor. Butyric acid, for example, is responsible for the odor of rancid butter. Note also the relationship, shown in their names, of the C_6 to C_{10} acids to goats and the implication of the name valeric acid (Latin *valere*, powerful) for the C_5 acid. Some of these acids are also found in human sweat. Valeric acid is a sex attractant for the sugar beet wire worm.[3]

A Direct Chemical Engine. Mankind has a large number of machines for converting the energy of a chemical reaction into mechanical work, but they effect this conversion only indirectly. A battery, for example, generates electricity and the electricity is used to run a motor. A combustion engine generates heat and the heat is used to expand a gas which then pushes against a piston or the vanes of a turbine. In contrast to these crude devices, a muscle converts chemical energy directly into mechanical work. Furthermore, it will work at room temperature, does not require a heavy container for the hot gas or working fluid, and can be made very light.

Can we design a machine as elegant as a mosquito? At present the answer is that we cannot, but we may have made a beginning. We still understand very little about the mechanism by which chemical energy causes an actual muscle fiber to contract and do work, but we do know how to make a much simpler and cruder device that, like muscle, **makes direct use of the energy of a**

---

[3] As we have seen in other chapters, compounds of many different types can function as insect pheromones. Other examples of acid pheromones are *trans*-9-oxo-2-decenoic acid, which attracts honeybees, and *trans*-3,*cis*-5-tetradecadienoic acid, a sex attractant for the female black carpet beetle.

$$\underset{H}{\overset{HOOC}{}}C=C\underset{CH_2CH_2CH_2CH_2CH_2-\overset{O}{\overset{\|}{C}}-CH_3}{\overset{H}{}}$$

trans-9-oxo-2-decenoic acid

$$\underset{}{\overset{HOOC-CH_2}{}}C=C\underset{H}{\overset{H}{}}\;\;\underset{C=C}{\overset{H\;\;H}{}}C-CH_2CH_2CH_2CH_2CH_2CH_3$$

trans-3,*cis*-5-tetradecadienoic acid

chemical reaction. It should be emphasized, however, that this simple device and an actual muscle have very little else in common.

Natural and synthetic fibers are made of extremely large molecules called polymers (Greek: *poly*, many; *meros*, parts). These fiber polymers are long-chain molecules containing thousands of atoms in the chain, and the particular one in which we are interested also has thousands of carboxyl groups attached to the chain (Figure 10-5). When the COOH groups are in the acid (COOH)

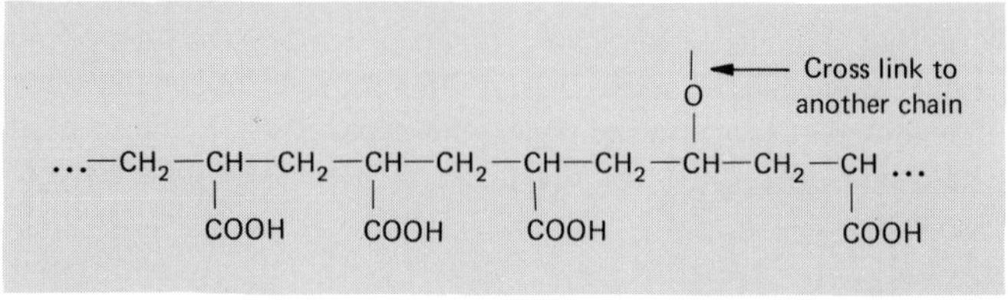

FIGURE 10-5. Part of the acrylic acid–vinyl alcohol copolymer.

form rather than in the carboxylate (COO^-) form, the molecular chains tend to coil up and the fiber constructed from them is short. When the polymer is in the COO^- form, however, the chains tend to stretch out. There are two reasons for this: the COO^- groups repel one another and try to get further apart, and the carboxylate form of the fiber absorbs water which also causes it to swell and become longer. This very large carboxylate salt molecule is too large to dissolve in water, so the water dissolves in it instead and causes the swelling. To summarize, we have short fibers with COOH groups and long fibers with COO^- groups.

The conversion of the long fiber (COO^-) to the short fiber (COOH) is simply a matter of bathing the fiber in a strong acid such as HCl (Figure 10-6). The energy that is used to lift the weight comes essentially from the reaction

$$NaOH + HCl \rightarrow NaCl + H_2O + energy$$

Although lifting a weight may not seem like much of an accomplishment, it

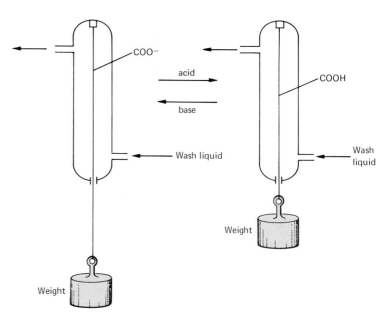

FIGURE 10-6. Conversion of the $-COO^-$ functional groups to $-COOH$ functional groups shrinks the fiber and lifts the weight.

represents exerting a force on something and moving it; this is all that any machine does, whether man-made or natural. Some day a practical direct chemical engine will be developed and, when it is, a lot of pistons, gears, and cog wheels are going to be out of work. The human heart, for example, is a pump that works simply by contractions of its fibers.

Preparation of Carboxylic Acids

OXIDATION

Carboxylic acids can be made by oxidation of several different functional groups. Successful use of this method for preparing a carboxylic acid depends on finding an oxidizing agent that will attack the part of the molecule you wish to oxidize while leaving the rest of the molecule unchanged. Almost any organic molecule can be oxidized, but the reaction is not particularly useful if the only products are CO_2 and H_2O.

In primary alcohols, RCH_2OH, the alkyl group R is usually sufficiently resistant to oxidation so that it remains unchanged while the CH_2OH group is converted to COOH (equations 10-4a and 10-4b). The aldehyde, RCHO, is probably an intermediate in

$$3\ RCH_2OH + 16\ H^+ + 2\ Cr_2O_7^{2-} \xrightarrow{H_2SO_4} 3\ R-\underset{\underset{O}{\|}}{C}-OH + 4\ Cr^{3+} + 11\ H_2O$$

(sodium dichromate)   (chromic sulfate)

(10-4a)

reaction 10-4, but is itself rapidly oxidized the rest of the way to the carboxylic acid.

$$RCHO \xrightarrow[H_2SO_4]{Cr_2O_7^{2-}} RCOOH \qquad (10\text{-}4b)$$

Benzene rings are sufficiently inert so that almost any side chain can be oxidized to the carboxylic acid without affecting the ring (equation 10-5). Notice that the side chain in equation 10-5 is

Ph–CH_2CH_2–OH $\xrightarrow[H_2SO_4]{Na_2Cr_2O_7}$ Ph–C(=O)–OH + CO_2 + 2 H_2O

benzoic acid

(10-5)

oxidized all the way back to the carbon next to the ring. This procedure will also work with alkyl groups (equation 10-6). For reactions such as 10-5 and 10-6, a higher temperature or a higher concentration of oxidant is used than in the case of simpler oxidations such as in equation 10-4.

Ph–$CH_2CH_2CH_2CH_3$ $\xrightarrow[H_2SO_4]{Na_2Cr_2O_7}$ Ph–C(=O)–OH + 3 CO_2 + 4 H_2O

(10-6)

HYDROLYSIS

Another way of making carboxylic acids is by hydrolysis of related compounds such as amides, acid chlorides, anhydrides, or esters. In all of these reactions an OH group from water or OH⁻ displaces another group such as the amide (NH_2), acyloxy ($-O-\overset{O}{\underset{\|}{C}}-R$), or alkoxy ($-O-R$) group.

$$CH_3-\overset{O}{\underset{\|}{C}}-Cl + 2NaOH \rightarrow NaCl + CH_3-\overset{O}{\underset{\|}{C}}-O^{-\,+}Na \quad (10\text{-}7)$$

acetyl chloride
colorless liquid with an irritating odor; bp 51°C

sodium acetate

$$CH_3-\overset{O}{\underset{\|}{C}}-Cl + H_2O \xrightarrow{H^+,\,\text{catalyst}} CH_3-\overset{O}{\underset{\|}{C}}-OH + HCl \quad (10\text{-}8)$$

In reaction 10-7, OH⁻ bonds to the carbon of the C=O and displaces Cl⁻. This gives acetic acid, but the alkali present immediately converts the acid to the sodium salt.

$$CH_3-\overset{O}{\underset{\|}{C}}-NH_2 + H_2O + HCl \rightarrow CH_3-\overset{O}{\underset{\|}{C}}-OH + NH_4Cl \quad (10\text{-}9)$$

acetamide
colorless solid; mp 81°C

$$CH_3-\overset{O}{\underset{\|}{C}}-NH_2 + NaOH \rightarrow CH_3-\overset{O}{\underset{\|}{C}}-O^{-\,+}Na + NH_3 \quad (10\text{-}10)$$

Again, in equation 10-10, the alkali converts the acid into the sodium salt. In equation 10-9, the excess acid used as a catalyst converts the NH_3 that would otherwise be formed into the ammonium salt NH_4Cl.

The formation of acids by hydrolysis of esters will be discussed in a later section.

THE CARBONATION OF GRIGNARD REAGENTS

A Grignard reagent is a magnesium compound made by the reaction of an alkyl or aryl halide with metallic magnesium. As we have seen in Chapter 8, the alkyl or aryl group of a Grignard reagent will react with various positively charged or electrophilic reagents such as H^+ or the positive end of the carbonyl group, $\overset{\delta+}{C}-\overset{\delta-}{O}$, of an aldehyde or ketone (equation 10-11). It will come as no surprise, therefore, to learn that Grignard reagents also add to carbon dioxide.

$$CH_3CH_2MgBr + \underset{H_3C\quad CH_3}{\overset{O}{\underset{\|}{C}}} \rightarrow CH_3CH_2-\underset{H_3C\quad CH_3}{\overset{OMgBr}{\underset{|}{C}}} \quad (10\text{-}11)$$

$$CH_3CH_2MgBr + \overset{O}{\underset{\underset{O}{\|}}{C}} \rightarrow CH_3CH_2-\overset{OMgBr}{\underset{\underset{O}{\|}}{\underset{|}{C}}} \quad (10\text{-}12)$$

The product of the carbonation reaction (equation 10-12) is a magnesium halide salt of the carboxylic acid. It is converted to the free carboxylic acid on treatment with a strong inorganic acid such as HCl (equation 10-13). The carbonation reaction will work with

$$CH_3CH_2-\overset{O}{\underset{\|}{C}}-O^{-\,+}MgBr + HCl \rightarrow CH_3CH_2-\overset{O}{\underset{\|}{C}}-OH + MgBrCl$$

(10-13)

any Grignard reagent. For example, benzoic acid can be made in this way from bromobenzene (equation 10-14). Although the magnesium halide salt of the acid still contains a carbonyl group, it is present mostly as an insoluble precipitate in the ether solution and has little tendency to react with a second molecule of Grignard reagent.

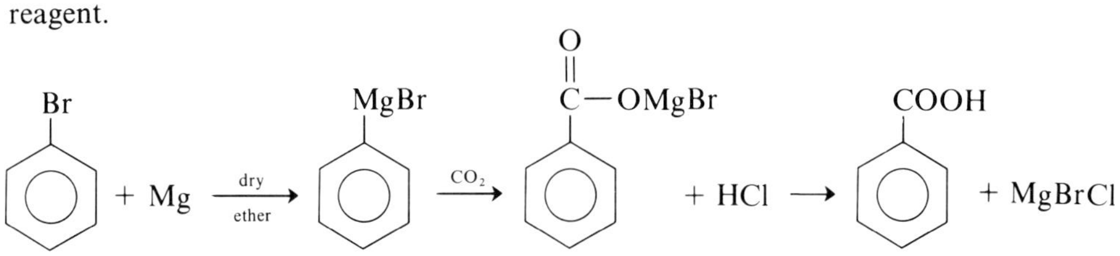

$$C_6H_5Br + Mg \xrightarrow[\text{ether}]{\text{dry}} C_6H_5MgBr \xrightarrow{CO_2} C_6H_5-\overset{O}{\underset{\|}{C}}-OMgBr + HCl \rightarrow C_6H_5COOH + MgBrCl$$

(10-14)

Formation of Acid Chlorides and Anhydrides

ACID CHLORIDES

The conversion of a carboxylic acid to an acid chloride can be accomplished by treating the acid with the acid chloride of a strong inorganic acid. One such reagent, illustrated in equations 10-15 and 10-16, is thionyl chloride ($SOCl_2$); another is phosphorus pentachloride (PCl_5).

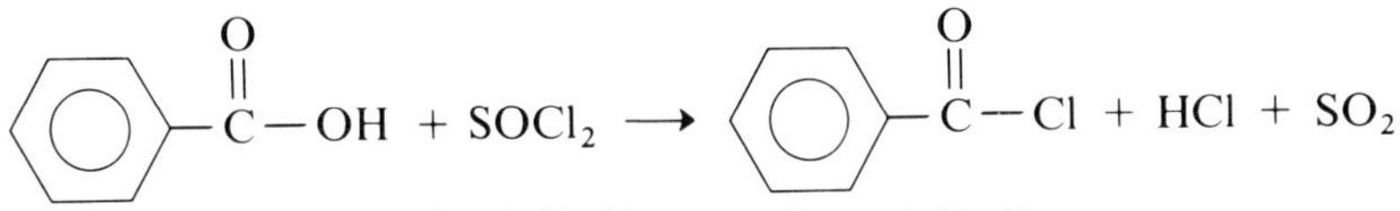

$$C_6H_5-\overset{O}{\underset{\|}{C}}-OH + SOCl_2 \rightarrow C_6H_5-\overset{O}{\underset{\|}{C}}-Cl + HCl + SO_2$$

thionyl chloride — colorless; bp 78.8°C

benzoyl chloride — colorless; mp 42°C; bp 245°C

(10-15)

$$CH_3-\overset{O}{\underset{\|}{C}}-OH + SOCl_2 \rightarrow CH_3-\overset{O}{\underset{\|}{C}}-Cl + HCl + SO_2$$

(10-16)

Acid chlorides are mainly used in chemical syntheses in indirect but efficient ways of replacing the OH of a carboxyl group with NH_2, NHR, or OR groups. Thus acid chlorides react with ammonia or with amines (substituted ammonias) to give **amides** in high yields.

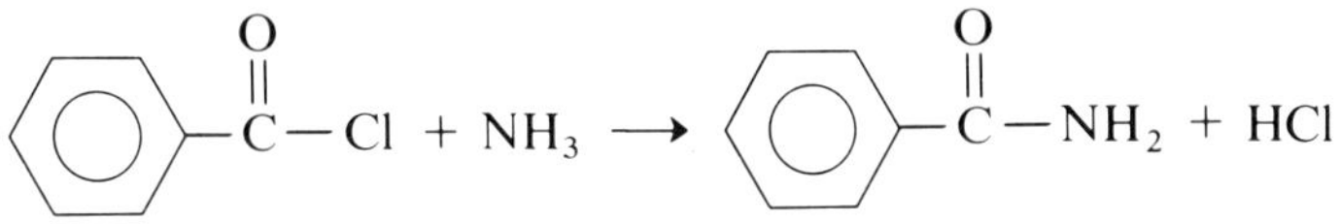

$$C_6H_5-\overset{O}{\underset{\|}{C}}-Cl + NH_3 \rightarrow C_6H_5-\overset{O}{\underset{\|}{C}}-NH_2 + HCl$$

benzamide
colorless; mp 130°C

(10-17)

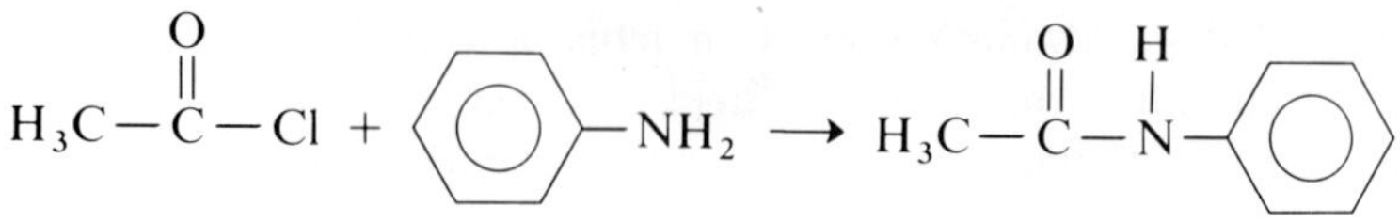

$$H_3C-\overset{O}{\underset{\|}{C}}-Cl + \langle\bigcirc\rangle-NH_2 \rightarrow H_3C-\overset{O}{\underset{\|}{C}}-\overset{H}{\underset{|}{N}}-\langle\bigcirc\rangle \quad (10\text{-}18)$$

aniline   acetanilide or
colorless; bp 184°C   N-phenylacetamide
colorless; mp 114°C

Acid chlorides will also react with water to regenerate the acid, as we have seen in equation 10-8. The hydrolysis reaction is very similar to the reaction of an acid chloride with an alcohol or "substituted water."

$$CH_3-\overset{O}{\underset{\|}{C}}-Cl + CH_3CH_2OH \rightarrow CH_3-\overset{O}{\underset{\|}{C}}-O-CH_2CH_3 + HCl \quad (10\text{-}19)$$

acetyl chloride   ethyl acetate
colorless liquid; fruity odor;
bp 77.1°C

The electrophilic substitution reaction of acid chlorides with aromatic compounds (Chapter 8), in the presence of $AlCl_3$ as a catalyst, is a way of making aromatic ketones (equations 10-20 and 10-21). The reaction is known as the Friedel-Crafts ketone synthesis.

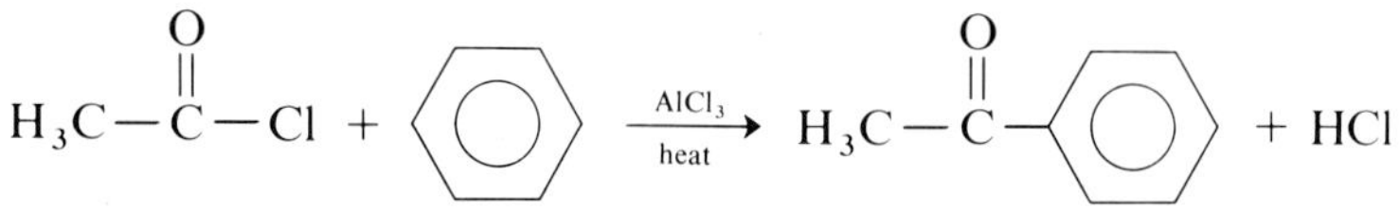

$$H_3C-\overset{O}{\underset{\|}{C}}-Cl + \langle\bigcirc\rangle \xrightarrow[\text{heat}]{AlCl_3} H_3C-\overset{O}{\underset{\|}{C}}-\langle\bigcirc\rangle + HCl \quad (10\text{-}20)$$

$$\langle\bigcirc\rangle-\overset{O}{\underset{\|}{C}}-Cl + \langle\bigcirc\rangle-CH_3 \xrightarrow[\text{heat}]{AlCl_3}$$

ortho-methylbenzophenone and para-methylbenzophenone + HCl (10-21)

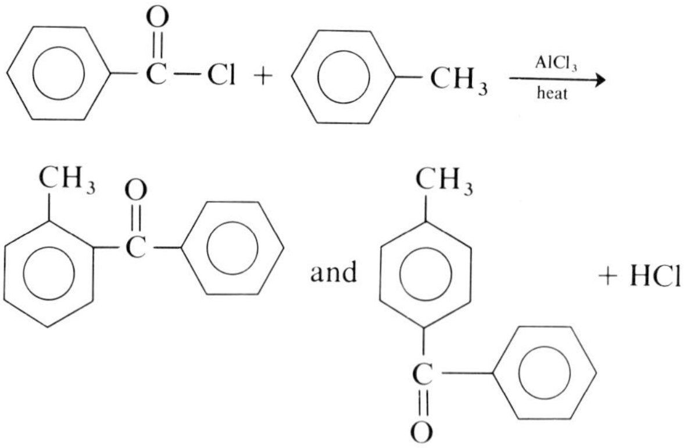

ACID ANHYDRIDES

The reaction of an acid chloride with the sodium salt of an acid gives an acid anhydride (equation 10-22).

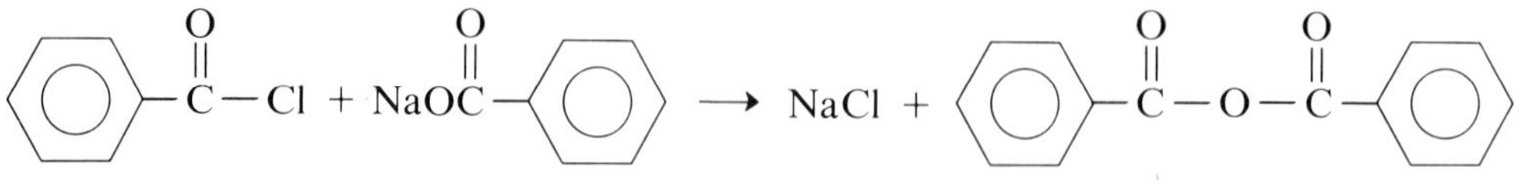

$$\langle\bigcirc\rangle-\overset{O}{\underset{\|}{C}}-Cl + NaO\overset{O}{\underset{\|}{C}}-\langle\bigcirc\rangle \rightarrow NaCl + \langle\bigcirc\rangle-\overset{O}{\underset{\|}{C}}-O-\overset{O}{\underset{\|}{C}}-\langle\bigcirc\rangle$$

benzoic anhydride
colorless solid; mp 42°C

(10-22)

Acetic anhydride can be made in the same way from acetyl chloride and sodium acetate (equation 10-23); however, this compound is made industrially by a different route, shown in equation 10-24.

$$CH_3-\overset{\overset{O}{\|}}{C}-Cl + NaO\overset{\overset{O}{\|}}{C}CH_3 \rightarrow NaCl + CH_3-\overset{\overset{O}{\|}}{C}-O-\overset{\overset{O}{\|}}{C}-CH_3$$

acetic anhydride

liquid with irritating odor; bp 140°C; slightly soluble in water

(10-23)

$$CH_2=C=O + CH_3-\overset{\overset{O}{\|}}{C}-OH \rightarrow CH_3-\overset{\overset{O}{\|}}{C}-O-\overset{\overset{O}{\|}}{C}-CH_3$$

ketene, a highly reactive unsaturated ketone
colorless gas; bp −56°C

(10-24)

Cyclic acid anhydrides are formed readily by heating a dicarboxylic acid, but only if the two carboxyl groups are close enough together. For example, phthalic acid, succinic acid, and maleic acid readily form the corresponding cyclic anhydrides on heating (equations 10-25, 10-26, 10-27), but isophthalic acid and fumaric acid will not do so. This property is often used in determining the structure of new carboxylic acids.

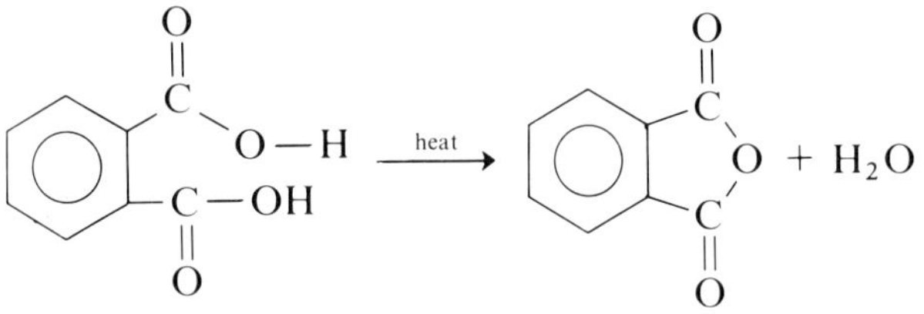

phthalic acid
colorless; mp 206°C

phthalic anhydride
colorless; mp 131°C

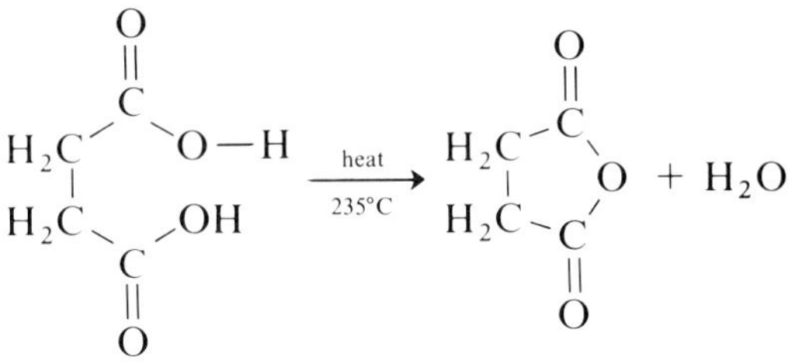

(10-25)

succinic acid
mp 185°C

succinic anhydride
mp 119.6°C

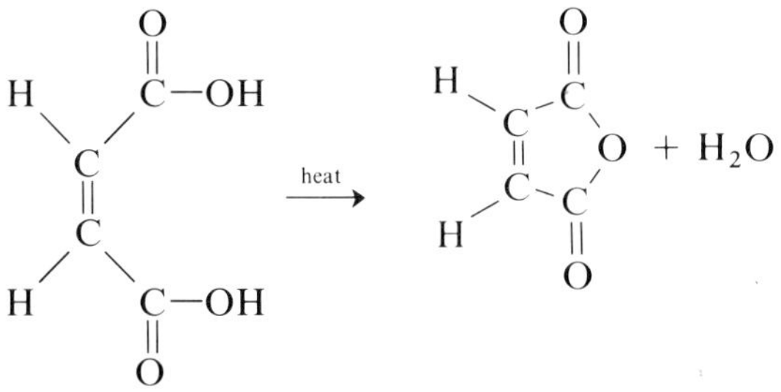

(10-26)

maleic acid
colorless solid; mp 130.5°C

maleic anhydride
colorless solid; mp 53°C

(10-27)

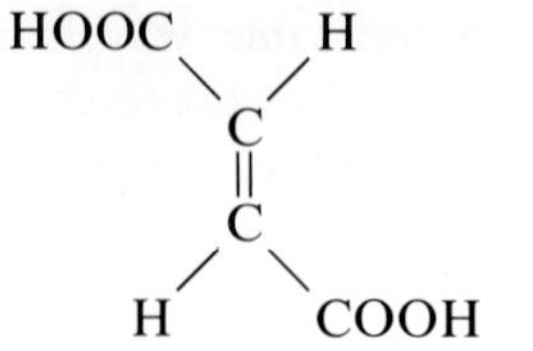

fumaric acid
mp 287°C

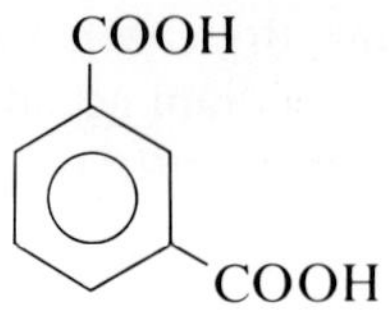

isophthalic acid
mp 330°C

Strong heating of fumaric acid slowly converts it into its cis isomer, maleic acid, which then loses water to give maleic anhydride.

Succinic acid occurs naturally in amber, which is a kind of fossilized pine resin, as well as in various fungi and lichens. The alchemist Agricola obtained it by distilling amber in 1546.

Phthalic anhydride is used in the manufacture of **phenolphthalein**, a laxative and also an acid-base indicator.

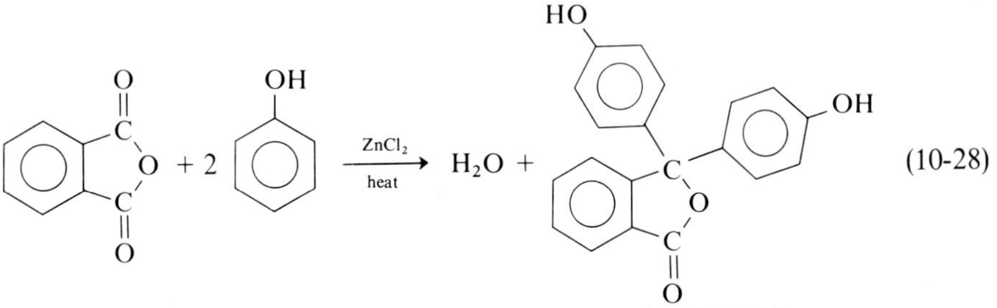

$$\text{phthalic anhydride} + 2\ \text{phenol} \xrightarrow[\text{heat}]{ZnCl_2} H_2O + \text{phenolphthalein} \qquad (10\text{-}28)$$

phenolphthalein
colorless; mp 258°C

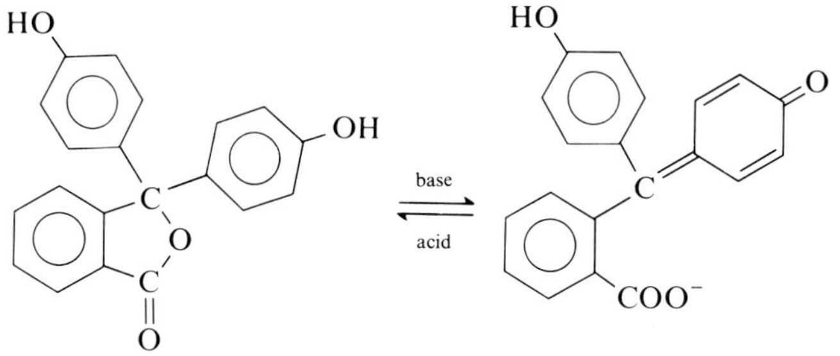

colorless $\underset{\text{acid}}{\overset{\text{base}}{\rightleftharpoons}}$ red

Maleic anhydride is a powerful skin irritant.

Cantharidine is a naturally occurring anhydride secreted by the blistering beetle or "Spanish fly," *Cantharis vesicatoria*. In spite of the fact that it blisters the skin and is toxic to the point of being fatal, it has been used as an aphrodisiac (Greek, *aphrodite*, goddess of love).

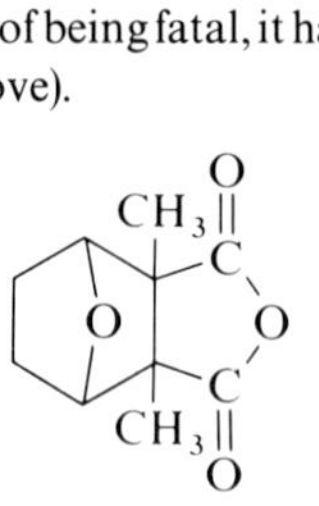

cantharidine

Reactions of Anhydrides. Anhydrides give reactions very similar to those of acid chlorides, and indeed acid chlorides might be considered to be anhydrides of carboxylic acids and HCl even though they are not made in that way.

$$R-\overset{O}{\underset{\|}{C}}-Cl + H_2O \rightarrow R-\overset{O}{\underset{\|}{C}}-OH + HCl \quad (10\text{-}29)$$

$$R-\overset{O}{\underset{\|}{C}}-O-\overset{O}{\underset{\|}{C}}-R + H_2O \rightarrow R-\overset{O}{\underset{\|}{C}}-OH + HO-\overset{O}{\underset{\|}{C}}-R \quad (10\text{-}30)$$

Like acid chlorides, anhydrides can be hydrolyzed or used to make amides or esters. For example, **aspirin**, or sodium acetylsalicylate, is made from **salicylic acid** and acetic anhydride. The reaction amounts to the formation of an acetate ester from the phenolic hydroxyl group of salicylic acid.

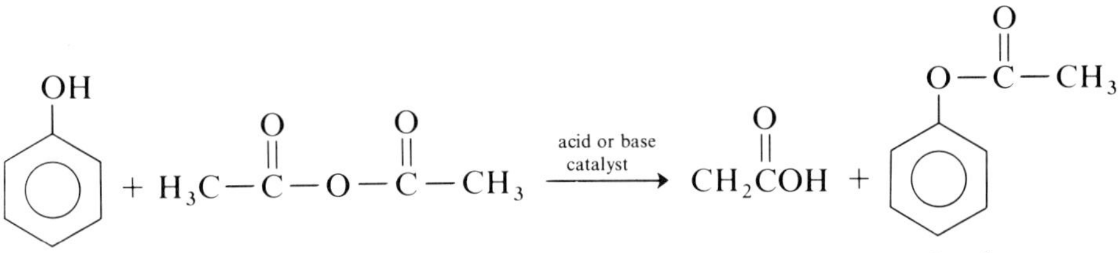

$$\text{C}_6\text{H}_5\text{OH} + H_3C-\overset{O}{\underset{\|}{C}}-O-\overset{O}{\underset{\|}{C}}-CH_3 \xrightarrow{\text{acid or base catalyst}} CH_2COH + \text{phenyl acetate} \quad (10\text{-}31)$$

phenyl acetate
bp 195°C

$$\text{salicylic acid (COOH, OH)} + H_3C-\overset{O}{\underset{\|}{C}}-O-\overset{O}{\underset{\|}{C}}-CH_3 \xrightarrow{\text{acid or base catalyst}}$$

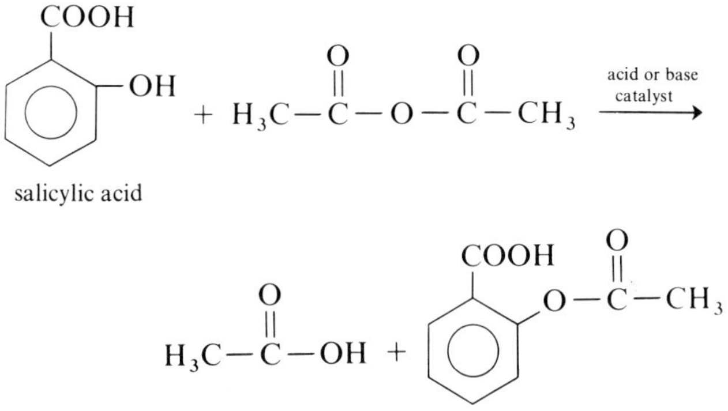

salicylic acid

$$H_3C-\overset{O}{\underset{\|}{C}}-OH + \text{acetylsalicylic acid (COOH, O-C(O)-CH}_3\text{)} \quad (10\text{-}32)$$

acetylsalicylic acid
mp 135°C

The name salicylic acid for *o*-hydroxybenzoic acid comes from the occurrence of structurally related compounds in willow bark (*Salix*). Many salicylates and compounds related to salicylic acid possess medicinal properties, and some of them have been used for a very long time. For example, both Hippocrates (circa 400 B.C.) and the early American Indians used willow bark infusions to relieve pain and fever. Another example is oil of wintergreen, which contains methyl salicylate. The use of oil of wintergreen as an external medication for muscles and joints makes it a sort of athlete's perfume.

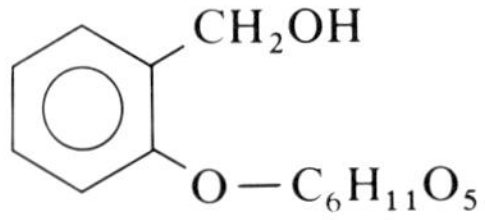

salicin
an analgesic found in willow bark
(CH₂OH, O—C₆H₁₁O₅)

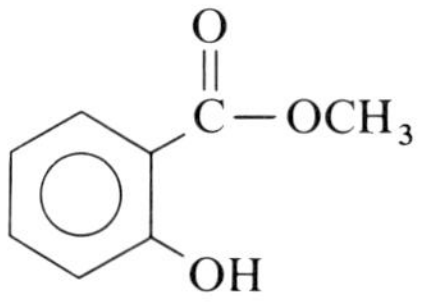

methyl salicylate
wintergreen odor, in oil of wintergreen
(C(O)—OCH₃, OH)

Although the older salicylates were known to relieve the pain and stiffness of arthritis, this relief was often accompanied by nausea and stomach upsets. Aspirin, which is relatively free from these side effects, was first synthesized in 1853, but its value as an analgesic was not realized until forty years later when it was put on the market by the Baeyer Company in Germany.

The sharp odor sometimes noticed when an aspirin bottle is opened is due to traces of acetic acid, formed by hydrolysis of the aspirin.

Esters

Esters are derivatives of carboxylic acids in which the OH group has been replaced by an alkoxy (OR) or aryloxy (OAr) group. An example of an ester with an aryloxy group is phenyl acetate which, as we have just seen, can be made from phenol and acetic anhydride. An example of an ester made from an alcohol rather than from a phenol is ethyl acetate.

$$\underset{R-C-O-R}{\overset{O}{\|}} \quad \text{or} \quad \underset{R-C-O-Ar}{\overset{O}{\|}}$$

The presence of an ester functional group in a molecule causes it to have a strong infrared absorption band near 6000 nm.

Nomenclature of Esters

Esters are usually named after the parent acid, using a system like that used for naming salts. Thus benzoic acid plus sodium hydroxide react to give the salt, sodium benzoate. Benzoic acid plus methyl alcohol react to give the ester methyl benzoate.

sodium benzoate, a **salt**
solid; decomposes before it melts; soluble in water

methyl benzoate, an **ester**
liquid; bp 199.6°C; insoluble in water

Despite the similarity of the names, salts and esters are very different types of compound. Thus, sodium benzoate is an ionic compound with properties typical of ionic compounds (very high melting points, soluble in water), whereas methyl benzoate is a typical covalent compound. Note also that the reaction of benzoic acid with sodium hydroxide to give sodium benzoate occurs instantaneously, whereas the reaction of benzoic acid with methyl alcohol to give methyl benzoate requires refluxing for several hours in the presence of a catalyst, usually a trace of H_2SO_4.

When the ester functional group is part of a complicated molecule containing other functional groups as well, it is sometimes more convenient to deal with the ester part of the molecule by means of a prefix. For example, acetylsalicylic acid is an ester related to the phenol, salicylic acid, and acetic acid. In acetylsalicylic acid, a hydrogen atom of salicylic acid has been replaced by an acetyl group, CH_3CO. This compound could also be called *o*-acetoxybenzoic acid, regarding it as benzoic acid in which one hydrogen has been replaced by an acetoxy group, CH_3COO.

Preparation of Esters

In addition to the indirect preparation of esters from the alcohol (or phenol) and an acid chloride or anhydride, they can be made directly from the alcohol and the acid. One might reasonably inquire why esters are ever made by the indirect route, using a separate reaction to make the acid chloride or anhydride, when a one-step direct route exists. The reason is that the direct reaction produces an equilibrium mixture which may contain only a low yield of the desired ester. Take reaction 10-33 for example. If we

$$CH_3-\overset{O}{\underset{\|}{C}}-OH + CH_3CH_2OH \underset{}{\overset{H^+ \text{ catalyst}}{\rightleftharpoons}} H_2O + CH_3CH_2-O-\overset{O}{\underset{\|}{C}}-CH_3$$

(10-33)

start with 1 mole each of acetic acid and ethyl alcohol, the reaction stops when there is still 0.34 mole of unreacted acid and alcohol and only 0.66 mole of ester. When the alcohol is a cheap one like ethyl alcohol and the acid is expensive, the conversion of the acid to ester is increased by using a large excess of the alcohol.

The acid catalyst in the esterification reaction makes the carbonyl group of the acid more reactive to the electrophilic alcohol, as suggested in the **reaction mechanism** below.

$$CH_3-\overset{O}{\underset{\|}{C}}-OH + H^+ \text{ (catalyst)} \rightleftharpoons CH_3-\overset{\overset{+}{O}H}{\underset{\|}{C}}-OH + CH_3CH_2OH \rightleftharpoons \left[CH_3-\overset{OH}{\underset{|}{\underset{H-\overset{+}{O}-CH_2CH_3}{C}}}-OH \right] \rightleftharpoons$$

$$H^+ + H_2O + CH_3-\overset{O}{\underset{\|}{C}}-OCH_2CH_3$$

Note that this reaction mechanism predicts that the water formed in the reaction gets its oxygen atom from the acid, not from the OH group of the alcohol. This has been shown by using the ^{18}O isotope of oxygen to "label" the oxygen atom.

$$CH_3-\overset{O}{\underset{\|}{C}}-\boxed{OH} \quad H\boxed{O}-CH_2CH_3 \quad \text{and not} \quad CH_3-\overset{O}{\underset{\|}{C}}-O\boxed{H} \quad HO\boxed{}-CH_2CH_3$$

Reactions of Esters

HYDROLYSIS

The hydrolysis of an ester to give the carboxylic acid and the alcohol is catalyzed by strong acids (H^+). It is just the reverse of the acid-catalyzed esterification reaction, and also does not go to completion but instead gives an equilibrium mixture. Thus if we start with 1 mole of ethyl acetate in equation 10-33, there will be 0.66 moles of ester remaining when the reaction reaches equilibrium. Just as excess alcohol can be used to drive the reaction to the ester-plus-water side, excess water can be used to drive the reaction to the acid-plus-alcohol side.

Another way of hydrolyzing an ester is to use sodium hydroxide instead of an acid catalyst. The sodium hydroxide not only "catalyzes" the hydrolysis but also reacts with the acid, thus ensuring complete reaction (equation 10-34). This hydrolysis with aqueous

$$CH_3-\overset{O}{\underset{\|}{C}}-OCH_2CH_3 + NaOH \rightarrow CH_3-\overset{O}{\underset{\|}{C}}-O^-Na^+ + HOCH_2CH_3 \quad (10\text{-}34)$$

alkali is known as **saponification** (Latin, *sapon*, soap) because soap is the product of hydrolysis of esters of glycerol and long-chain "fatty" acids.

Triglycerides and Soap. Animal fat consists mostly of a mixture of esters of the trihydric alcohol, glycerol, and various long-chain carboxylic acids.

$$\begin{array}{l} CH_2O\overset{O}{\underset{\|}{C}}CH_2CH_2CH_2CH_2CH_2CH_2CH_2CH_2CH_2CH_2CH_2CH_2CH_2CH_2CH_2CH_3 \\ | \\ CHO\overset{O}{\underset{\|}{C}}CH_2CH_2CH_2CH_2CH_2CH_2CH_2CH_2CH_2CH_2CH_2CH_2CH_2CH_2CH_2CH_3 \\ | \\ CH_2O\overset{O}{\underset{\|}{C}}CH_2CH_2CH_2CH_2CH_2CH_2CH_2CH_2CH_2CH_2CH_2CH_2CH_2CH_2CH_2CH_3 \end{array}$$

tristearin, the triglyceride of stearic acid

The fat can be either a relatively hard substance, such as tallow, or soft and oily. The softer fats and the oils have one or more double bonds in the acid part of the triglyceride. For comparison, tallow contains more stearic and palmitic acids (as triglycerides) and corn oil contains more oleic and linoleic acids.

$$HO-\overset{O}{\underset{\|}{C}}-(CH_2)_7-CH=CH-(CH_2)_7CH_3$$
oleic acid

$$HO-\overset{O}{\underset{\|}{C}}-(CH_2)_7-CH=CH-CH_2-CH=CH-(CH_2)_4-CH_3$$
linoleic acid

Soap can easily be made at home from fat saved in the kitchen. The fat is simply boiled for a while with "lye" or alkali (NaOH) in an iron kettle until it is converted into a concentrated solution of the sodium salts of the acids and a layer of glycerol (equation 10-35). On cooling, the soap solution solidifies into a cake. This soap will usually have a dark yellow color from

TABLE 10-2. *Saturated and Unsaturated Acids in Typical Fats and Oils.*

Fat or oil	Acid group			
	Stearate	Palmitate	Oleate	Lineoleate
Tallow	23	28	40	3
Butter	12	25	30	2
Corn oil	3	10	29	48
Peanut oil	3	9	57	25

$$
\begin{array}{c}
\text{CH}_2\text{—O—}\overset{\overset{\displaystyle O}{\|}}{\text{C}}\text{—R} \\
| \\
\text{CH—O—}\overset{\overset{\displaystyle O}{\|}}{\text{C}}\text{—R} \;+\; 3\,\text{NaOH} \\
| \\
\text{CH}_2\text{—O—}\overset{\overset{\displaystyle O}{\|}}{\text{C}}\text{—R} \\
\text{fat}
\end{array}
\;\rightarrow\;
\begin{array}{c}
\text{CH}_2\text{OH} \\
| \\
\text{CHOH} \\
| \\
\text{CH}_2\text{OH} \\
\text{glycerol}
\end{array}
\;+\; 3\,\text{Na}^+{}^-\text{O—}\overset{\overset{\displaystyle O}{\|}}{\text{C}}\text{—R} \quad (10\text{-}35)
$$
                                                                                                      soap

impurities in the fat and some excess alkali which makes it rather harsh. A softer soap can be made by dispersing the glycerol in the soap solution as it cools. If you wish to be completely self-sufficient, you can make your own alkali. Save the wood ashes from your stove and strain hot water through them. The filtrate contains large amounts of Na_2CO_3 and K_2CO_3 and will serve quite well to hydrolyze fat.

Waxes are esters of long-chain acids with long-chain alcohols. Beeswax, for example, is mostly ceryl myristate, $C_{26}H_{53}OOCC_{13}H_{27}$, and carnauba wax is mostly myricyl cerotate, $C_{31}H_{63}OOCC_{25}H_{51}$. Carnauba wax, from the leaves of a Brazilian palm, is used in polishes.

FIGURE 10-7. Backyard soap manufacture. [From A. M. Earl, *Home Life in Colonial Days*, Macmillan, New York, 1937]

AMMONOLYSIS OF ESTERS

Many of the reactions of water (H_2O) and ammonia (NH_3) are alike, as they should be for hydrides of adjacent elements in the periodic table which both have unshared pairs of electrons to make them nucleophilic. Thus, both water and ammonia are bases, though ammonia is a stronger base than water. Another similarity is that both water and ammonia can cleave ester linkages. The reactions are known as hydrolysis and ammonolysis.

$$CH_3-\overset{O}{\overset{\|}{C}}-O-CH_2CH_3 \qquad CH_3-\overset{O}{\overset{\|}{C}}-O-CH_2CH_3$$

$$CH_3-\overset{O}{\overset{\|}{C}}-OCH_2CH_3 + H_2O \rightleftarrows CH_3-\overset{O}{\overset{\|}{C}}-OH + HOCH_2CH_3 \qquad (10\text{-}36)$$

$$CH_3-\overset{O}{\overset{\|}{C}}-OCH_2CH_3 + NH_3 \rightleftarrows CH_3-\overset{O}{\overset{\|}{C}}-NH_2 + HOCH_2CH_3 \qquad (10\text{-}37)$$

$$Ph-\overset{O}{\overset{\|}{C}}-O-CH_3 + NH_3 \rightleftarrows Ph-\overset{O}{\overset{\|}{C}}-NH_2 + HOCH_3 \qquad (10\text{-}38)$$

benzamide
mp 130°C

REDUCTION OF ESTERS

Esters are reduced by lithium aluminum hydride to derivatives of two alcohols, one corresponding to the "acid" part of the ester and the other to the "alcohol" part. Note that the first product in

$$4\ CH_3CH_2-\overset{O}{\overset{\|}{C}}-O-CH_3 + 2\ LiAlH_4 \xrightarrow{\text{dry ether}} LiAl(OCH_2CH_2CH_3)_4 + LiAl(OCH_3)_4 \qquad (10\text{-}39)$$

equation 10-39 is a derivative of 1-propanol, not ethanol. The COO group of the ester is reduced by $LiAlH_4$ to a CH_2O group. The alcohols are liberated from the lithium aluminum derivatives by treatment with acid (equations 10-40 and 10-41). Applied to fats,

$$LiAl(OCH_2CH_2CH_3)_4 + 4\ HCl \rightarrow LiCl + AlCl_3 + 4\ HOCH_2CH_2CH_3 \qquad (10\text{-}40)$$

$$LiAl(OCH_3)_4 + 4\ HCl \rightarrow LiCl + AlCl_3 + 4\ HOCH_3 \qquad (10\text{-}41)$$

this sequence of reactions is a way of making glycerol plus long-chain alcohols.

ESTERS AND GRIGNARD REAGENTS

The carbonyl group of an ester, like that of an aldehyde or a ketone, can add a Grignard reagent. However, the addition product

$$R-\overset{O}{\overset{\|}{C}}-O-R' + R''-MgCl \rightarrow \left[R-\underset{R''}{\overset{O-MgCl}{\overset{|}{C}}}-O-R' \right] \qquad (10\text{-}42)$$

of equation 10-42, unlike the product from an aldehyde or a ketone, is unstable and decomposes to give a ketone (equation 10-43).

$$\begin{bmatrix} \text{O}-\text{MgCl} \\ | \\ \text{R}-\text{C}-\text{O}-\text{R}' \\ | \\ \text{R}'' \end{bmatrix} \rightarrow \begin{matrix} \text{O} \\ \| \\ \text{R}-\text{C} \\ | \\ \text{R}'' \end{matrix} + \text{ClMgOR}' \quad (10\text{-}43)$$

The ketone, of course, reacts with more Grignard reagent to give the usual product.

$$\begin{matrix} \text{O} \\ \| \\ \text{R}-\text{C} \\ | \\ \text{R}'' \end{matrix} + \text{ClMg}-\text{R}'' \rightarrow \begin{matrix} \text{O}-\text{MgCl} \\ | \\ \text{R}-\text{C}-\text{R}'' \\ | \\ \text{R}'' \end{matrix} \quad (10\text{-}44)$$

If equations 10-42 through 10-44 are combined, the overall result can be written as equation 10-45. Hydrolysis of the reaction product

$$\begin{matrix} \text{O} \\ \| \\ \text{R}-\text{C}-\text{OR}' \end{matrix} + 2\,\text{R}''-\text{MgCl} \rightarrow \begin{matrix} \text{O}-\text{MgCl} \\ | \\ \text{R}-\text{C}-\text{R}'' \\ | \\ \text{R}'' \end{matrix} + \text{R}'\text{O}-\text{MgCl} \quad (10\text{-}45)$$

therefore gives an alcohol in which two of the alkyl groups are alike and come from the Grignard reagent, while the third comes from the "acid" part of the ester.

$$\begin{matrix} \text{O}-\text{MgCl} \\ | \\ \text{R}-\text{C}-\text{R}'' \\ | \\ \text{R}'' \end{matrix} + \text{HCl} \rightarrow \begin{matrix} \text{OH} \\ | \\ \text{R}-\text{C}-\text{R}'' \\ | \\ \text{R}'' \end{matrix} + \text{MgCl}_2 \quad (10\text{-}46)$$

Occurrence and Properties of Esters

In previous sections a few naturally occurring esters have been used to illustrate reactions, but perhaps not enough emphasis has been placed on how widespread the ester functional group really is.

Low molecular weight esters are liquids, usually with a pleasant fruity odor, and indeed esters are often responsible for plant and fruit odors. Ethyl acetate is used as a solvent for lacquers and as a paint thinner. Octyl acetate is responsible for the odor of oranges, and pentyl acetate is responsible for the odor of bananas. The beautiful odor of new mown hay and sweet clover is due to coumarin, a cyclic ester.

coumarin
mp 68°C

The catnip plant (*Nepeta cataria* L.) also contains a cyclic ester, nepetalactone. Nepetalactone is very attractive to cats of all kinds, including lions.

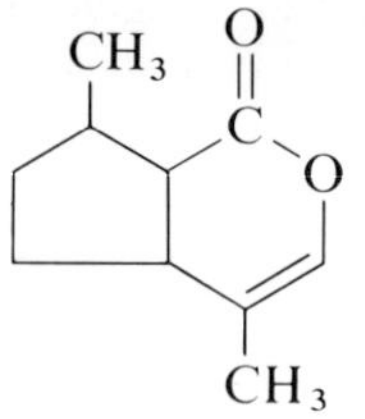

nepetalactone
bp (at 0.05 mm Hg) 71°C

Boll weevils can be attracted (to their destruction) by the sex attractant or pheromone propylure. It is interesting that the cis isomer not only fails to attract the male of the pink bollworm moth but also offsets the attractive effect of the trans isomer.

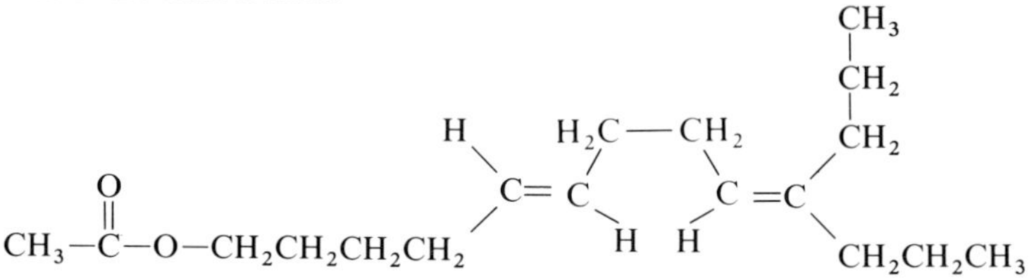

10-propyl-*trans*-5,9-tridecadien-1-ol acetate or propylure

A great deal of work is being done in a search for easily manufactured specific lures to control harmful insects without damaging useful ones. In the absence of any detailed theory to predict which compounds will be effective, a broad screening approach is used. That is to say, a large number of compounds are tested on the insect in question, and a few are found to have some weak activity. Then the structures of the weakly active compounds are changed systematically, and the structural features that make the compound more effective are noted. Similar procedures are used in other fields in which the theoretical background is insufficient, for example, in the search for better antimalarial drugs.

Some plants such as the pyrethrin flower, an African relative of the chrysanthemum, produce their own insecticide. Pyrethrin is used in insect sprays for its quick knockdown effect.

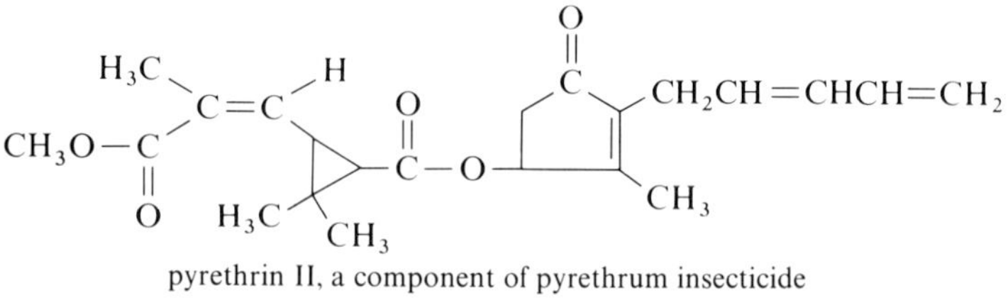

pyrethrin II, a component of pyrethrum insecticide

Another example of a naturally occurring insecticide, and one that has the advantage of being quite specific for just certain insects, is juvabione. Juvabione occurs in balsam fir and in paper made from the pulped wood. It is a juvenile hormone for the insect *Pyrrhocoris apterus*. Nymphs of this insect, exposed to juvabione, never become sexually mature and die without reproducing themselves.

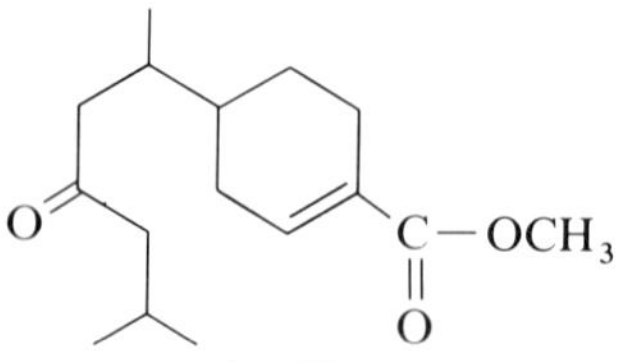

juvabione

The discovery of juvabione came about as the result of an interesting accident. A research group was trying to raise batches of Pyrrocoris apterus for some now-forgotten biological purpose and encountered mysterious difficulties. The eggs would not hatch and the nymphs never developed into adults, although other laboratories had not experienced any difficulty. Eventually, the cause was traced to the paper used in lining the cages and from there to the balsam fir from which the paper was made.

The compound aflatoxin B_1 is one of a group of complex cyclic esters produced by the fungus *Aspergillus flavus*. These substances were first investigated in 1960 after 100,000 young turkeys died from eating peanut meal contaminated with the fungus. The aflatoxins are among the most potent carcinogens (cancer producing compounds) known, as little as 1 microgram per day being sufficient to cause cancer in rats.

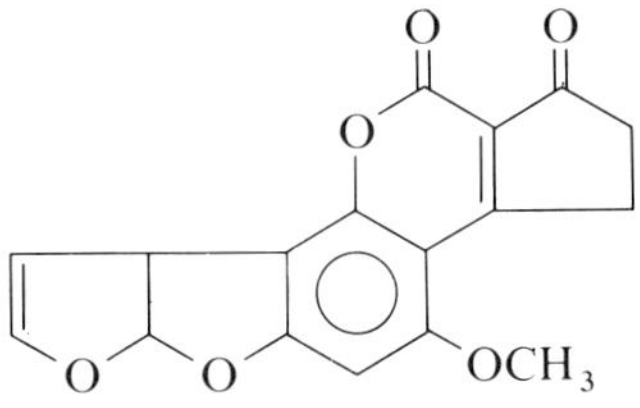

aflatoxin B_1

Amides

An amide is a derivative of a carboxylic acid in which the hydroxyl group has been replaced by an NH_2 group or substituted NH_2 group.

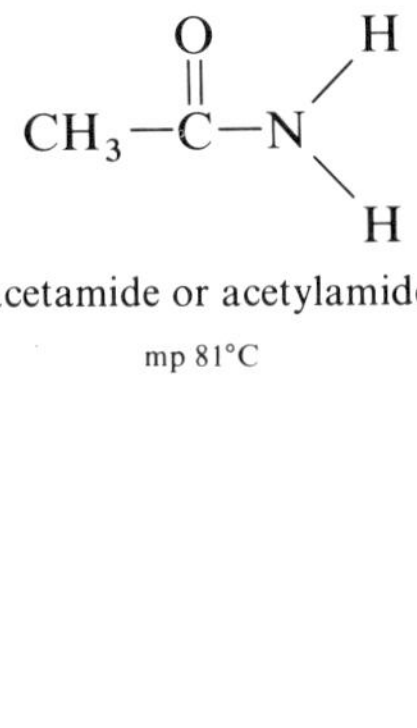

acetamide or acetylamide
mp 81°C

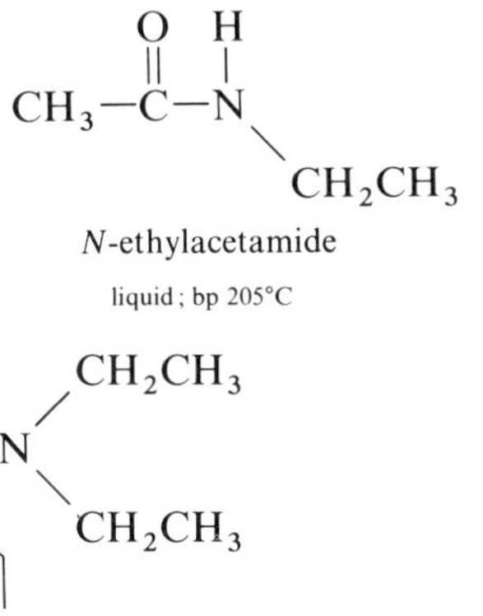

N-ethylacetamide
liquid; bp 205°C

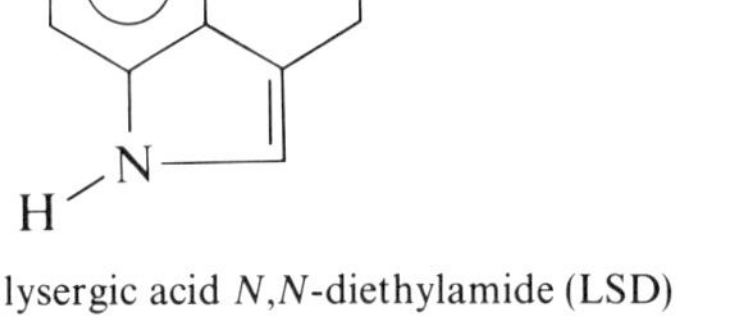

lysergic acid N,N-diethylamide (LSD)
pointed prisms from benzene; mp 80–85°C

The psychomimetic properties of LSD were discovered accidentally one afternoon in 1943 by A. Hofmann in his laboratory in Switzerland. Hofmann went home and to bed and the symptoms wore off in about 2 hr. In general LSD symptoms somewhat resemble those of schizophrenia and can be brought on by an oral dose as small as 50 µg. Lysergic acid itself and a number of derivatives are found in rye grain that has been contaminated with the fungus

Claviceps purpurea, also known as ergot. Outbreaks of ergot-induced mass insanity have been reported from pre-Christian times and as recently as 1952. In the middle ages ergotism was called St. Anthony's fire. It should be noted that the alkaloids of ergot are a complex mixture which probably has more drastic effects than those of LSD alone. Besides hallucinations, the symptoms of ergotism can include peripheral pains, vomiting, diarrhea, gangrene, convulsion, and death.

The amide functional group is neither very acidic nor very basic. The greater electronegativity of nitrogen as compared to oxygen prevents amides from being acidic like the closely related carboxylic acids, and the presence of the electron-withdrawing carbonyl group prevents them from being as basic as ammonia or a substituted ammonia such as aniline. For example, a mixture of a carboxylic acid, an amide, and aniline could be separated by extraction first with sodium bicarbonate and then with dilute hydrochloric acid.

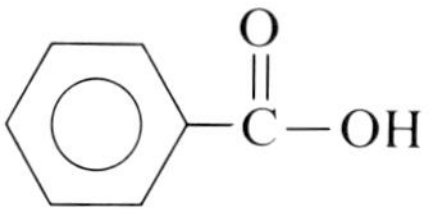

$Ph-C(=O)-OH$ → extracted by $NaHCO_3$

benzoic acid

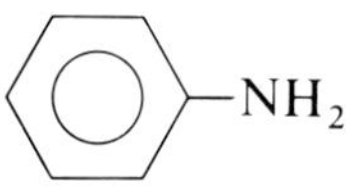

$Ph-NH_2$ → extracted by dil. HCl

aniline

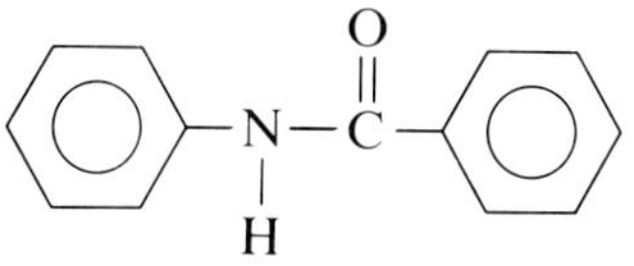

$Ph-N(H)-C(=O)-Ph$ → left behind

benzamide

The compound **carbamic acid** is unstable and decomposes immediately into ammonia and carbon dioxide. It is represented by

$$[H_2NCOOH] \rightarrow NH_3 + CO_2$$
carbamic acid

a number of esters, including the tranquilizer meprobamate (Miltown).

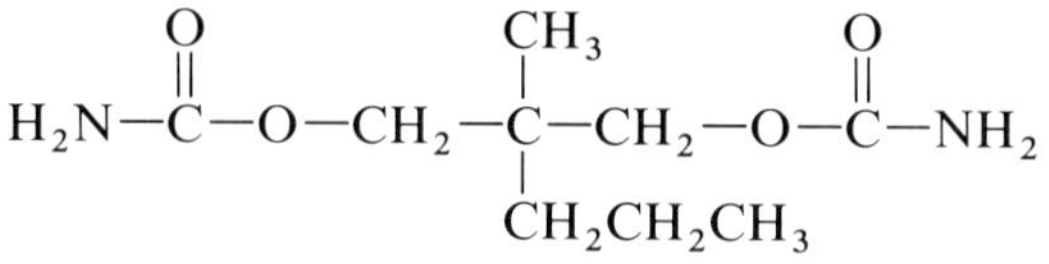

$$H_2N-\overset{O}{\overset{\|}{C}}-O-CH_2-\underset{CH_2CH_2CH_3}{\overset{CH_3}{\underset{|}{\overset{|}{C}}}}-CH_2-O-\overset{O}{\overset{\|}{C}}-NH_2$$

meprobamate
mp 104°C

The use of 2.4 g of meprobamate per day is said to have induced a physical dependence on this drug, the withdrawal symptoms including vomiting, tremors, anxiety, hallucinations, and epileptic fits.

The most important naturally occurring amides are proteins. These compounds will be discussed in detail in Chapter 14, but for the time being we will note that it is possible to connect together molecules of a compound containing both an amino group (NH$_2$) and a carboxyl group by means of an amide linkage. Such linkages are the main structural feature of proteins, in which they are known as **peptide bonds** (equation 10-47).

$$R-\underset{\underset{NH_2}{|}}{\overset{\overset{H}{|}}{C}}-\overset{O}{\overset{\|}{C}}-OH + H_2N-\underset{\underset{R}{|}}{\overset{\overset{H}{|}}{C}}-\overset{O}{\overset{\|}{C}}-OH \rightarrow H_2O + R-\underset{\underset{NH_2}{|}}{\overset{\overset{H}{|}}{C}}-\overset{O}{\overset{\|}{C}}-\underset{\underset{}{|}}{\overset{\overset{H}{|}}{N}}-\underset{\underset{R}{|}}{\overset{\overset{H}{|}}{C}}-COOH$$

(10-47)

Relationships Between Reactions

Most of the reactions covered in this chapter involve the addition of a nucleophile, or electron-rich reagent, to the positive end of a carbonyl group. This first step, which is like that of many of the reactions of aldehydes and ketones (Chapter 8), is followed by the departure of a second nucleophile (equation 10-48). The overall

$$R-\underset{:Y}{\overset{\overset{O}{\|}}{C}}-X \rightarrow \left[R-\underset{+Y}{\overset{\overset{\bar{O}}{|}}{C}}-X \right] \rightarrow R-\overset{\overset{O}{\|}}{\underset{+Y}{C}} + X^- \quad (10\text{-}48)$$

result is the displacement of one nucleophile attached to carbonyl by another. In the reaction of an acid chloride with hydroxide ion, for example, Ÿ in equation 10-48 is HÖ:$^-$, X is Cl, and the products are RCOOH and Cl$^-$ (actually RCOO$^-$ and H$_2$O and Cl$^-$).

The reaction of an acid chloride with ammonia is essentially the same process. The nucleophilic electrons are the uncharged pair on NH$_3$ and the displaced nucleophile is again Cl$^-$. A minor difference in detail is that an extra proton is removed from the nitrogen in a fast second step. The reaction of an acid chloride with the sodium salt of a carboxylic acid is still another example of essentially the same process of addition followed by nucleophilic displacement.

$$CH_3-\overset{\overset{O}{\|}}{C}-Cl \xrightarrow{\dot{Y}} CH_3-\underset{+Y}{\overset{\overset{\bar{O}}{|}}{C}}-Cl \rightarrow CH_3-\underset{+Y}{\overset{\overset{O}{\|}}{C}} + Cl^-$$

$$CH_3-\overset{\overset{O}{\|}}{C}-Cl \xrightarrow{H\ddot{O}:^-} CH_3-\underset{\underset{H}{\overset{|}{O}}}{\overset{\overset{-O}{|}}{C}}-Cl \rightarrow CH_3-\underset{\underset{H}{\overset{|}{O}}}{\overset{\overset{O}{\|}}{C}} + Cl^- \rightarrow CH_3COO^- + H_2O$$

$$CH_3-\overset{\overset{O}{\|}}{C}-Cl \xrightarrow{:NH_3} CH_3-\underset{\underset{H\ H\ H}{+N}}{\overset{\overset{-O}{|}}{C}}-Cl \rightarrow CH_3-\underset{\underset{H\ \ H}{N}}{\overset{\overset{O}{\|}}{C}} + H^+ + Cl^-\ \ \searrow NH_4^+$$

$$CH_3-\overset{\overset{O}{\|}}{C}-Cl \xrightarrow{\overset{..}{:}\overset{..}{O}-\overset{\overset{O}{\|}}{C}-\text{Ph}} CH_3-\overset{\overset{^-O}{|}}{\underset{\underset{\underset{\text{Ph}}{|}}{\underset{C=O}{|}}}{\overset{|}{C}}}-Cl \rightarrow CH_3-\overset{\overset{O}{\|}}{\underset{\underset{\underset{\text{Ph}}{|}}{\underset{C=O}{|}}}{C}}\overset{O}{\diagdown} + Cl^-$$

Table 10-3 shows a set of similarly related reactions in which both the entering nucleophile and the departing nucleophile are varied. The intermediates are not shown and, to facilitate comparisons, products such as the acids formed by reaction of ammonia are shown in the free state rather than as the ammonium salts.

TABLE 10-3. *Products of the Reactions of Various Nucleophiles and Substrates*

Nucleophile	Substrate			
	$R-\overset{\overset{O}{\|}}{C}-Cl$	$R-\overset{\overset{O}{\|}}{C}-O-R'$	$R-\overset{\overset{O}{\|}}{C}-O-\overset{\overset{O}{\|}}{C}-R$	$R-\overset{\overset{O}{\|}}{C}-NH_2$
H_2O	$R-\overset{\overset{O}{\|}}{C}-OH$ and HCl	$R-\overset{\overset{O}{\|}}{C}-OH$ and HOR'	$R-\overset{\overset{O}{\|}}{C}-OH$ and $HO-\overset{\overset{O}{\|}}{C}-R$	$R-\overset{\overset{O}{\|}}{C}-OH$ and NH_3
CH_3OH	$R-\overset{\overset{O}{\|}}{C}-OCH_3$ and HCl	$R-\overset{\overset{O}{\|}}{C}-OCH_3$ and HOR'	$R-\overset{\overset{O}{\|}}{C}-OCH_3$ and $HO-\overset{\overset{O}{\|}}{C}-R$	$R-\overset{\overset{O}{\|}}{C}-OCH_3$ and NH_3
NH_3	$R-\overset{\overset{O}{\|}}{C}-NH_2$ and HCl	$R-\overset{\overset{O}{\|}}{C}-NH_2$ and HOR'	$R-\overset{\overset{O}{\|}}{C}-NH_2$ and $HO-\overset{\overset{O}{\|}}{C}-R$	$R-\overset{\overset{O}{\|}}{C}-NH_2$ and NH_3

SUMMARY

A. Carboxylic acids
   1. Nomenclature
      (a) The systematic names are based on the name of the alkane having the same number of carbon atoms in the main chain (including the carboxyl group).
      (b) They take the form alkanoic acid.
      (c) For naming substituted acids, the carboxyl carbon is designated as carbon 1. In an older system, the *next* carbon is designated as α.

2. Carboxylic acids are more acidic than alcohols because of
   (a) the effect of the C=O group on the electronegativity of the OH oxygen
   (b) the delocalization of charge in the carboxylate ion.
3. After butanoic acid, carboxylic acids become progressively less soluble
   (a) Differences in solubility in bases (NaOH for phenols, NaHCO$_3$ for acids) can be used to separate mixtures of phenols and acids.
   (b) Mixtures of acids can be separated by paper chromatography.
4. The boiling points increase by about 20°C for each CH$_2$ group.
5. Preparation
   (a) By oxidation of alcohols, aldehydes, or aromatic side chains.
   (b) By hydrolysis of amides, acid chlorides, anhydrides, or esters.

B. Acid chlorides
   1. Acid chlorides may be prepared by the reaction of the acid with thionyl chloride.
   2. They react (by displacement of Cl$^-$) with NH$_2$, NHR, or OH functional groups, giving amides, substituted amides, and esters.
   3. They react with aromatic compounds in the presence of AlCl$_3$ to give ketones (electrophilic substitution).

C. Acid anhydrides
   1. May be prepared from acid chlorides and the sodium salt of an acid, another nucleophilic displacement reaction.
   2. The reactions of anhydrides are like those of acid chlorides.
   3. The formation of cyclic anhydrides shows the relative locations of pairs of carboxyl groups in the same molecule.

D. Esters
   1. Esters are named like salts, but are strictly covalent molecules.
   2. They can be made from the alcohol and acid chloride or anhydride or from the alcohol and the acid.
   3. The equilibrium reaction between acid and alcohol to give ester and water is catalyzed by acids, which increase the electrophilicity of the carbonyl group.
   4. Esters can be hydrolyzed either by means of an acid catalyst or by means of alkali.
   5. Fats are esters of glycerol with long-chain acids; waxes are esters of long-chain alcohols with long-chain acids.
   6. The reaction of esters with ammonia resembles the reaction with water, but gives amides rather than acids.
   7. Esters can be reduced with LiAlH$_4$ to two alcohols.
   8. Esters react with Grignard reagents to give tertiary alcohols.

E. Amides
   1. Amides are nearly neutral, in contrast to the carboxylic acids and ammonia or substituted ammonias from which they are derived.
   2. The amide functional group serves to link the amino acid units of proteins.
   3. Amides may be prepared from ammonia (or amines, substituted ammonias) and acids, acid chlorides, acid anhydrides, or esters.

EXERCISES

1. Give the common and systematic names for

   (a) H—C(=O)—OH

   (b) C$_6$H$_5$—C(=O)—OH

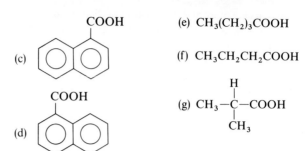

(c) [1-naphthoic acid structure]

(d) [2-naphthoic acid structure, COOH on other ring position]

(e) $CH_3(CH_2)_3COOH$

(f) $CH_3CH_2CH_2COOH$

(g) $CH_3-\underset{\underset{CH_3}{|}}{\overset{\overset{H}{|}}{C}}-COOH$

2. Draw structures for
   (a) propanoic acid
   (b) β-methylbutyric acid
   (c) 3-methylbutanoic acid
   (d) capric acid
   (e) β-naphthoic acid
   (f) formic acid

3. Give an example of (a) an acid found in red ants and (b) an acid found in vinegar.

4. Explain why acetic acid is more acidic than ethyl alcohol.

5. Why is it that ethyl acetate (bp 77°C) has a lower boiling point than either ethyl alcohol (bp 78.50°C) or acetic acid (bp 118°C), even though the latter two compounds consist of smaller molecules? *Hint:* See page 118 of Chapter 7.

6. Trichloroacetic acid, Cl_3CCOOH, is a stronger acid than acetic acid. Why? *Hint:* See Chapter 2.

7. Explain why acetamide is less basic than ammonia.

8. Give a method for separating the following. Assume insolubility in water.
   (a) A mixture of a carboxylic acid RCOOH and a phenol, ArOH.
   (b) A mixture of a carboxylic acid, a phenol, and an alcohol.
   (c) A mixture of carboxylic acids.

9. (Open book) Give an example of an acid that acts as an insect sex attractant. Does your compound attract all insects or just one kind?

10. Write equations showing how to convert each of the following substances into benzoic acid.

    (a) 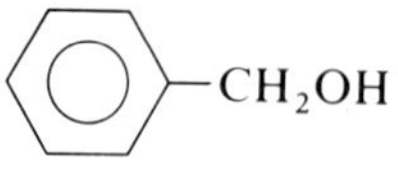 Ph–CH_2OH
    (b) 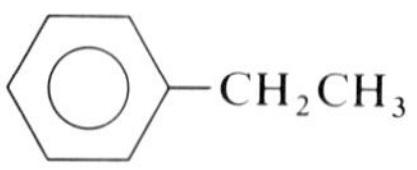 Ph–CH_2CH_3
    (c) 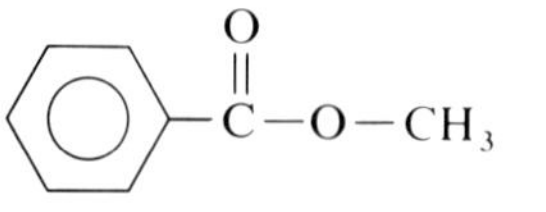 Ph–C(=O)–O–CH_3
    (d) Ph–C(=O)–Cl
    (e) Ph–C(=O)–NH_2
    (f) 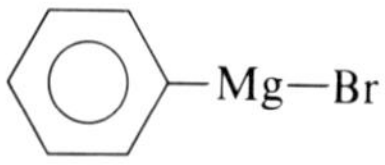 Ph–Mg–Br

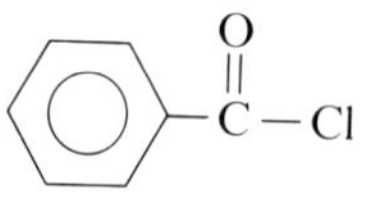

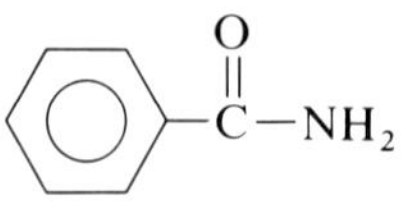

11. Complete the following reactions

(a) $CH_3CH_2CH_2CH_2-\underset{\underset{O}{\|}}{C}-OH + SOCl_2 \rightarrow$

(b) 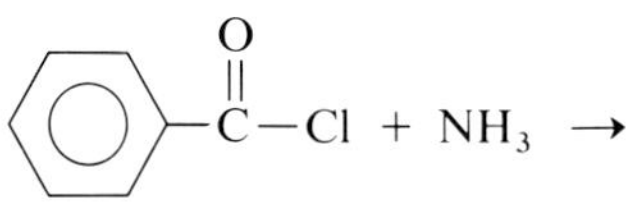 $C_6H_5-\underset{\underset{O}{\|}}{C}-Cl + NH_3 \rightarrow$

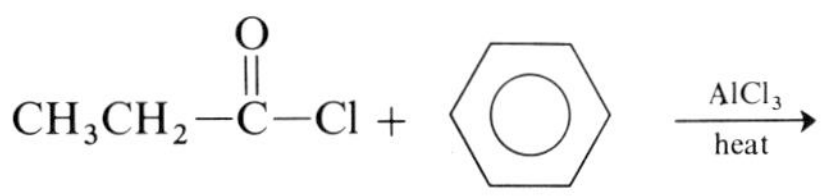

(c) $CH_3CH_2-\underset{\underset{O}{\|}}{C}-Cl + C_6H_6 \xrightarrow[\text{heat}]{AlCl_3}$

(d) $CH_3CH_2-\underset{\underset{O}{\|}}{C}-Cl + Na^+{}^-O-\underset{\underset{O}{\|}}{C}-CH_3 \rightarrow$

(e) 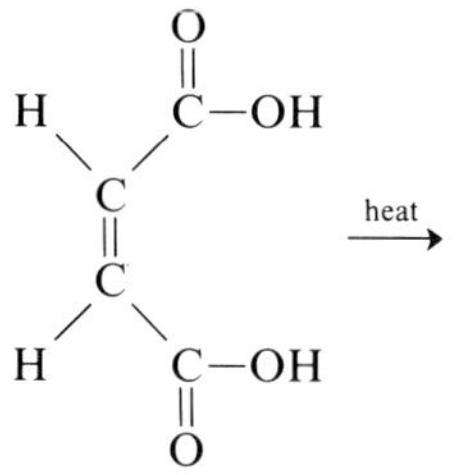
$$\begin{array}{c} H \\ \diagdown \\ C \\ \| \\ C \\ \diagup \\ H \end{array} \begin{array}{c} C-OH \\ \| \\ O \\ \\ C-OH \\ \| \\ O \end{array} \xrightarrow{\text{heat}}$$

(f) 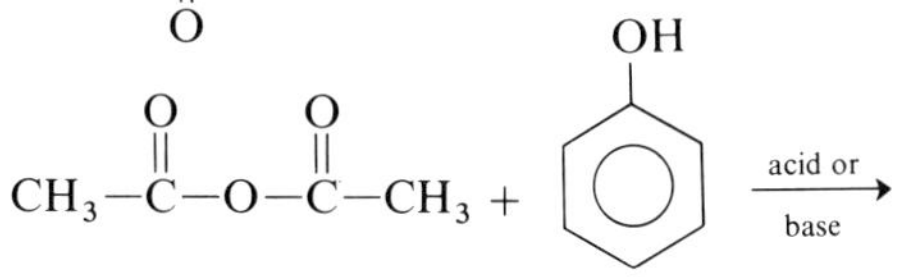 $CH_3-\underset{\underset{O}{\|}}{C}-O-\underset{\underset{O}{\|}}{C}-CH_3 + C_6H_5OH \xrightarrow[\text{base}]{\text{acid or}}$

12. (Open book) Give an example of (a) an acid-base indicator; (b) a naturally occurring acid anhydride; (c) a well-known analgesic; and (d) a substance with a wintergreen odor.

13. Give the names, formulas, and natural occurrence or some use or interesting property of three esters.

14. (Open book) Write the equation for a reaction that would convert juvabione (page 210) into (a) methyl alcohol and an acid; (b) two alcohols; (c) an amide and methyl alcohol.

15. Explain how to make soap from fat. Use equations.

16. What is the main structural difference between the molecules of a typical vegetable oil and an animal fat such as tallow?

17. Write the structural formula for (a) tristearin and (b) tricaprylin.

18. Write an equation showing how to make long-chain alcohols from fats.

19. Complete the following reactions

(a) 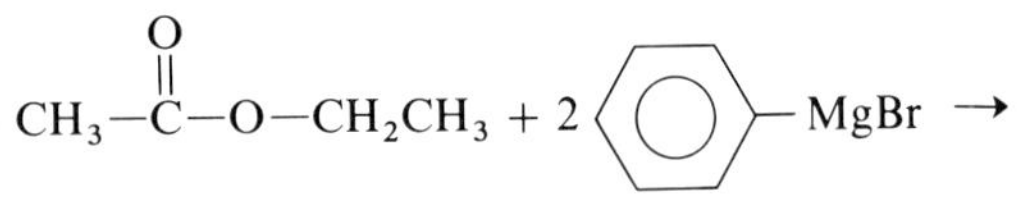 $CH_3-\underset{\underset{O}{\|}}{C}-O-CH_2CH_3 + 2\, C_6H_5-MgBr \rightarrow$

(b) $4CH_3-\underset{\underset{O}{\|}}{C}-OCH_2CH_3 + 2\, LiAlH_4 \rightarrow$

Carboxylic Acids, Esters, and Related Compounds

20. What ester could be used to make each of the following alcohols by reaction with ethyl magnesium bromide?

(a) $(CH_3CH_2)_3COH$

(b) 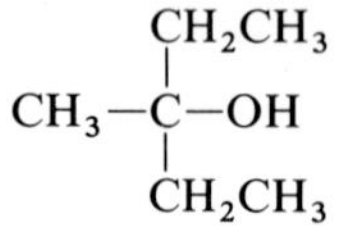 $CH_3-\underset{\underset{CH_2CH_3}{|}}{\overset{\overset{CH_2CH_3}{|}}{C}}-OH$

(c) $CH_3CH_2\underset{\underset{}{}}{\overset{\overset{OH}{|}}{C}}HCH_2CH_3$

(d) 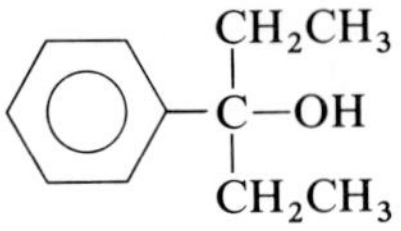 $C_6H_5-\underset{\underset{CH_2CH_3}{|}}{\overset{\overset{CH_2CH_3}{|}}{C}}-OH$

21. Explain with equations and sketches of separatory funnels how you could separate a mixture of N,N-diethyl benzamide, benzoic acid, and N,N-diethylaniline.

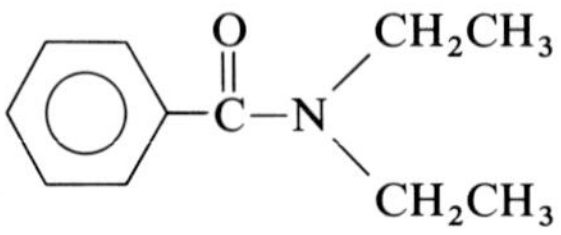

N,N-diethylbenzamide

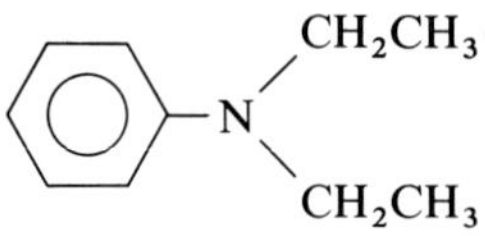

N,N-diethylaniline

22. What structural feature is most responsible for the color of phenolphthalein (page 202) in basic solution? *Hint:* See Chapter 8, page 157.

Chapter 11
Nitrogen Compounds

In earlier chapters we have encountered quite a few nitrogen compounds, including nitrobenzene in Chapter 6 and amides in Chapter 10. The purine and pyrimidine bases used for genetic coding will be taken up again in Chapter 15 and the peptide linkage of proteins in Chapter 14. In this chapter we will discuss some new types of nitrogen compound and also survey the occurrence of nitrogen compounds in nature.

Nitrogen Fixation

The first thing that anyone notices about the element nitrogen is the stability and apparent unreactivity of molecular nitrogen, N_2. Air, for example, consists of about 78% nitrogen by volume and, when the highly reactive oxygen is exhausted by combustion, the nitrogen remains unchanged. Conversely, in the violent decomposition of a number of nitrogen-containing compounds, molecular nitrogen is a major product. On the other hand, molecular nitrogen cannot be entirely unreactive or else it would all be present merely as atmospheric nitrogen.

Strictly organic reactions that consume molecular nitrogen are unknown, but there are a number of inorganic reactions that do so and nitrogen is also "fixed" or converted into nitrogen compounds by certain bacteria. In nature, the only nonbiochemical nitrogen-fixing reaction is the conversion of molecular nitrogen into oxides of nitrogen by lightning bolts in the atmosphere. In the laboratory, nitrogen is observed to react with metallic lithium at room temperature to give Li_3N, and it has also been reduced to ammonia (NH_3) in a process catalyzed by certain titanium compounds. By far the greatest amount of nitrogen is fixed biochemically by bacteria living on the root nodules of plants such as clover or peas. The reaction is of unknown mechanism but produces ammonia which is converted into more complex nitrogen-containing molecules by the plant (Figure 11-1). Plants that do not themselves harbor nitrogen-fixing bacteria need to be fertilized with ammonia or nitrates.

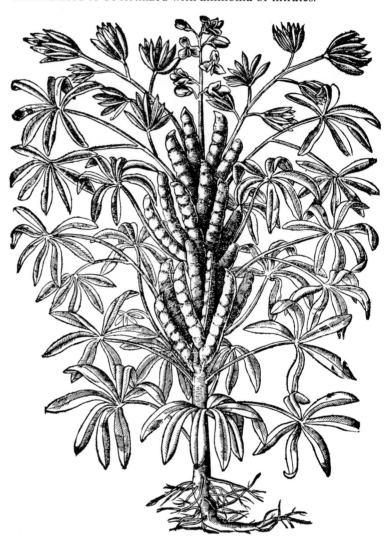

LUPINE

FIGURE 11-1. Ammonia is synthesized in the roots of this plant at ordinary temperatures and 1 atm pressure. The commercial synthesis from nitrogen and hydrogen uses an iron catalyst, temperatures of 450–600°C, and pressures of 200–600 atm. [From J. W. Krutch, *Herbal*, Putnam, N.Y., 1965 and Pierandrea Mattioli, *Commentaries on the Six Books of Dioscorides*, Prague, 1563]

Ammonia and Amines

Ammonia is a colorless, water-soluble gas with a pungent odor. It boils at $-33°C$ and freezes at $-78°C$. In spite of its low boiling point, liquid ammonia is easily handled in the laboratory because the heat of vaporization is so high that it evaporates only slowly. Like water, ammonia is a hydrogen-bonded liquid.

There is a whole inorganic chemistry of ammonia and solutions of compounds in liquid ammonia, paralleling the chemistry of water. Thus an acid in ammonia is a compound that gives a proton to the solvent to give NH_4^+, just as an acid in water is a compound that gives a proton to H_2O to form H_3O^+. Conversely, a strong base in ammonia solution removes a proton from the solvent to give amide ion, NH_2^-.

An amine is an ammonia molecule substituted with from one to three alkyl or aryl groups.

CH_3-NH_2
methylamine
bp −6.5°C; soluble in water

$CH_3CH_2-NH_2$
ethylamine
bp 16.6°C; soluble in water

$CH_3CH_2-\overset{\overset{H}{|}}{N}-CH_2CH_3$
diethylamine
bp 55.5°C; soluble in water

$CH_3CH_2-\overset{\overset{CH_2CH_3}{|}}{N}-CH_2CH_3$
triethylamine
bp 89.5°C; slightly soluble in water

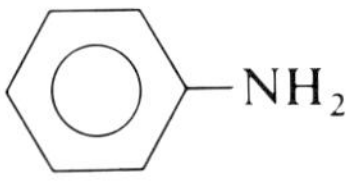

aniline, phenylamine, or aminobenzene
bp 184°C; slightly soluble in water

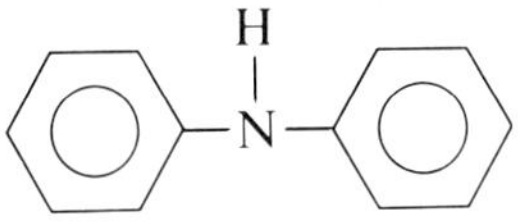

diphenylamine
bp 302°C; insoluble in water

Although amines are not basic enough to remove protons from ammonia, they do react as bases in water (equations 11-1). Of course,

$$R-\overset{..}{N}H_2 + H_2O \rightleftharpoons R-\overset{\overset{H}{+|}}{\underset{\underset{H}{|}}{N}}-H + HO^-$$

$$R_3N: + H_2O \rightleftharpoons R_3\overset{+}{N}H + HO^- \quad (11\text{-}1)$$

they will also react with stronger acids than water to form salts (equation 11-2). In both cases, a proton is added to the nitrogen,

$$CH_3-\overset{..}{\underset{\underset{H}{|}}{N}}-H + HO\overset{\overset{O}{\|}}{C}CH_3 \rightleftharpoons CH_3-\overset{\overset{H}{+|}}{\underset{\underset{H}{|}}{N}}-H \quad O-\overset{\overset{O}{\|}}{C}-CH_3$$

methylammonium acetate (11-2)

using the lone pair of electrons to establish the new bond. Although ammonia and the amines are pyramidal, with the unshared pair of electrons at the apex of the pyramid, ammonium salts and quaternary ammonium and alkylammonium compounds are tetrahedral, like saturated carbon compounds.

$$\underset{\text{an amine; pyramidal}}{\overset{..}{\underset{H\ H\ R}{N}}} \qquad \underset{\text{an alkylammonium salt; tetrahedral}}{\overset{H}{\underset{H\ H\ R}{\overset{+}{N}}}} \qquad \underset{\text{a quaternary ammonium ion; tetrahedral}}{\overset{R}{\underset{R\ R\ R}{\overset{+}{N}}}}$$

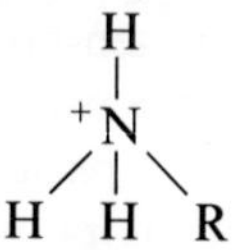

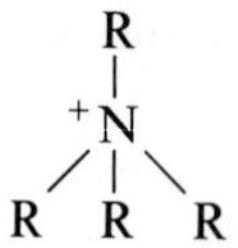

A strong base such as NaOH will liberate an amine from its salts (equation 11-3).

$$\underset{\text{methylammonium chloride}}{CH_3\overset{+}{N}H_3Cl^-} + NaOH \rightarrow \underset{\text{methylamine}}{CH_3NH_2} + NaCl + H_2O \qquad (11\text{-}3)$$

Dimethylamine and trimethylamine occur in herring brine and contribute to the odor downwind from Norwegian fish-oil factories.

Unlike carbon, nitrogen has little tendency to form long chains or rings consisting solely of nitrogen atoms; therefore, the chemistry of nitrogen is largely the chemistry of carbon *and* nitrogen compounds. The nitrogen analog of ethane is hydrazine, and even this compound is highly reactive.

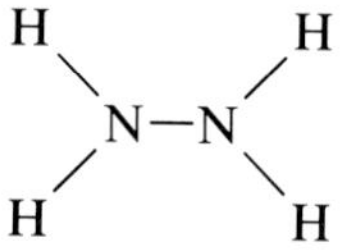

hydrazine
mp 2°C; bp 114°C;
flammable, used as rocket fuel

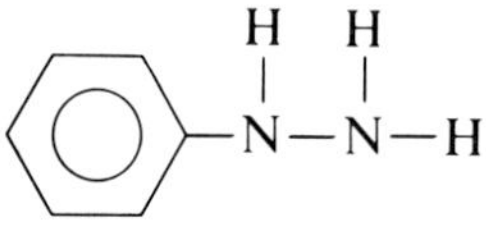

phenylhydrazine
mp 19.6°C

CLASSIFICATION OF AMINES

In the chemistry of "substituted water" we found it worth while to use separate names for alcohols and ethers, but substituted ammonias are all known as amines whether they have one, two, or three substituents. However, the three kinds of amine do differ in some of their reactions, and they are referred to as **primary amines** (one alkyl or aryl group), **secondary amines** (two alkyl or aryl groups), and **tertiary amines** (three alkyl or aryl groups). Amines of all three kinds are bases and form salts with acids.

A possible source of confusion lies in the fact that *t*-butylamine

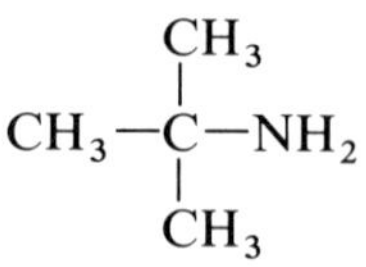

$$CH_3-\underset{\underset{CH_3}{|}}{\overset{\overset{CH_3}{|}}{C}}-NH_2$$

is a primary amine even though it has a tertiary alkyl substituent. An example of a tertiary amine is trimethylamine

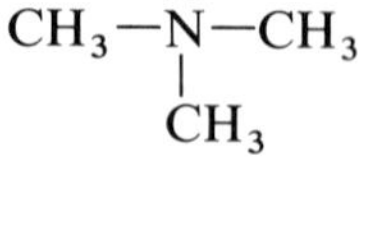

$$CH_3-\underset{\underset{CH_3}{|}}{\overset{..}{N}}-CH_3$$

PREPARATION

Just as the hydrolysis of an alkyl halide gives an alcohol (equation 11-4), the reaction with ammonia gives an amine (equation 11-5).

$$H_2O: + CH_3Cl \rightarrow H-O-CH_3 + H^+ + Cl^- \quad (11\text{-}4)$$

$$H_3N: + CH_3Cl \rightarrow H_3\overset{+}{N}-CH_3 + Cl^- \quad (11\text{-}5)$$

Both reactions are examples of nucleophilic displacement of halide ion by a reagent having an unshared pair of electrons. The main difference is that methylamine is basic enough to hold on to the extra proton, and methanol is not. To produce free methylamine we have to treat the methylammonium chloride with strong base (equation 11-6).

$$CH_3\overset{+}{N}H_3\ Cl^- + NaOH \rightarrow CH_3-NH_2 + NaCl + H_2O \quad (11\text{-}6)$$

Secondary amines can be made by treating alkyl halides with primary amines instead of with ammonia (equations 11-7 and 11-8).

$$R-\ddot{N}H_2 + R'-Cl \rightarrow R-\overset{+}{N}H_2-R'\ Cl^- \quad (11\text{-}7)$$

$$R-\overset{+}{N}H_2-R'\ Cl^- + NaOH \rightarrow R-NH-R' + NaCl + H_2O \quad (11\text{-}8)$$

Tertiary amines can be made similarly from secondary amines (equations 11-9 and 11-10).

$$R-NH-R' + R''-Cl \rightarrow R-\overset{+}{N}H(R')(R'')\ Cl^- \quad (11\text{-}9)$$

$$R-\overset{+}{N}H(R')(R'')\ Cl^- + NaOH \rightarrow R-N(R')(R'') + NaCl + H_2O \quad (11\text{-}10)$$

Reaction of a tertiary amine with an alkyl halide gives a **quaternary ammonium salt**, there being no such thing as a quaternary amine (equation 11-11).

$$R-N(R')(R'') + R'''-Cl \rightarrow R-\overset{+}{N}(R')(R'')(R''')\ Cl^- \quad (11\text{-}11)$$

The reaction of ammonia with an alkyl halide usually gives primary and secondary amines as byproducts. To avoid this, it is better to make primary amines from amides by the Hofmann reaction, equation 11-12.

$$R-\underset{\underset{O}{\|}}{C}-NH_2 + NaOBr \rightarrow R-NH_2 + NaBr + CO_2 \quad (11\text{-}12)$$

$$\underset{\text{caproamide}}{CH_3CH_2CH_2CH_2CH_2\underset{\underset{O}{\|}}{C}NH_2} + NaOBr \rightarrow \underset{\text{1-aminopentane}}{CH_3CH_2CH_2CH_2CH_2NH_2} + CO_2 + NaBr \quad (11\text{-}13)$$

Aniline is prepared by reduction of nitrobenzene (equation 11-14).

$$C_6H_5-NO_2 + [H] \rightarrow H_2O + C_6H_5-NH_2 \quad (11\text{-}14)$$

Cyano compounds, otherwise known as nitriles, can also be reduced to give primary amines. An example is shown in equation 11-15. The symbol [H] in equations 11-14 and 11-15 stands for a

$$CH_3CH_2CH_2C\equiv N + [H] \rightarrow CH_3CH_2CH_2CH_2NH_2 \quad (11\text{-}15)$$

reducing agent such as $SnCl_2$ plus acid or $LiAlH_4$.

ADDITION TO ALDEHYDES AND KETONES

Primary amines (RNH_2) and monosubstituted ammonias such as hydroxylamine ($HONH_2$) react with aldehydes and ketones to give compounds with C=N double bonds. These products, called Schiff bases, are used to identify the aldehyde or ketone by means of the melting point of the derivative.

$$\underset{\text{hydroxylamine}}{HO-NH_2} + CH_3-\underset{\underset{O}{\|}}{C}-CH_3 \xrightarrow{H^+} H_2O + \underset{\underset{\text{mp 61°C}}{\text{acetone oxime}}}{\underset{CH_3}{\overset{HO}{\diagdown}}N=\underset{CH_3}{\overset{}{C}}} \quad (11\text{-}16)$$

$$\underset{\text{phenylhydrazine}}{C_6H_5-NH-NH_2} + CH_3-\underset{\underset{O}{\|}}{C}-CH_3 \xrightarrow{H^+} H_2O + \underset{\underset{\text{mp 42°C}}{\text{acetone phenylhydrazone}}}{C_6H_5-\underset{H}{\overset{}{N}}-\underset{CH_3}{\overset{CH_3}{\diagup}}N=C} \quad (11\text{-}17)$$

An oxime can be reduced with sodium and alcohol to make a primary amine, and this is another way of making such compounds (equation 11-18).

$$\underset{\substack{\text{cyclopentanone} \\ \text{oxime}}}{\underset{H_2C-CH_2}{\overset{H_2C-CH_2}{\diagdown}}C=N\diagup^{OH}} \xrightarrow{H_2,\,Ni,\,90°C} H_2O + \underset{\text{cyclopentylamine}}{\underset{H_2C-CH_2}{\overset{H_2C-CH_2}{\diagdown}}C\diagup^{H}_{NH_2}} \quad (11\text{-}18)$$

Secondary amines can be made by the reduction of the addition products of primary amines and aldehydes or ketones (equation 11-19).

$$\underset{R}{\overset{R}{\diagdown}}C=O + R'NH_2 \xrightarrow{H^+} \underset{R}{\overset{R}{\diagdown}}C=N-R' \xrightarrow[HCOOH]{[H]} \underset{R}{\overset{R}{\diagdown}}CH-\underset{}{\overset{H}{\underset{|}{N}}}-R'$$

(11-19)

AMINES AND NITROUS ACID

Nitrous acid ($HO-N=O$ or HNO_2) reacts with secondary amines to give nitrosamines; in these compounds the hydrogen on the nitrogen of the secondary amine has been replaced by a nitroso ($-N=O$) group (equation 11-20).

$$\underset{H_3C}{\overset{H_3C}{\diagdown}}N-H + HONO \rightarrow \underset{H_3C}{\overset{H_3C}{\diagdown}}N-N=O + H_2O \quad (11\text{-}20)$$

N-nitrosodimethylamine
yellow oil; bp 153°C

More interesting is the reaction of primary amines with nitrous acid. Aniline and other primary *aromatic* amines give diazonium salts, which can be prepared in solution at 0°C (equation 11-21).

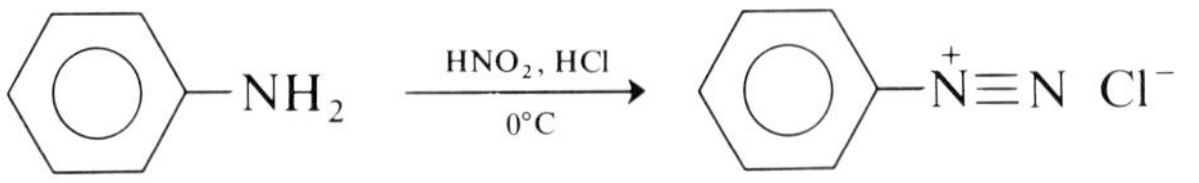

$$\text{C}_6\text{H}_5-NH_2 \xrightarrow[0°C]{HNO_2, HCl} \text{C}_6\text{H}_5-\overset{+}{N}\equiv N \ Cl^- \quad (11\text{-}21)^1$$

benzenediazonium chloride

On heating a diazonium salt in aqueous acid, nitrogen is evolved and the corresponding phenol is formed (equation 11-22). This reaction thus provides a convenient way of converting an amine, $ArNH_2$, into the corresponding phenol, $ArOH$ (equation 11-23).

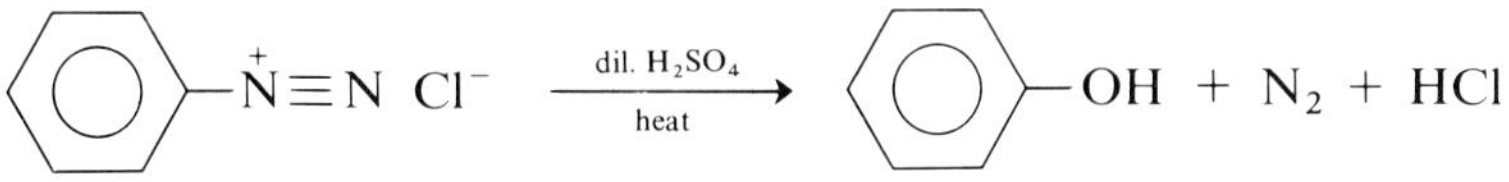

$$\text{C}_6\text{H}_5-\overset{+}{N}\equiv N \ Cl^- \xrightarrow[\text{heat}]{\text{dil. } H_2SO_4} \text{C}_6\text{H}_5-OH + N_2 + HCl$$

(11-22)

$$ArNH_2 \xrightarrow[HCl]{HNO_2} ArN_2Cl \xrightarrow[\text{heat}]{\text{dil. } H_2SO_4} ArOH \quad (11\text{-}23)$$

[1] A nitroso derivative

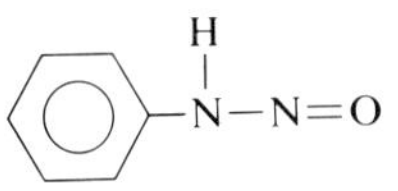

$$\text{C}_6\text{H}_5-\underset{|}{\overset{H}{N}}-N=O$$

is probably formed first in this case also and then reacts with HCl to give the diazonium salt and water.

The reaction of *aliphatic* primary amines with nitrous acid does not give an isolable diazonium salt. Instead, hydrolysis products are obtained

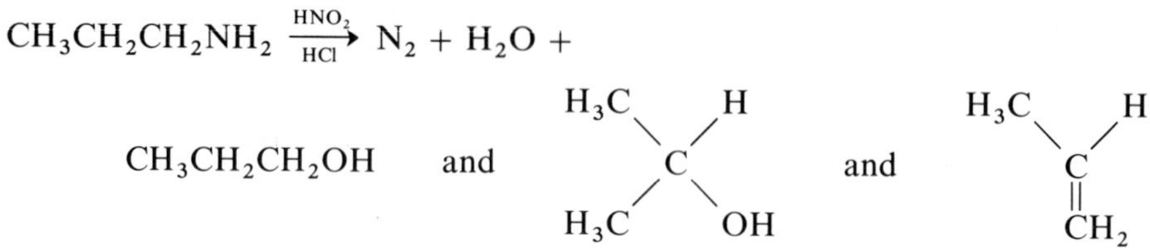

$$CH_3CH_2CH_2NH_2 \xrightarrow[HCl]{HNO_2} N_2 + H_2O +$$

$$CH_3CH_2CH_2OH \quad \text{and} \quad \begin{array}{c} H_3C \\ \diagdown \\ H_3C \end{array} \!\!\! C \!\!\! \begin{array}{c} H \\ \diagup \\ OH \end{array} \quad \text{and} \quad \begin{array}{c} H_3C \\ \diagdown \\ \end{array} \!\!\! C \!\!\! \begin{array}{c} H \\ \diagup \\ \end{array} \quad (11\text{-}24)^2$$
$$ \overset{\parallel}{CH_2}$$

Diazonium Salts and Azo Compounds

Compounds with a double bond between nitrogen atoms are known as **azo** compounds. Aromatic azo compounds are colored, the precise color depending on the presence of other functional groups in the molecule.

Ph—N=N—Ph

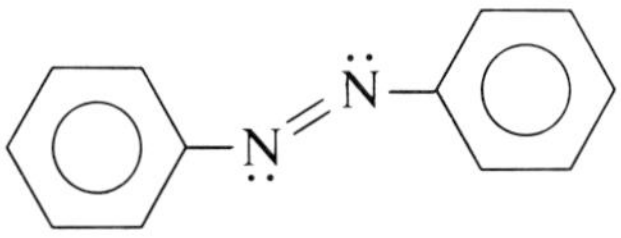

trans-azobenzene

orange-red; mp 68°C

Although azo compounds are not found in naturally occurring dyestuffs, they are the basis of many synthetic dyes. Azo dyes invariably have one or more additional functional groups in the molecule, both to modify the color and to aid in the process of making the dye stick to the fabric. The azo dyes are usually synthesized from diazonium salts using the diazo-coupling reaction, an example of electrophilic aromatic substitution. For comparison and review, equations for the diazo-coupling reaction (equation 11-25) and an equation for another aromatic substitution reaction, nitration (equation 11-26) are shown next to each other.

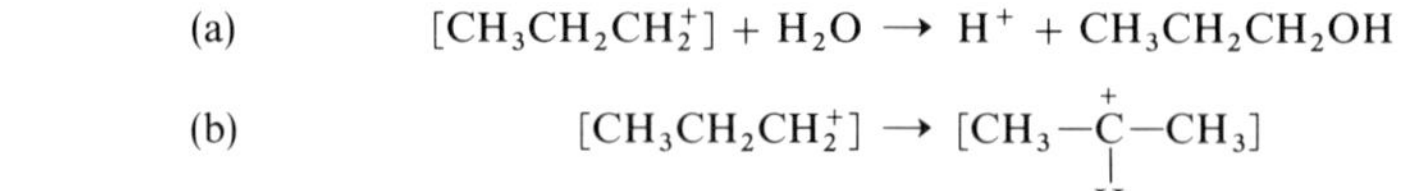

$$O_2N^+ + \underset{}{\bigcirc}\text{—OH} \rightarrow O_2N\text{—}\underset{}{\bigcirc}\text{—OH} + H^+ \qquad (11\text{-}25)$$

electrophilic
reagent

---

[2] The diazonium salt $CH_3CH_2CH_2N_2^+Cl^-$ loses N_2 and forms a carbonium ion, $[CH_3CH_2CH_2^+]$. This intermediate is not isolable either, but reacts in several ways.

(a) $\quad [CH_3CH_2CH_2^+] + H_2O \rightarrow H^+ + CH_3CH_2CH_2OH$

(b) $\quad [CH_3CH_2CH_2^+] \rightarrow [CH_3-\overset{+}{\underset{H}{C}}-CH_3]$

The conversion of the primary carbonium ion to the secondary carbonium ion $[CH_3\overset{+}{C}HCH_3]$ is an example of a rearrangement reaction. The secondary carbonium ion then reacts to give 2-propanol and propylene.

$$[CH_3-\overset{+}{C}HCH_3] \rightarrow H^+ + CH_3-CH=CH_2$$

$$[CH_3-\overset{+}{C}HCH_3] + H_2O \rightarrow H^+ + CH_3-\overset{OH}{\underset{}{C}H}-CH_3$$

A Short Course in
Modern Organic Chemistry

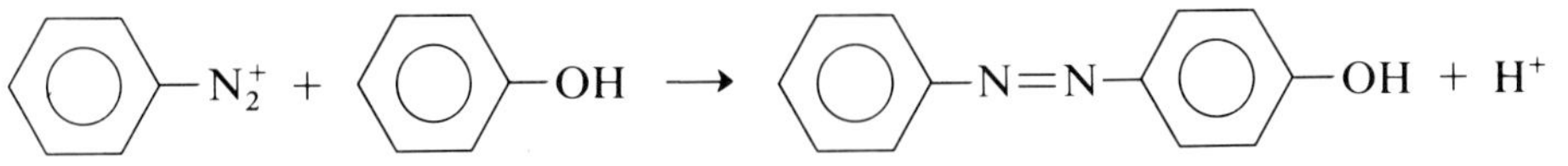

$$\text{C}_6\text{H}_5\text{-N}_2^+ + \text{C}_6\text{H}_5\text{-OH} \rightarrow \text{C}_6\text{H}_5\text{-N=N-C}_6\text{H}_4\text{-OH} + \text{H}^+$$

electrophilic reagent

p-hydroxyazobenzene
orange; mp 152°C

(11-26)

p-Hydroxyazobenzene, though colored, does not stick to fabrics very well and is therefore not used as a dye.

A typical azo dye is primulin red, made from a rather bulky diazonium salt and β-naphthol (equation 11-27). Primulin red is put

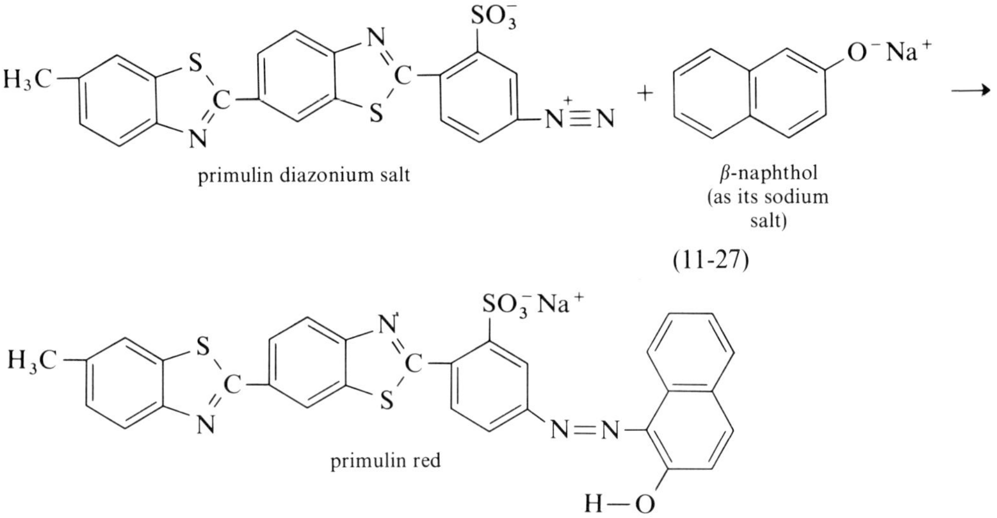

[primulin diazonium salt] + [β-naphthol (as its sodium salt)] →

(11-27)

[primulin red]

on the fabric by a process known as **ingrain dyeing**. You will notice that the diazo-coupling reaction makes a very large molecule out of two smaller, soluble molecules. This is the secret of ingrain dyeing: the cloth is dipped into the diazonium salt solution, to coat the fibers with that reagent, and then into the β-naphthol solution. Formation of the primulin red takes place right on the fiber; either the product is entangled in crevices of the fiber or it is too insoluble to wash off.

Since the intermediate primulin diazonium salt is somewhat sensitive to light, designs can be photographed onto cloth using primulin red. A cardboard stencil or a negative is placed over the cloth after it has been dipped in the diazonium salt solution. Exposure of the unshielded parts of the cloth to sunlight for a few minutes destroys the diazonium salt, and when the cloth is then developed by dipping into a β-naphthol solution, only the shielded parts turn red.

Nitrogen Heterocycles

A heterocyclic compound is a ring compound containing one or more atoms other than carbon as part of the ring. Many important natural products such as alkaloids, hemoglobin, chlorophyll, various drugs, vitamins, and the bases that form the genetic code of nucleic acids, are derivatives of simple five- or-six-membered nitrogen heterocycles.

FIVE-MEMBERED RINGS

The five-membered ring compound pyrrole is found in coal tar. Although the structural formula appears to be that of a secondary amine, pyrrole is only very weakly basic. The reason for this is that the "unshared" pair of electrons on nitrogen plus the four π electrons of the two "double bonds" constitute an aromatic sextet of electrons like that in benzene. The electron pair is therefore not really available for bonding to a proton and forming a salt.

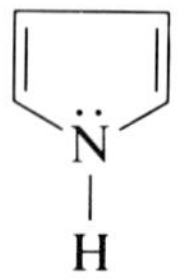

pyrrole
bp 131°C

The most important naturally occurring compounds related to pyrrole are porphyrins (Greek, *porphyra*, purple). Porphyrins are metal salts of a flat, 16-membered ring compound (protoporphyrin, Figure 11-2) made up of four pyrrole rings connected by a system of double and single bonds. Hemin (Figure 11-2) is an iron salt of

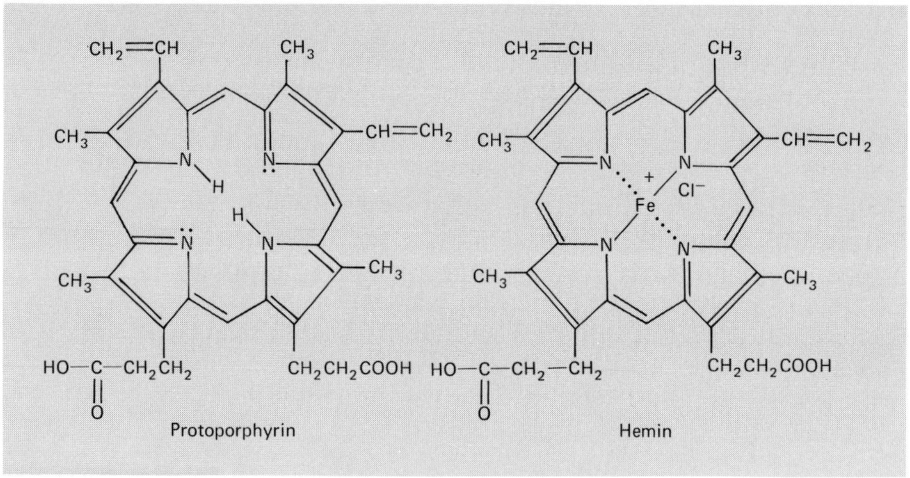

FIGURE 11-2. Protoporphyrin and hemin.

protoporphyrin: the iron is bonded to all four nitrogen atoms and replaces two hydrogens that would otherwise be attached to two of the nitrogen atoms. The respiratory pigment **hemoglobin**, which carries the oxygen in the blood of red-blooded animals, is a compound of hemin and a protein or polypeptide (Chapter 14), but it is the hemin part of the complex that carries the oxygen molecule, loosely bonded to the central iron atom.

The chlorophylls are magnesium salts remarkably similar to hemin in their structure (Figure 11-3). Their function in plants is still not completely understood, but chlorophyll molecules in the chloroplasts of green leaves absorb sunlight and make its energy available for the process of photosynthesis. Photosynthesis makes

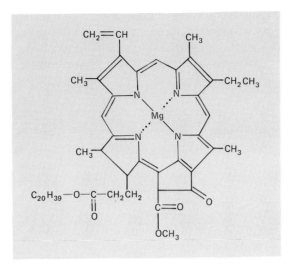

FIGURE 11-3. Chlorophyll a. Chlorophyll b has an aldehyde group in place of one of the methyl groups.

carbohydrates (Chapter 13) out of carbon dioxide and water and simultaneously liberates oxygen. The ordinary reaction, represented by the burning of paper, goes the other way, giving off energy and carbon dioxide and water and consuming oxygen.

It is striking that both hemin and chlorophyll are part of energy-supplying mechanisms, one for animals and one for plants. Their structural similarity may be derived from a stage in evolution before there was any distinction between plants and animals.

Indole, so called because it can be made from indigo (Chapter 8) has a benzene ring fused to a pyrrole ring. Indole and skatole are

indigo

indole
mp 52°C

skatole
mp 95°C

found in feces (characteristic odor!) and are formed by the action of bacteria on the amino acid tryptophan, formed by the digestion of protein.

tryptophan

Curiously enough, pure indole in low concentration has a flowery rather than a fecal odor. It is found in extracts of jasmine and orange blossoms.

The indole ring system is quite common in hallucinogens and other drugs that have mental or emotional effects. These compounds will be discussed in a later section of this chapter.

The imidazole ring is found in the amino acid L-histidine (Chapter 14) and in histamine. Histamine is apparently formed in the body by the decarboxylation of histidine and is found, combined with protein, in all tissues. In contrast to the protein-bound histamine,

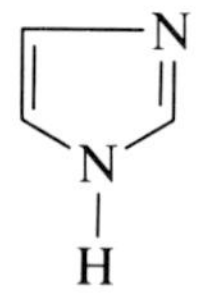

imidazole or
1,3-diazole
mp 90°C

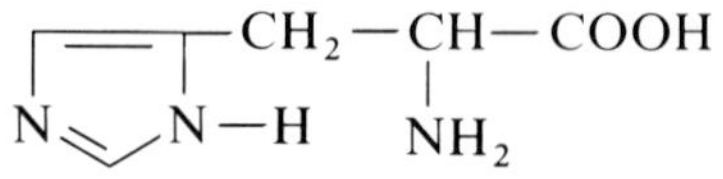

histidine

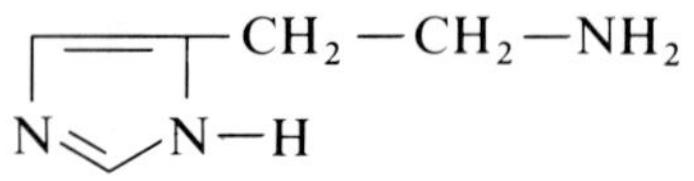

histamine
mp 83°C

free histamine is very poisonous. Many of the symptoms of allergies (hay fever, asthma) and of the common cold are caused by the release of histamine during allergic reactions.

Some of the symptoms of allergies and colds can be alleviated by means of antihistamines. Examples are Benadryl (also used as a sea-sickness remedy), Chlor-Trimeton, and Pyribenzamine. Notice that all three of these anti-

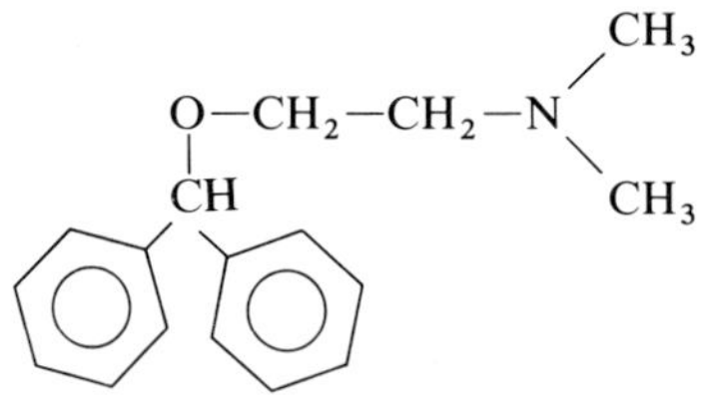

diphenhydramine or Benadryl
mp 166°C

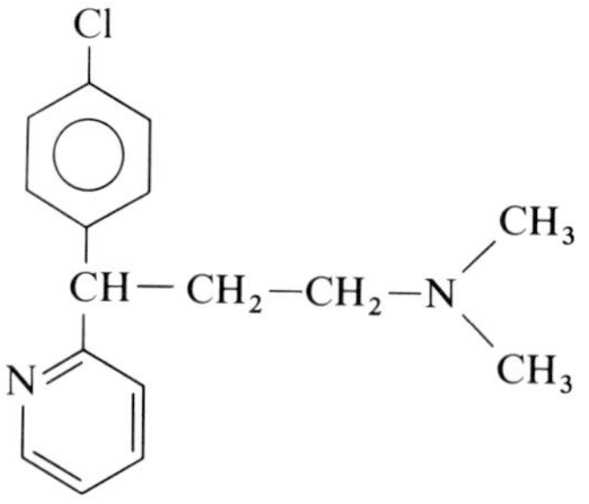

chlorpheniramine or Chlor-Trimeton,
maleic acid salt
mp 130°C

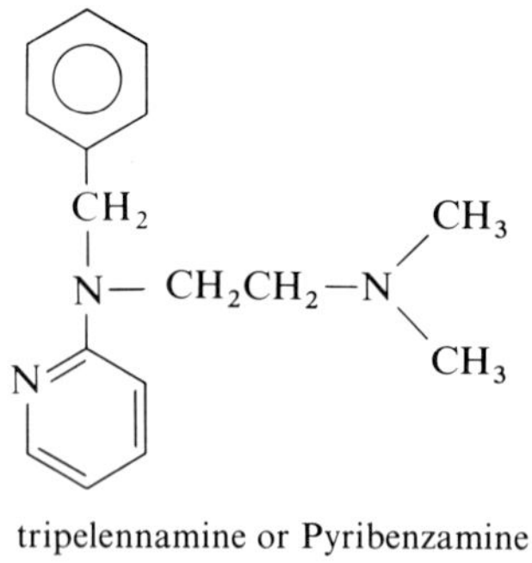

tripelennamine or Pyribenzamine

histamines resemble histamine slightly in having a CH_2CH_2N part structure, and that they resemble each other even more closely. Although we usually do not know how a given drug works, the general idea that similar structures will have similar effects is a useful guide in the search for new drugs.

SIX-MEMBERED RINGS

Pyridine (Figure 11-4) is a much more stable and more aromatic compound than pyrrole. It is a base and readily forms salts with strong acids, since the unshared pair of electrons on nitrogen is not a

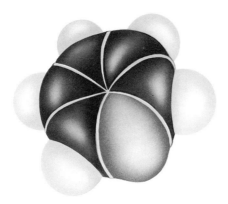

FIGURE 11-4. Model of a pyridine molecule.

part of the aromatic sextet. Pyridine occurs in coal tar and has an unpleasant pungent odor. It is used to "denature" ethyl alcohol.

pyridine
bp 115°C; soluble in water

Piperidine is the saturated compound obtained by adding six hydrogen atoms to pyridine. Its name comes from the presence of the same structure in piperine, the alkaloid[3] of black pepper (*Piper nigrum*).

piperidine

piperine
mp 130°C

Nicotinic acid or its amide must be present in the diet to prevent the deficiency disease pellagra (dermatitis, central nervous symptoms). The name nicotinic acid is derived from that of the alkaloid nicotine, which can be oxidized to nicotinic acid by means of nitric acid.

nicotinamide
mp 128°C

nicotine
oil; bp 123°C at 17 mm

---

[3] **Alkaloid** is a term rather loosely used to denote certain nitrogen compounds found in plants, usually basic and usually having some important physiological activity. It is derived from the word "alkali" for the hydroxides of the alkali metals, NaOH, KOH.

Nictotine occurs in tobacco leaves to the extent of 2–8 % as citric and malic acid salts. About 500 tons are used each year as an agricultural insecticide. It is moderately toxic to man, the fatal dose being about 40 mg. Count Bucarmé was killed by nicotine in 1850, the first known instance of its use by a murderer.

As can be seen from the carbon atom marked with an asterisk, nicotine is optically active. The naturally occurring form is levorotatory and has about twice the physiological impact of its enantiomer.

Coniine is an alkaloid with a saturated six-membered, or piperidine, ring. Found in the hemlock herb (not to be confused with the hemlock spruce of *Evangeline*), coniine was the poison used in the political execution of Socrates in 399 B.C. (Figure 11-5). Large doses of coniine paralyze the respiratory system. The free base has an odor like that of a foul pipe or a cigar butt.

coniine
bp 166°C

FIGURE 11-5. David's "The Death of Socrates." [Courtesy of the Metropolitan Museum of Art, Wolfe Fund, 1931]

The bark of the cinchona tree, native to the Andes and grown commercially in Java, contains several alkaloids, of which the most important is quinine. Quinine is still one of the best drugs for the treatment and prevention of malaria, although it is supplemented nowadays by numerous synthetic antimalarials which are cheaper. Malaria is still a serious problem, particularly since the appearance of strains of mosquitoes that are resistant to DDT. The ring systems found in the quinine molecule (structure page 179) are quinoline and quinuclidine. Most synthetic antimalarials incorporate the quinoline ring system in one way or another.

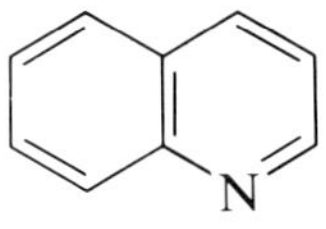

quinoline

quinuclidine

THE PURINE RING SYSTEM

Substituted purines occur not only in the nucleic acids (Chapter 15) but also in other important natural products. The purines of nucleic

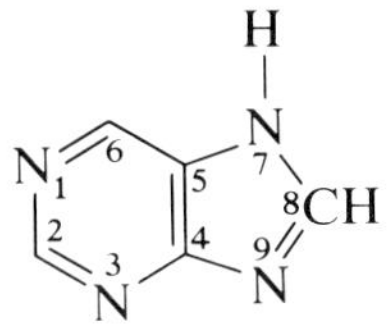

purine
mp 216°C

acid are adenine, or 6-aminopurine, and guanine, or 2-amino-6-hydroxypurine. The most stable forms of the purines may have carbonyl or $>C=N-H$ groups rather than hydroxyl or amino groups. However, they can react in either form, and the interconversion is so easy that the distinction is not important. Guanine is found in guano (bird manure).

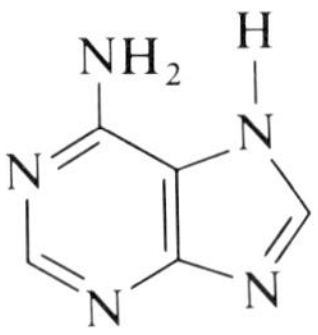

adenine
colorless solid; dec. 360°C

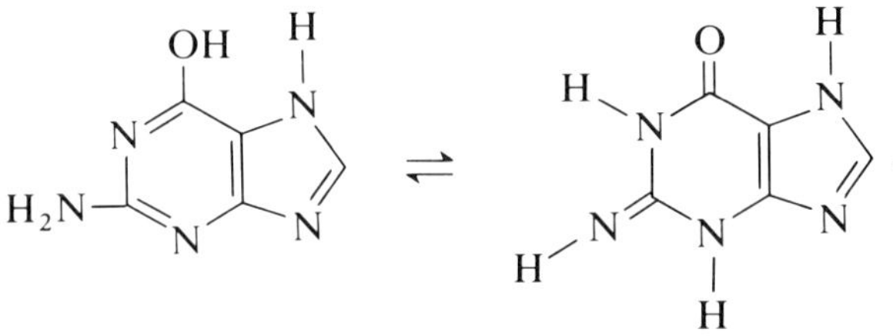

guanine
colorless solid; dec. above 360°C; insoluble in water; soluble in alkali

Uric acid, or 2,6,8-trihydroxypurine, is found in blood, in urine, and in guano (25%). Crystals of the sodium salt of uric acid deposit in the joints of persons suffering from gout.

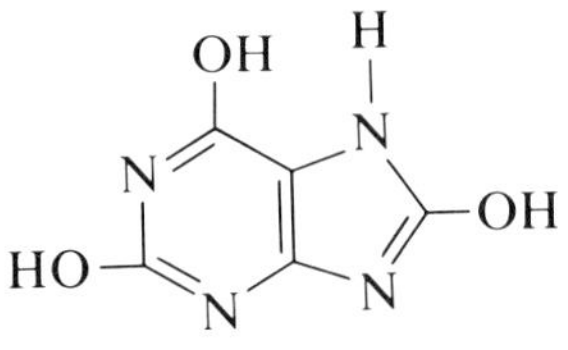

uric acid
dec. to give HCN on heating; soluble in alkali

Caffeine is an alkaloid present in coffee, tea, and cola nuts. It is a mild cardiac, central nervous system, and respiratory stimulant, with nervousness and insomnia as side effects. It has been estimated that a fatal dose might be about 10 g.

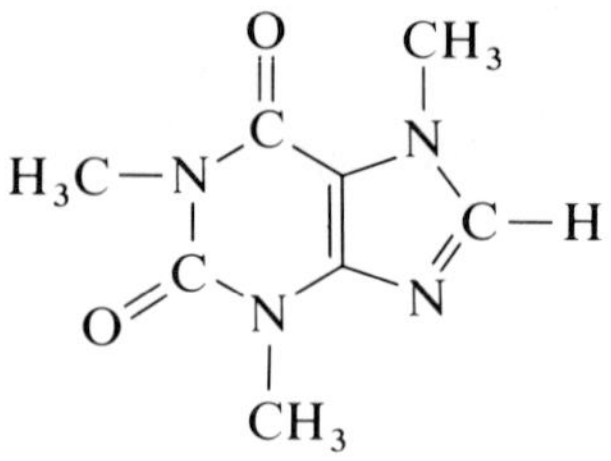

caffeine
mp 238°C

PTERINS

Pterins are compounds having a pyrimidine ring fused with a pyrazine or 1,4-diazine ring. The pterins take their name from their occurrence in butterfly wing pigments.[4] Xanthopterin is responsible for the yellow color both of butterfly wings and of urine. A physiologically important relative of xanthopterin is pteroylglutamic acid, a substance necessary for the growth of certain microorganisms and used in the treatment of some kinds of anemia.

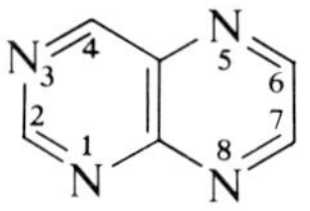

pteridine
yellow; mp 138°C

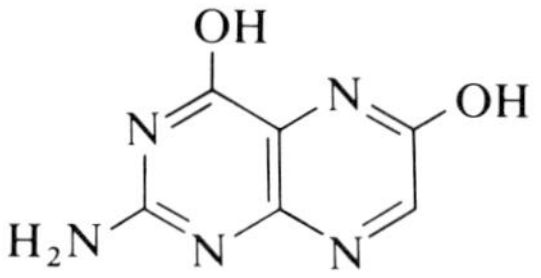

xanthopterin
orange-yellow; dec. above 410°C

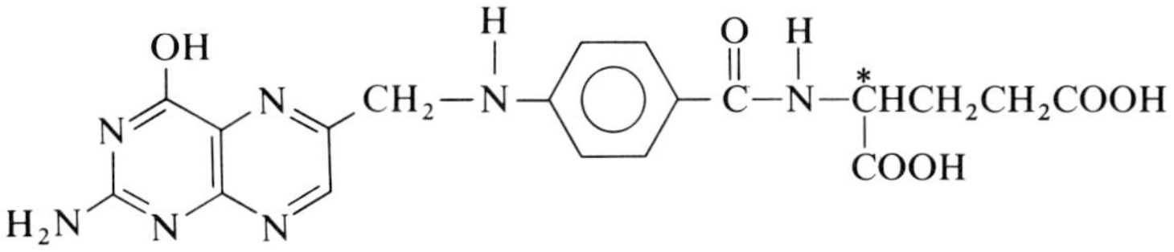

pteroylglutamic acid
yellow-orange; dec. above 250°C

Psychomimetic Drugs

Many compounds are known, both synthetic and naturally occurring, that have the effect of changing a person's mental or emotional state. The majority of these psychomimetic drugs are nitrogen compounds. Quite a few are naturally occurring alkaloids.

EPINEPHRINE AND RELATED COMPOUNDS

Fundamental to any eventual understanding of the mode of action of psychomimetic drugs is an understanding of the effect on mind and mood of compounds naturally present in the body. Epinephrine, or adrenalin, is a hormone released into the blood stream by the medulla of the adrenal gland. When a person is startled

[4] The same etymological root is found in such words as helicopter and pterodactyl.

or alarmed, the sympathetic nervous system triggers the release of adrenalin; the hormone, in turn, produces a number of physiological

$$HO-C_6H_3(OH)-\overset{*}{C}H(OH)CH_2NHCH_3$$

epinephrine or adrenalin
mp 211°C; $[\alpha]_D^{25} = -50°$

effects that are likely to have survival value in situations of sudden danger. The pupils of the eye dilate, the blood pressure increases, the heart beat is strengthened, the air channels leading to the lung become wider, and the metabolic rate and rate of oxygen consumption are increased. Epinephrine is used in the treatment of bronchial asthma because of its property of dilating the bronchial tubes.

Norepinephrine (the prefix **nor** indicates the absence of a methyl group) is also produced by the adrenal medulla. It is also formed, along with acetylcholine, at postganglionic nerve endings where it is part of the mechanism of nerve impulse transmission from the end of one nerve fiber to the beginning of the next one. Norepinephrine

$$HO-C_6H_3(OH)-\overset{*}{C}H(OH)-CH_2-NH_2$$

norepinephrine
dec. 216.5–218°C; $[\alpha]_D^{25} = -37.3°$ in HCl (aq)

also acts as a vasoconstrictor and raises the blood pressure. Surprisingly, it is found in bananas, particularly in the peel.

A **sympathomimetic drug** is one that has effects like those of epinephrine, that is, effects like those obtained by stimulating the sympathetic nervous system. The dextrorotatory enantiomer of epinephrine is an example; its effects are like those of (−)-epinephrine but only about one-twentieth as strong.

Table 11-1 lists a number of drugs related to epinephrine. The word **amphetamine** applied to these drugs refers both to a particular compound, 1-phenyl-2-aminopropane, and to a series of its derivatives. Benzedrine is used (and abused) to overcome sleepiness, increase alertness, counteract depression, and lessen fatigue. Unfortunately, the *eventual* reaction is a still greater fatigue and depression. Some amphetamines ("STP," "speed," mescaline) cause hallucinations. Apparently they mimic substances naturally present in the brain just well enough to interfere seriously with some important biochemical process.

The use of mescal buttons (cut from the flowering tops of the cactus) in American Indian religious ceremonies dates back to pre-Columbian times. Mescaline causes "a kaleidoscopic play of visual hallucinations in indescribably rich colors." The mescal buttons contain at least fifteen other alkaloids in addition to mescaline and produce auditory and tactile hallucinations as well as visions.

TABLE 11-1. *Amphetamines*

Compound	Formula
Amphetamine or benzedrine; bp 200°C	C₆H₅–CH₂–*CH(NH₂)–CH₃
Ephedrine or (−)-α(R), β(R)-N,α-dimethyl-β-hydroxyphenethylamine; mp 34°C	C₆H₅–*CH(OH)–*CH(CH₃)–NHCH₃
2,5-Dimethoxy-4-methyl-amphetamine or "STP"	2,5-(OCH₃)₂-4-CH₃-C₆H₂–CH₂–*CH(CH₃)–NH₂
Phenylephrine hydrochloride or Neo-Synephrine	3-HO-C₆H₄–*CH(OH)–CH₂–NHCH₃·HCl
N-Methylamphetamine, methamphetamine, or "speed"	C₆H₅–CH₂–CH(CH₃)–NH(H)–CH₃
Mescaline; mp 35°C; moderately soluble in water	3,4,5-(CH₃O)₃C₆H₂–CH₂CH₂NH₂

FIGURE 11-6. The mescal or peyote cactus. [From R. E. Schultes, *Science*, **163**:245 (1969)]

SEROTONIN AND RELATED COMPOUNDS

The indole derivative serotonin is a vasoconstrictive drug that is present in brain tissue. It has been suggested, on the basis of the bizarre effects of some related compounds, that the mental disease schizophrenia results from some abnormality in the metabolism of serotonin.

serotonin

bufotenine
mp 146°C

Bufotenine, an alkaloid isolated from toads and from toadstools (merely a coincidence), is a powerful hallucinogen. Note that it differs from serotonin, the normal brain component, only in having two methyl groups on nitrogen. Psilocybine and psilocine, two indole alkaloids from the Aztec sacred mushroom *Psilocybe mexicana*, have similar hallucinogenetic properties (Figure 11-7).

psilocybine

psilocine

Lysergic acid N,N-diethylamide (LSD, Chapter 10) is also an indole alkaloid.

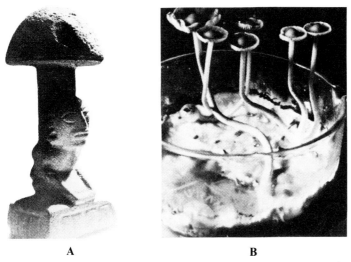

A   B

FIGURE 11-7. **(A)** A pre-Columbian carving representing the sacred mushroom. **(B)** *Psilocybe mexicana*. [From F. J. Ayd, Jr., and B. Blackwell (eds.), *Discoveries in Biological Psychiatry*, 1970, by permission of J. B. Lippincott Company]

THE TROPANE ALKALOIDS

The tropane alkaloids, which include various narcotics, hallucinogens, and violent poisons, take their name from the tropane ring system that they have in common as part of their structures.

tropane
bp 163°C

cocaine
mp 98°C

atropine, (±) form
mp 114°C; the (−)-form is called hyoscyamine

scopolamine or hyoscine
$[\alpha]_D^{20} = -28°$; monohydrate; mp 59°C; the (±) mixture is called atroscine

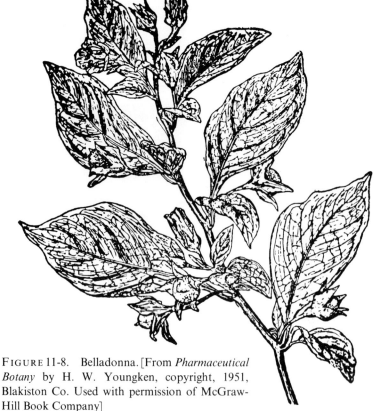

FIGURE 11-8. Belladonna. [From *Pharmaceutical Botany* by H. W. Youngken, copyright, 1951, Blakiston Co. Used with permission of McGraw-Hill Book Company]

Cocaine is found in the leaves of the coca plant, native to the Andes. Its hydrochloride is used as a surface anaesthetic, but it is an addictive drug and therefore dangerous. Several plants such as the deadly nightshade (*Atropa belladonna*, Figure 11-8), henbane, mandrake (Figure 11-9), and Jimson weed contain atropine, hyoscyamine, and scopolamine. Fatally poisonous in large doses, these compounds are hallucinogenic in lesser doses. They were used in medieval times in connection with the conjuration of demons and in prophecy and soothsaying.

Scopolamine has been used as a "truth drug" although the quality of the truths elicited may be doubted in view of the hallucinogenic properties of the drug. This compound also appears in the literature of forensic chemistry in the role of a poison; 25 mg was isolated from Mrs. Crippen's body. The name belladonna applied to the deadly nightshade refers to the fact that atropine applied to the eye (1 part in 130,000 parts of water) causes the pupil to dilate, giving the eye an attractive soft look.

FIGURE 11-9. Mandrake. [From J. W. Krutch, *Herbal*, Putnam, N.Y., 1965, and Pierandrea Mattioli, *Commentaries on the Six Books of Dioscorides*, Prague, 1563]

THE MORPHINE ALKALOIDS

Opium, the latex from the opium poppy, contains at least two dozen different alkaloids. About half of them are derivatives of benzylisoquinoline and are relatively innocuous. Papaverine is used as a smooth muscle relaxant and antispasmodic

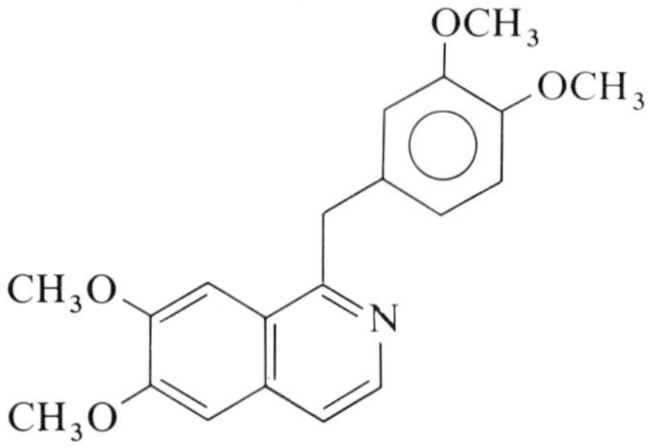

papaverine
mp 147°C

Morphine (Figures 11-10 and 11-11) is an addictive drug that produces drowsiness and sleep. An excellent pain reliever, its use is severely limited by the danger of addiction. Addiction to drugs such as morphine is not just the acquisition of a habit but rather a change in body chemistry, a physical dependence on the drug. A person who has become *physically dependent*[5] on a drug needs it just for the normal functioning of the body; for him it is like another vitamin or essential amino acid. This would not be so bad—after all, many people have to take medication all their lives—except that the "sustaining dose" increases steadily and so do the deleterious side effects. The addict does not have a long life expectancy.

Heroin, the diacetyl derivative of morphine, is more potent than the parent alkaloid and has similar effects. **Codeine** is a monomethyl ether of morphine in which the phenolic hydroxyl group has been

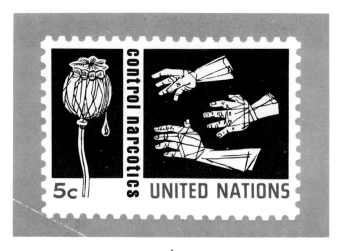

A

[5] Physical dependence can also occur with barbiturates (400 mg/day of pentobarbital) and even with alcohol.

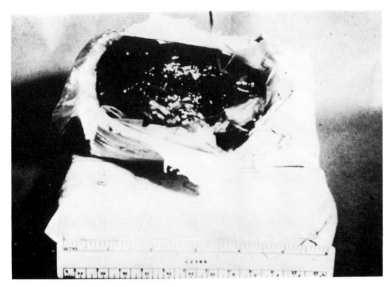
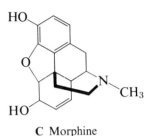

B Raw opium   **C** Morphine

FIGURE 11-10. Morphine is one of several alkaloids present in raw opium, the dried latex of the opium poppy, *Papaver somniferum*. Heroin, which does not occur naturally, is made by acetylating morphine. [**A**. Courtesy of the U.N. **B**. Courtesy of the Florida Crime Laboratory]

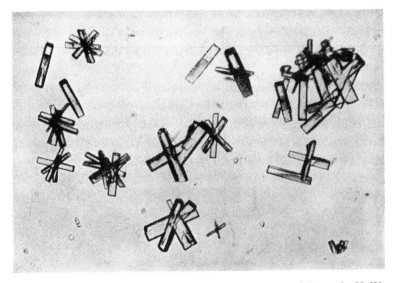

FIGURE 11-11. Morphine crystals. [From *Pharmaceutical Botany* by H. W. Youngken, copyright, 1951, Blakiston Co. Used with permission of McGraw-Hill Book Co.]

replaced by a methoxyl. **Thebaine** is the dimethyl derivative. Codeine is less addictive, and less euphoric, than either morphine or heroin. It is used in cough syrups.

Naturally, compounds that might have the analgesic and pain-relieving effects of morphine without its addictive effects have long been sought for. Heroin, which is now considered to be as bad as, or worse than, morphine, was once thought to be a useful way of weaning addicts from that narcotic. The search continues.

BARBITURATES

Barbituric acid is a pyrimidine derivative made by a reaction (equation 11-28) that you may recognize as being similar to the formation of an amide from an amine and an ester (equation 11-29).

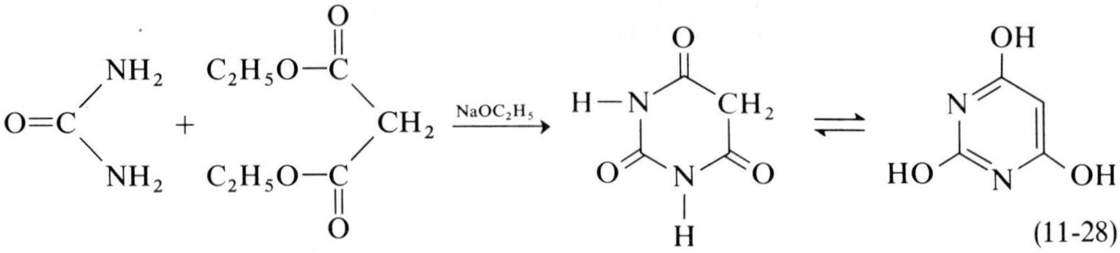

$$\text{urea} + \text{diethyl malonate} \xrightarrow{\text{NaOC}_2\text{H}_5} \text{barbituric acid (mp 248°C)} \rightleftharpoons \text{hydroxy form} \tag{11-28}$$

$$R-NH_2 + C_2H_5O-\overset{O}{\underset{\|}{C}}-R' \rightarrow R-\overset{H}{\underset{|}{N}}-\overset{O}{\underset{\|}{C}}-R' + C_2H_5OH \tag{11-29}$$

Barbituric acid exists partly in the amide form and partly in the hydroxy form. As an acid, it is somewhat stronger than acetic acid. The proton replaced when a salt is formed is one that is attached to the nitrogen in one isomeric form of the acid and to the oxygen in the other form.

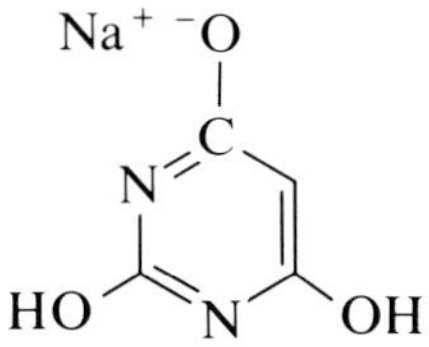

Derivatives of the amide form of barbituric acid in which one or both of the hydrogen atoms of the CH_2 group have been replaced by alkyl or aryl substituents are called barbiturates, a somewhat confusing practice that wrongly suggests that they might be salts or esters. The barbiturates are used as sedatives and soporifics in sleeping pills, but can be quite dangerous; overdoses lead to death and continual use leads to addiction. Phenobarbital is used to prevent epileptic seizures.

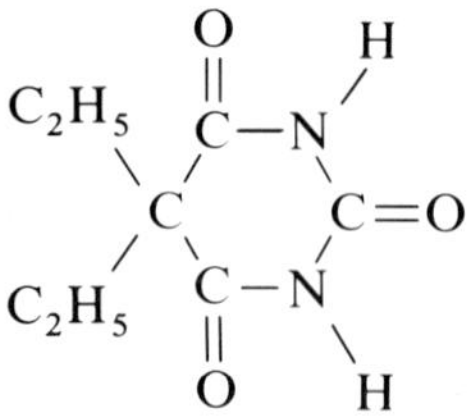

5,5-diethylbarbituric acid, barbital (generic name), or Veronal (proprietary name)

mp 188°C; moderately soluble in water; reacts with alkali to form a sodium salt

$$\text{phenobarbital or Luminal}$$

(structure: phenyl and CH₃CH₂ groups attached to C, linked via C–N(H)–C(=O)–N(H)–C(=O) ring)

phenobarbital or Luminal

In the language of the drug culture, barbiturates are "downers" and are used to counteract the effects of amphetamines or "uppers," and vice versa. The use of an amphetamine to get started in the morning and a barbiturate to get tranquilized at bedtime is a particularly dangerous and debilitating habit.

The Art of Synthesis

A considerable number of organic chemists have as their principal activity the laboratory synthesis of rather complicated natural products, such as the ones discussed in this chapter. There are several reasons for developing such syntheses:

1. If the synthetic product is made by well-known reactions and is identical in every way to the natural product, it proves the structure of the natural product. Of course, the choice of a structure to synthesize is not just a guess, but is guided by the reactions and spectroscopic properties of the natural product.
2. In some cases the laboratory synthesis can be made economical enough so that it is cheaper to make the natural product than it is to harvest it and extract it from a plant or animal.
3. Modifications of the synthesis might give products with properties more desirable than those of the natural product. For example, the natural product might have some medicinal property but accompanied by undesirable side effects. Variations on the structure would be tried in the hope of keeping the desired property and minimizing the side effects.
4. The same mysterious motive that causes some people to climb mountains or to make models of the Brooklyn bridge by glueing match sticks together causes others to undertake organic syntheses.

The people who synthesize complicated natural products have to be extraordinarily gifted chemists, both in planning the synthesis on paper and in the actual laboratory operations. In planning a synthesis, for example, it is necessary to choose the steps in a sequence such that incompatible groups are not present in the molecule at the same time. For example, the scheme

$$\text{HO}-\text{CH}_2\text{CH}_2\text{CH}_2-\text{Br} \xrightarrow{\text{Mg}} \text{HO}-\text{CH}_2\text{CH}_2\text{CH}_2-\text{MgBr} \xrightarrow{\text{CO}_2}$$
$$\text{HO}-\text{CH}_2\text{CH}_2\text{CH}_2-\text{COOH}$$

will not work because the hydroxyl group is acidic enough to destroy the Grignard reagent, just as water would.

Synthesis is a game in which all too often the card that turns up reads "Return to START, do not pass GO, do not collect two hundred dollars."

Once a feasible sequence of reactions has been invented, considerable skill in the actual laboratory work is also necessary because of the importance of having high yields in every step. For example, if a synthetic scheme has twenty steps, each with an 80% yield, the overall yield is $(0.8)^{20}$ which works out to only about 1%.

Figure 11-12 shows the sequence of steps used by R. B. Woodward and W. E. Doering in the synthesis of the antimalarial drug quinine. Although some of the reactions may be unfamiliar, the reader will recognize such processes as the formation of a C=N bond from the reaction of an aldehyde and an amine, catalytic hydrogenation, formation of an amide from an amine and an anhydride or acid chloride, oxidation of an alcohol to a ketone, formation of a quaternary ammonium salt, esterification and ester hydrolysis, resolution of a ($\pm$) mixture, and reduction of a ketone to an alcohol.

SUMMARY

A. Ammonia and amines
   1. Amines are classified as primary (one alkyl or aryl group), secondary (two such groups), or tertiary (three such groups).
   2. They are given names ending in -amine, i.e., methylamine, diethylamine.
   3. Amines can be prepared by
      (a) Reactions of ammonia with an alkyl halide or by the hydrolysis of an amide (Chapter 10). Secondary and tertiary amines can be made by reaction of an alkyl halide with a primary or secondary amine. Addition of a fourth alkyl group produces a quaternary salt.
      (b) The Hofmann reaction of an amide.
      (c) Reduction of a nitro compound, RNO_2, or a Schiff base (R_2C=NR'), or a nitrile, RC≡N.
   4. Primary amines add to aldehydes and ketones to give Schiff bases R_2C=NR'.
   5. Amines react with nitrous acid
      (a) Tertiary amines just give nitrite salts.
      (b) Secondary amines give nitrosamines, R_2NN=O.
      (c) Primary aliphatic amines give alcohols and alkenes (by way of unstable diazonium intermediates).
      (d) Primary aromatic amines give diazonium salts, $Ar\overset{+}{N}$≡N. These react with phenols to give azo compounds ArN=NAr' useful as dyes. This is an aromatic electrophilic substitution reaction.
B. Nitrogen heterocycles
   1. Five-membered rings
      (a) Pyrrole is an unsaturated, five-membered ring compound with one NH. It is not very basic. Pyrrole rings are found in the respiratory pigment hemin and in chlorophyll.

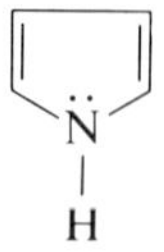

[OPPOSITE] FIGURE 11-12. The Woodward and Doering route to quinine.

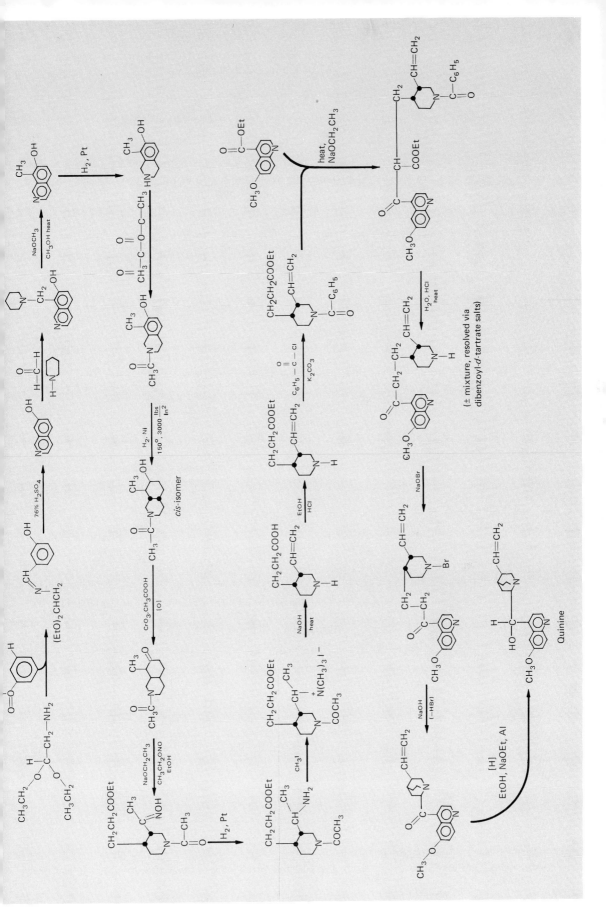

(b) Indole is a two-ring compound related to pyrrole. This ring system is found in various hallucinogens.

(c) Imidazole has two nitrogens in a five-membered ring. This ring system is found in physiologically active compounds such as histamine.

imidazole

2. Six-membered rings
   (a) Pyridine is a stable, aromatic compound. It is a base and forms salts with strong acids. The ring system is found in various alkaloids and in a vitamin nicotinic acid.

   (b) The saturated ring compound corresponding to pyridine is called piperidine. The ring system is found in the alkaloid coniine.

   (c) The fused ring compound with one benzene ring and a pyridine ring is called quinoline, from the occurrence of this ring system in the antimalarial drug quinine.

   (d) The purine ring system is found in nucleic acids and in alkaloids such as caffeine. It has four nitrogen atoms in two fused rings, one six-membered and the other five-membered.

   (e) The pteridine ring system has four nitrogen atoms in two fused six-membered rings. This ring system is found in butterfly wing pigments and in a vitamin.

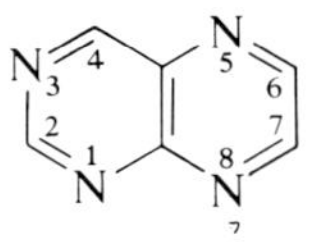

C. Nitrogen ring compounds or amines of the structural types already discussed are found in psychomimetic drugs, alkaloids, and barbiturates.

EXERCISES

1. Give the structures of triethylamine and triethylammonium ion and explain what makes triethylamine a base.

2. Classify the following as primary, secondary, or tertiary amines (or alcohols).

   (a) 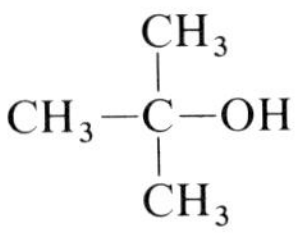 $CH_3-\underset{\underset{CH_3}{|}}{\overset{\overset{CH_3}{|}}{C}}-OH$

   (b) $CH_3-\underset{\underset{H}{|}}{\overset{\overset{CH_3}{|}}{C}}-\underset{\underset{H}{|}}{N}-H$

   (c) $CH_3-\underset{\underset{CH_3}{|}}{N}-CH_2-\underset{\underset{CH_3}{|}}{\overset{\overset{CH_3}{|}}{C}}-CH_3$

   (d) $CH_3-\underset{\underset{CH_3}{|}}{\overset{\overset{CH_3}{|}}{C}}-\ddot{N}H_2$

   (e) $CH_3-\underset{\underset{CH_3}{|}}{\overset{\overset{CH_3}{|}}{C}}-\underset{\underset{H}{|}}{\ddot{N}}-CH_3$

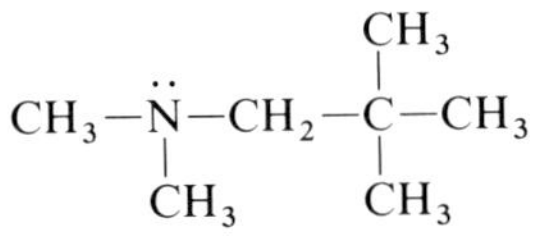

3. Complete the following reactions:

   (a) 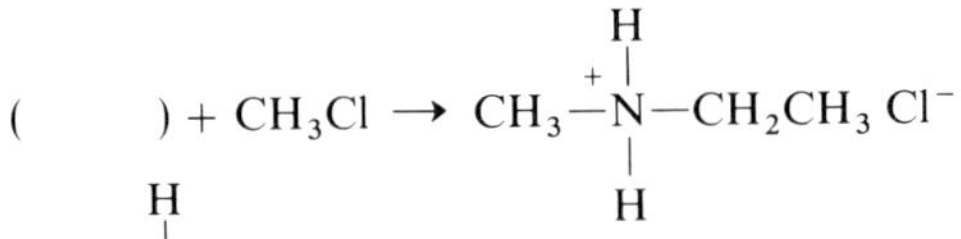 ( ) $+ CH_3Cl \rightarrow CH_3-\overset{+}{\underset{\underset{H}{|}}{\overset{\overset{H}{|}}{N}}}-CH_2CH_3\ Cl^-$

   (b) 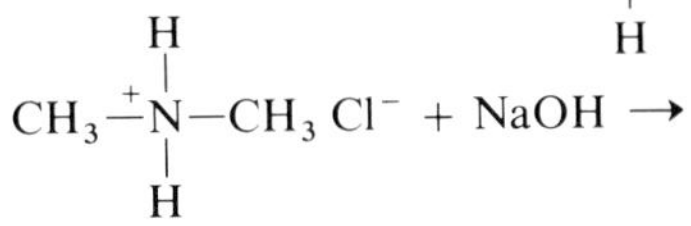 $CH_3-\overset{+}{\underset{\underset{H}{|}}{\overset{\overset{H}{|}}{N}}}-CH_3\ Cl^- + NaOH \rightarrow$

   (c) 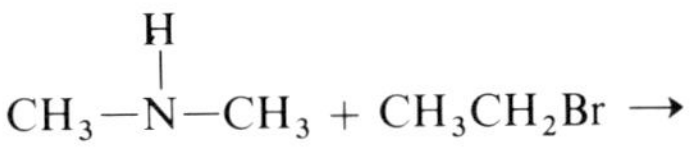 $CH_3-\underset{\underset{}{|}}{\overset{\overset{H}{|}}{N}}-CH_3 + CH_3CH_2Br \rightarrow$

   (d) 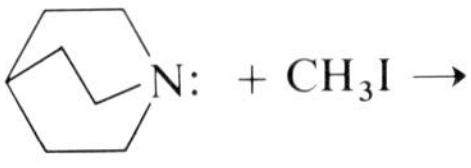 ⌬N: $+ CH_3I \rightarrow$

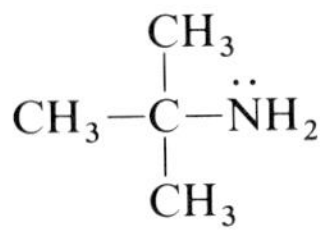

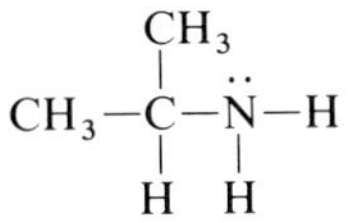

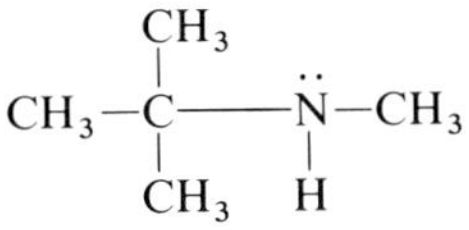

   (e) 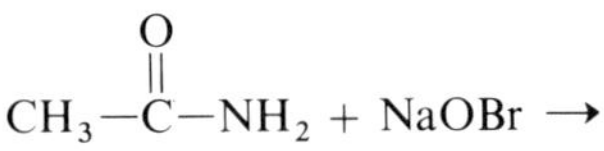 $CH_3-\overset{\overset{O}{\|}}{C}-NH_2 + NaOBr \rightarrow$

   (f) 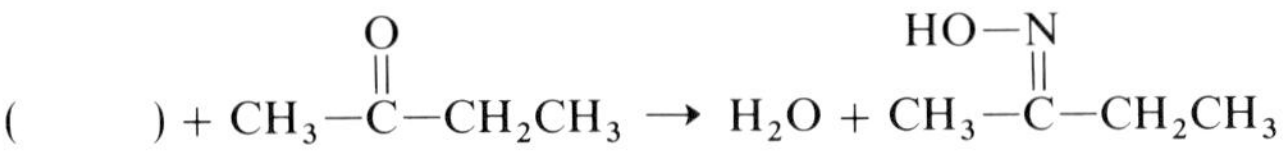 ( ) $+ CH_3-\overset{\overset{O}{\|}}{C}-CH_2CH_3 \rightarrow H_2O + CH_3-\overset{\overset{HO-N}{\|}}{C}-CH_2CH_3$

   (g) 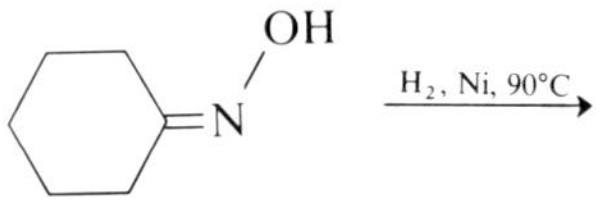 ⬡=N−OH $\xrightarrow{H_2,\ Ni,\ 90°C}$

(h)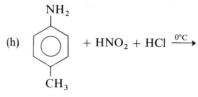

(i) 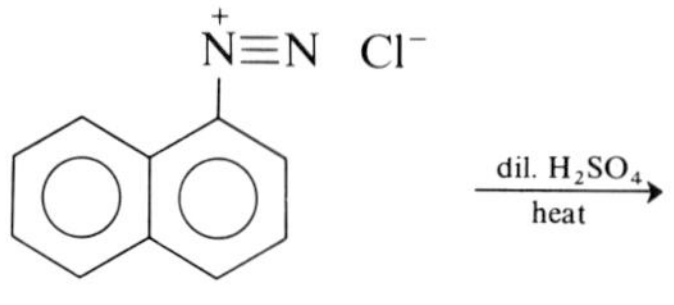 [naphthalene-N≡N⁺ Cl⁻] $\xrightarrow{\text{dil. H}_2\text{SO}_4, \text{ heat}}$

(j) $CH_3CH_2NH_2 + HNO_2 \rightarrow$

4. Write structures for
   (a) azobenzene
   (b) triethylamine
   (c) pyridine
   (d) piperidine
   (e) indole
   (f) pyrrole
   (g) 1,3-diazole
   (h) imidazole
   (i) nicotinic acid
   (j) quinoline
   (k) purine

5. Which is the stronger base, pyrrole or pyridine? Explain your answer.

6. Suggest a reason why pyridine is soluble in water whereas benzene is not.

7. (Open book) Give an example of
   (a) an alkaloid in pepper
   (b) a naturally occurring substance incorporating four pyrrole rings. What is its role in nature?
   (c) a naturally occurring derivative of imidazole
   (d) an antihistamine

8. (Open book) Identify any five of the following and either tell where it occurs, give a physiological property, or give its function in nature.

(a) 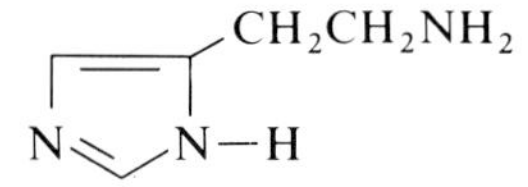 imidazole-CH₂CH₂NH₂

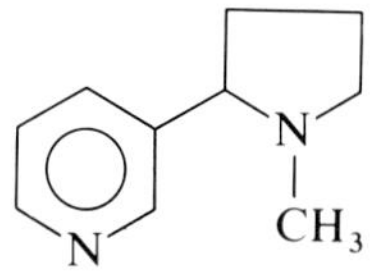

(b) 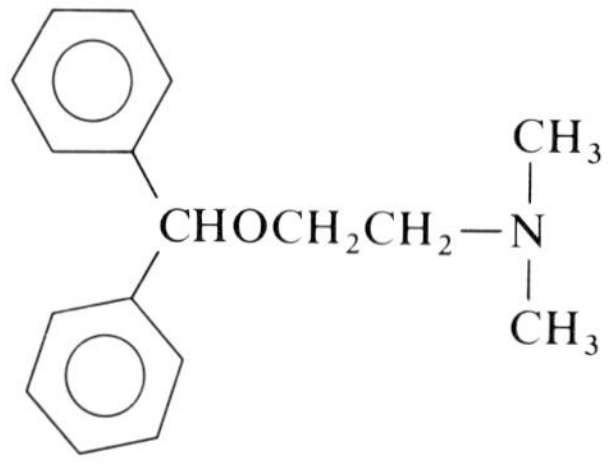 (C₆H₅)₂CHOCH₂CH₂—N(CH₃)₂

(c) nicotine (pyridine–N-methylpyrrolidine)

(d) 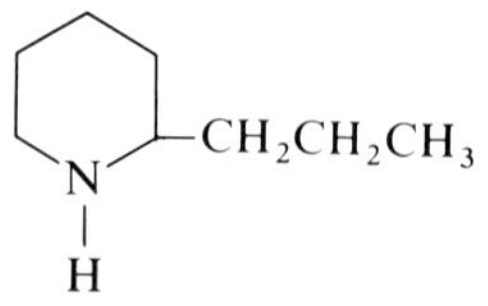 2-propylpiperidine

(e) 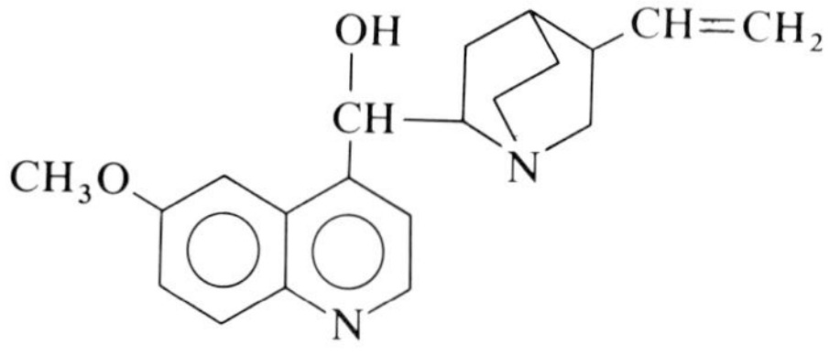 quinine-type structure (CH₃O-quinoline with CH(OH)–quinuclidine–CH=CH₂)

(f) [purine structure]   (g) [caffeine structure]

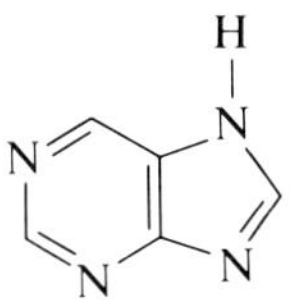

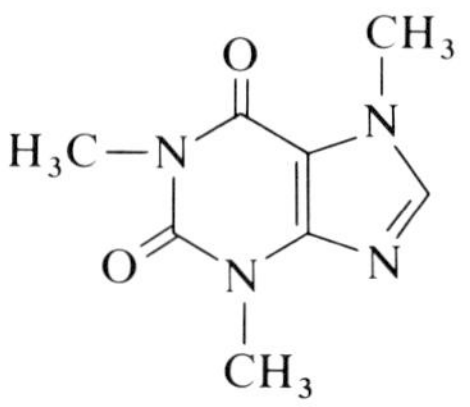

9. (Open book) Give the names, structures, and physiological properties of five substances that have emotional or mental effects.

10. Define (a) alkaloid, (b) sympathomimetic drug, and (c) amphetamine.

11. (Open book) Give three examples of the occurrence of the indole ring system in hallucinogenic drugs.

12. Match the items in column A with the appropriate item in column B.

A	B
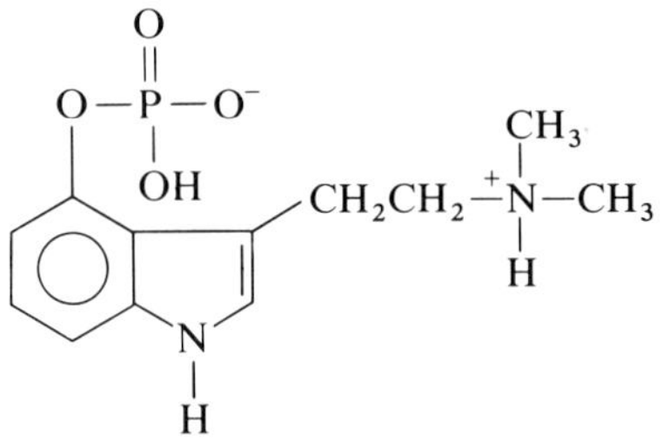 [psilocybin structure with phosphate group on indole and CH₂CH₂–N⁺(CH₃)₂H side chain]	LSD
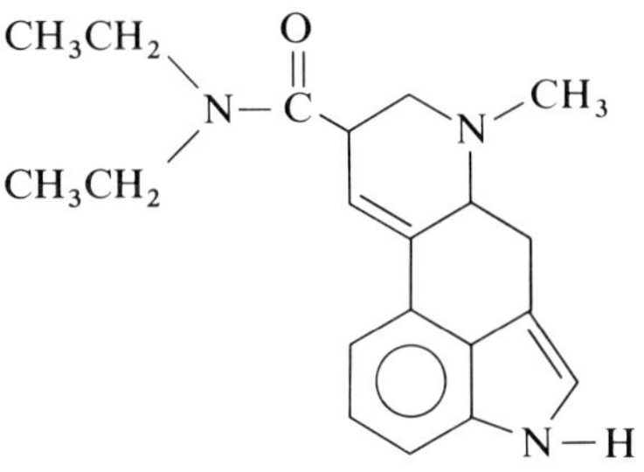 [lysergic acid diethylamide structure with (CH₃CH₂)₂N–C(=O)– group attached to ergoline ring system with N–CH₃ and N–H]	found in a mushroom
barbituric acid	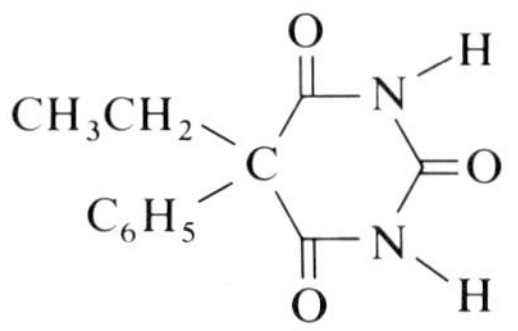 [phenobarbital structure: barbiturate ring with CH₃CH₂ and C₆H₅ substituents]
phenobarbital	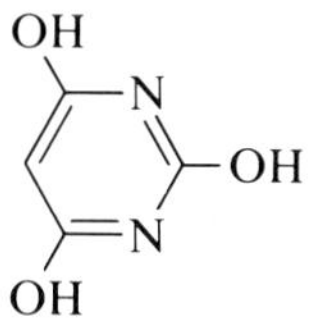 [2,4,6-trihydroxypyrimidine structure with OH groups]

Nitrogen Compounds

Chapter 12
Introduction to Large Molecules

The remaining chapters of this book will be concerned mainly with very large molecules, usually called macromolecules or polymers.[1] Polymer or macromolecular chemistry is important not only because of its industrial applications, to such things as paints and synthetic rubber, but even more because biological substances, such as proteins and nucleic acids are macromolecules.

For the most part, the chemistry of polymers and macromolecules is not fundamentally different from that of ordinary molecules, and it is possible to understand polymers on the basis of well-known general chemical principles, adding only a few new concepts and terms. Most macromolecules are actually fairly simple; they are big and composed of many parts, but the parts of a given macromolecule are usually very much alike.

Some New Terms

POLYMERS AND MONOMERS

The "poly" in **polymer** means many and the "mer" means part. The small molecules from which polymers are made are known as **monomers**. Thus polystyrene is a substance made by linking many styrene monomers together. A section of a polystyrene chain might be represented as in Figure 12-1, and the entire molecule by a

---

[1] For previous examples, see Chapter 4 (diamond), Chapter 6 (graphite), and Chapter 10, polyacrylic acid.

structural formula like the one shown below.[2]

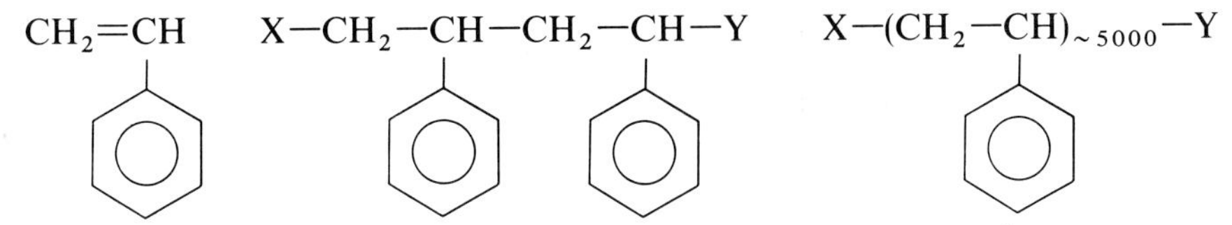

styrene
colorless liquid; bp 146°C

a dimer

polystyrene
colorless, highly transparent;
softens at about 90°C

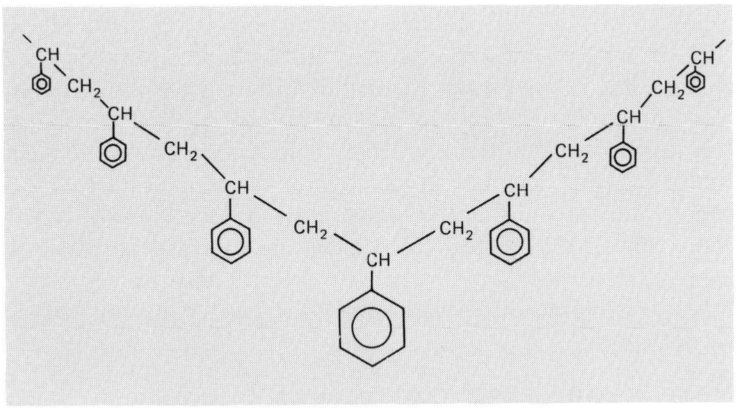

FIGURE 12-1. Part of a polystyrene chain.

A part structure that resembles the monomer

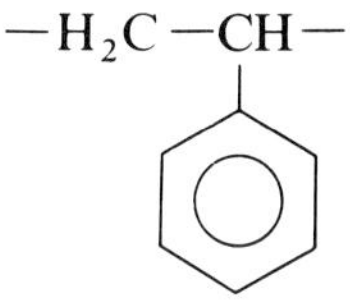

$$-H_2C-CH-$$

is called a **monomer unit**, in this example a styrene unit. The number of monomer units is called the **degree of polymerization**; in a typical polystyrene molecule the degree of polymerization might be about 5000 or so. Dimers and trimers are small molecules containing only two and three monomer units, whereas an **oligomer** contains some small but unspecified number of units.

Polystyrene occurs in nature as a component of styrax, the balsam of the sweet gum tree. Synthetic polystyrene is an example of a **thermoplastic** substance; that is, it softens when heated and can change its shape to conform to a mold, then when cooled it retains that shape. The process is reversible, so polystyrene powder or even fragments of discarded plastic can be used in molding.

Expanded polystyrene, a white, opaque material, is molded from especially prepared, minute polystyrene beads. The beads are made by polymerizing a mixture of styrene and pentane emulsified in water, and each bead contains some pentane trapped in its interior. When the beads are heated to the softening-point of polystyrene, about 90°C, the pentane vaporizes and the resulting

[2] The X and Y in these formulas are called end groups and are discussed on page 252.

pressure causes the beads to expand enormously. At the same time, the softened beads stick together to form the molded object. Because the beads are mostly just empty space, such objects are extremely light. For the same reason, expanded polystyrene is a good insulator for ice boxes. A dramatic demonstration of how little actual matter is in an object made of expanded polystyrene can be obtained by spilling some acetone on it; the parts hit by the acetone simply vanish as the polystyrene dissolves.

MOLECULAR WEIGHT

For a substance composed of small molecules to be considered pure, all the molecules must be alike or at least of the same size. Necessity has forced polymer chemists to have a more relaxed attitude: a pure sample of a polymer will invariably have molecules of more than one size. After all, what difference could it make if some of the molecules in a sample of polystyrene contained 5010 monomer units instead of 5000? On the other hand, it is worthwhile to differentiate between a sample in which most of the molecules contain about 2500 units and one in which most of the molecules contain about 5000 units. This is done by mentioning the **average molecular weight**. Solutions of polymers tend to be thick and viscous, and the solution is more viscous and syrupy the greater the average molecular weight.

END GROUPS AND CROSSLINKS

In our structural formula for polystyrene the ends of the chain were bonded to **end groups** X and Y whose nature was not specified. When the degree of polymerization or molecular weight is high, the nature of X and Y makes very little difference to the properties of the polymer since these end groups constitute such a tiny fraction of the molecule. In the case of polystyrene they are likely to be fragments of a molecule, called an **initiator**, that is added to the styrene to make it polymerize.

Benzoyl peroxide is a typical initiator used to polymerize monomers of the double bond or "vinyl" type. When the monomer, containing a trace of the peroxide, is heated or exposed to light, the peroxide decomposes into free radicals. In equation 12-1, the

$$\text{Ph-C(=O)-O-O-C(=O)-Ph} \\ \text{Ph-C(=O)-O:O-C(=O)-Ph} \xrightarrow{\text{heat or light}} 2\ \text{Ph-C(=O)-O}\cdot$$

(12-1)

formula for benzoyl peroxide was written twice, once in the usual way and once with the weak bond between the two oxygen atoms indicated by two dots. This is to serve as a reminder that a covalent bond consists of two electrons. Decomposition of the peroxide cleaves the O—O bond, and each fragment, or **benzoyloxy free radical**, has one of the original pair of electrons.

Free radicals are extremely reactive and, surrounded by styrene molecules, will add to form new radicals (equation 12-2). The new

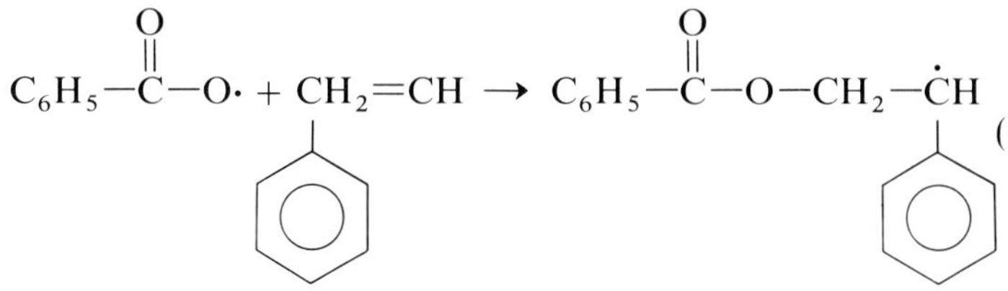

$$C_6H_5-\overset{O}{\underset{\|}{C}}-O\cdot + CH_2=\underset{\underset{C_6H_5}{|}}{CH} \rightarrow C_6H_5-\overset{O}{\underset{\|}{C}}-O-CH_2-\underset{\underset{C_6H_5}{|}}{\dot{C}H} \quad (12\text{-}2)$$

radical is still just as reactive, and is also surrounded by styrene molecules; therefore, it adds again (equation 12-3). This process

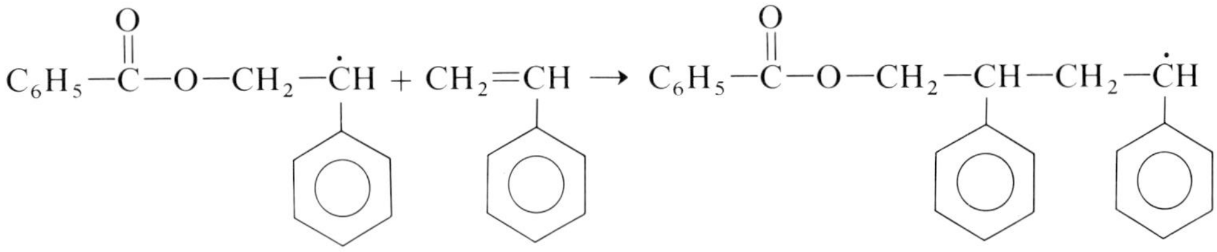

$$C_6H_5-\overset{O}{\underset{\|}{C}}-O-CH_2-\underset{\underset{C_6H_5}{|}}{\dot{C}H} + CH_2=\underset{\underset{C_6H_5}{|}}{CH} \rightarrow C_6H_5-\overset{O}{\underset{\|}{C}}-O-CH_2-\underset{\underset{C_6H_5}{|}}{CH}-CH_2-\underset{\underset{C_6H_5}{|}}{\dot{C}H}$$
$$(12\text{-}3)$$

continues and forms a long chain until two of the radicals (having used up most of the styrene) come close together. If they react by combination to form a new C—C bond, the result is a polystyrene molecule in which the end groups are both benzoic acid esters (equation 12-4).

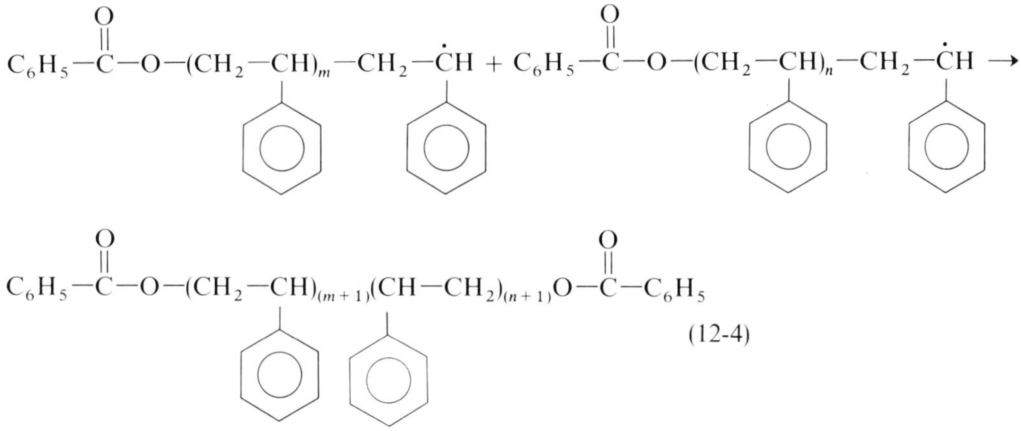

$$C_6H_5-\overset{O}{\underset{\|}{C}}-O-(CH_2-\underset{\underset{C_6H_5}{|}}{CH})_m-CH_2-\underset{\underset{C_6H_5}{|}}{\dot{C}H} + C_6H_5-\overset{O}{\underset{\|}{C}}-O-(CH_2-\underset{\underset{C_6H_5}{|}}{CH})_n-CH_2-\underset{\underset{C_6H_5}{|}}{\dot{C}H} \rightarrow$$

$$C_6H_5-\overset{O}{\underset{\|}{C}}-O-(CH_2-\underset{\underset{C_6H_5}{|}}{CH})_{(m+1)}(CH-CH_2)_{(n+1)}O-\overset{O}{\underset{\|}{C}}-C_6H_5$$
$$(12\text{-}4)$$

A **crosslink** in a polymer is a bond or a branch of a polymer chain that links two polymer chains together (Figure 12-2). Polystyrene, for example, can be crosslinked by mixing the original styrene with a little divinylbenzene. When this mixture polymerizes, some of the benzene rings of the polymer have vinyl groups that can grow on their own; these produce branches of the chain or are incorporated into another chain, linking the two together.

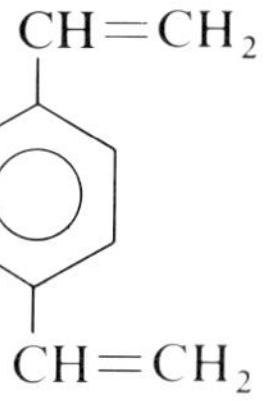

divinylbenzene

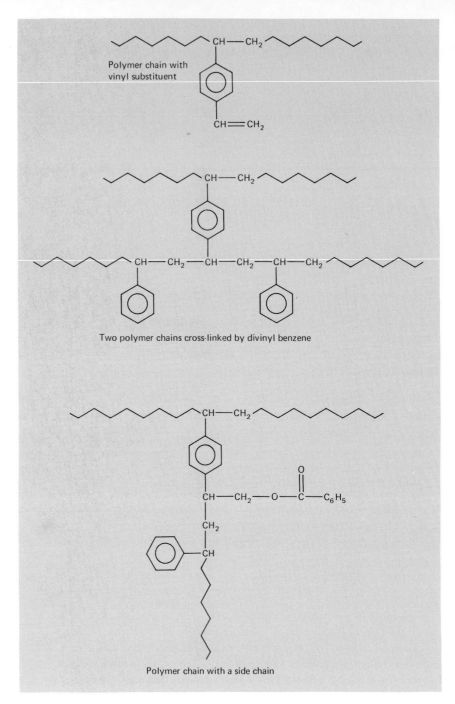

FIGURE 12-2. Crosslinking and branching in a styrene-divinyl-benzene copolymer.

As little as 0.01 % of divinylbenzene will change polystyrene from a substance that is readily soluble in benzene to a substance that is totally insoluble. In effect, the crosslinking has converted the polymer molecules into a single molecule that just sits there. If the amount of crosslinking is small enough, benzene will diffuse into the polymer and make it swell, but it still does not dissolve. Other examples of crosslinking will be encountered in connection with the vulcanization of rubber and in connection with DNA (Chapter 15).

A **thermosetting** plastic is one that can be made to soften and flow when it is first heated but which solidifies on prolonged heating (curing). It consists of a partially polymerized material still containing some unreacted monomer and also some unreacted crosslinking agent.

ASYMMETRY IN POLYMERS

In Chapter 9 it was stated that compounds with a single asymmetric carbon atom can occur in isomeric R and S forms called enantiomers. It was also noted that the presence of more than one asymmetric carbon atom in a molecule gives rise not only to enantiomeric forms but also to other isomers called diastereomers. Although most of the physical properties of a pair of enantiomers are identical, the physical properties of diastereomers can be quite different. The same is true in polymer chemistry.

Consider, for example, the polystyrene chain whose structure is given below.

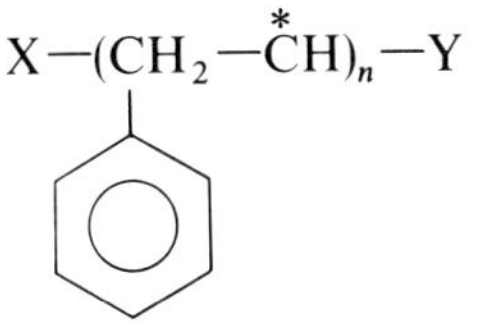

$$X-(CH_2-\overset{*}{C}H)_n-Y$$
$$|$$
$$\phi$$

If the chain has n asymmetric carbon atoms there will be 2^n optical isomers of which $2^n/2$ are diastereomers, that is, about 1 followed by 1500 zeroes when $n = 5000$. Of these 10^{1500} isomers, most will resemble each other rather closely in their physical properties and it is not sensible to expect to separate such a horrible mixture in any case. However it is worthwhile to distinguish between several *classes* of polymer that differ only in the configuration of their optically active carbons. This kind of difference between two polymers is known as a difference in **tacticity**.

The most complicated kind of asymmetric polymer is an **atactic** polymer (Figure 12-3). In an atactic polymer the sequence of R and S configurations is completely random, and a sample of the polymer will be a mixture of all the possible isomers.[3]

The simplest kind is an **isotactic** polymer (Figure 12-4). In any given chain of an isotactic polymer, the asymmetric carbon atoms are either all R or all S. There are big differences in physical properties between isotactic and atactic polymers. The isotactic polymers tend to crystallize more readily, have higher softening points, and

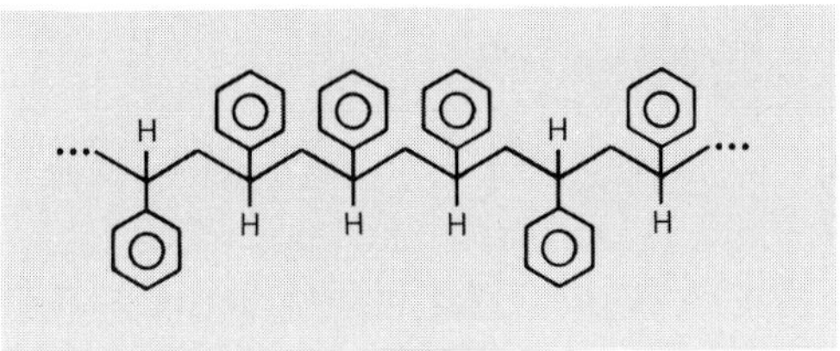

FIGURE 12-3. Atactic polystyrene, amorphous liquefies a little above 100°C.

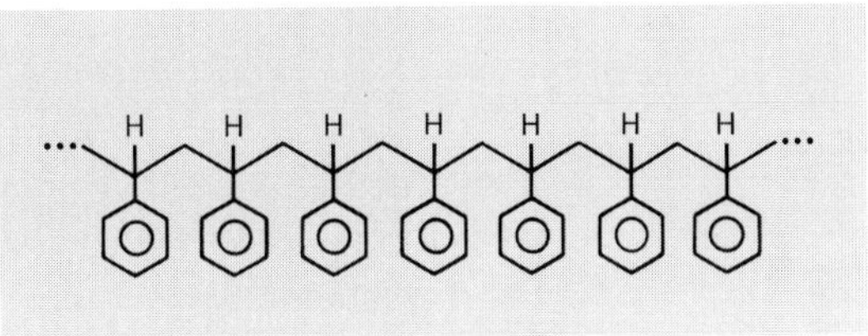

FIGURE 12-4. Isotactic polystyrene, crystalline, melts above 200°C.

[3] Provided of course that it contains enough molecules!

are mechanically much stronger. The polypeptide and nucleic acid macromolecules found in living organisms are all isotactic. As will be seen in later chapters, the molecules of these substances readily adopt orderly conformations, such as the α-helix of some proteins and the double helix of chromosomes, that appear to be essential to the mechanics of life.

Atactic polystyrene is made by means of free radical initiators in reactions like the one already described. Isotactic polystyrene is made using a **Ziegler-Natta catalyst**, such as a mixture of $TiCl_4$ and $Al(CH_2CH_3)_3$. It is generally believed that the polymer chain grows on the surfaces of these catalysts and that the surface serves to orient each new monomer unit in such a way that if the last one

FIGURE 12-5. Models of chain-segments of (**A**) atactic and (**B**) isotactic polypropylene.

took up an R configuration the new one will also. The process may be compared to the growth of a crystal in which each molecule has to be oriented in a particular way in order to fit into the crystal lattice.

A syndiotactic polymer is one in which the R and S configurations alternate.

Vinyl Polymers

POLYETHYLENE

Ethylene can be polymerized to a long-chain polymer, polyethylene, by several different catalysts; peroxides like benzoyl peroxide and more complicated catalysts made from aluminum alkyls and titanium tetrachloride are all effective. The polymer is a waxy-looking, translucent rather than transparent solid familiar to everyone from its use in kitchenware. The infrared spectrum of polyethylene is essentially that of the CH_2 group and is almost the same as the spectrum of cyclooctane.

$$\diagdown CH_2 \diagup CH_2 \diagdown CH_2 \diagup CH_2 \diagdown CH_2 \diagup CH_2 \diagdown CH_2 \cdots$$
polyethylene

$$\diagdown CH_2 \diagup CH_2 \diagdown CH_2 \diagup CH_2 \diagdown CH_2 \diagup CH_2 \diagdown CH \cdots$$
substituted polyethylene
(with R substituents)

POLYPROPYLENE

A number of substituted ethylenes, $CH_2{=}CHR$, can be polymerized. One example, styrene or vinyl benzene, has already been discussed. Others are polypropylene, polyvinyl chloride, polyvinyl acetate, and polymethacrylate.

Polypropylene is made from propylene using special catalysts (equation 12-5) so as to produce the isotactic form of the polymer. It is more flexible than polyethylene, and a strip of polypropylene can be used as a hinge. Some articles that were previously made of

$$CH_2{=}CH \;\; \longrightarrow \;\; (-CH_2-CH-)_n \quad\quad (12\text{-}5)$$
$$\;\;\;\;\;|\;|$$
$$\;\;\;CH_3\;\;\;\;\;\;\;\;\;\;\;\;\;\;\;\;CH_3$$

two pieces joined by a hinge are now molded from polypropylene as a single piece. A typical sample of polypropylene starts to soften and melt at about 120°C in contrast to polyethylene which begins to soften at about 85°C (very hot dish water).

POLYVINYL CHLORIDE

Polyvinyl chloride (PVC) is hard and tough and is extensively used for sewer, water, and gas pipes, raincoats, and a lot of other things. It is usually opaque because of the presence of fillers or of barium salts added to stabilize it against sunlight. PVC can be welded by means of a stream of hot gas and a PVC welding rod. Soft objects can be molded by heating a **plastisol** consisting of a paste of polyvinyl chloride particles mixed with a liquid **plasticizer** such as dimethyl phthalate.

POLYVINYL ACETATE

Polyvinyl acetate softens at too low a temperature to be used in the manufacture of molded objects, but it is used as an adhesive and a "permanent starch" for cloth. Vinyl acetate can be made by adding acetic acid to acetylene.

$$H-C\equiv C-H + HO-\overset{O}{\underset{\|}{C}}-CH_3 \xrightarrow{HgSO_4} H-\overset{H}{\underset{|}{C}}=\overset{H}{\underset{|}{C}}-O-\overset{O}{\underset{\|}{C}}-CH_3$$

vinyl acetate
bp 71°C at 728 mm

$$CH_2=CH \rightarrow X-(CH_2-CH)_n-Y$$
$$\underset{\underset{\underset{CH_3}{|}}{\underset{C=O}{|}}}{O} \qquad \underset{\underset{\underset{CH_3}{|}}{\underset{C=O}{|}}}{O}$$

polyvinyl acetate

A solution of polyvinyl acetate in acetone can be used to blow bubbles like soap bubbles, except that they can be larger (up to 130 cm) and last longer. As the acetone evaporates, the polyvinyl acetate forms a plastic film.

Polyvinyl alcohol cannot be made by polymerizing vinyl alcohol, because that compound does not exist.[4] However, it can be made by methanolysis of polyvinyl acetate (equation 12-6). Unlike the

$$X-(CH_2-CH)_n-Y + CH_3OH \xrightarrow{NaOCH_3} X-(CH_2-CH)_n-Y + CH_3-\overset{O}{\underset{\|}{C}}-OCH_3$$
$$\underset{\underset{\underset{CH_3}{|}}{\underset{C=O}{|}}}{O} \qquad \qquad \underset{OH}{}$$

polyvinyl alcohol

(12-6)

polymers described so far, polyvinyl alcohol is hydrophilic because of its many hydroxyl groups. Thus, it will dissolve in water but repel gasoline. It is used to make water-soluble packages, as a thickener for aqueous solutions, and as an adhesive.

Since polyvinyl alcohol is an alcohol, it will react with aldehydes to form polyvinyl acetals. An important example is polyvinyl butyral, used as the plastic between the layers of glass in shatterproof

[4] See Chapter 7. It is the enol form of acetaldehyde.

windshields. It is made from butyraldehyde and polyvinyl alcohol.

$$\sim CH_2-CH-CH_2-CH-CH_2-CH-CH_2-CH-CH_2-CH\sim$$

polyvinyl butyral

POLYACRYLATES

"Acrylic" polymers are made from esters of acrylic acid and methacrylic acid, and from acrylonitrile.

$$\begin{array}{ccc} CH_2=CH & CH_3 & CH_2=CH \\ | & | & | \\ C=O & CH_2=C & CN \\ | & | \\ OCH_3 & C=O \\ & | \\ & OCH_3 \end{array}$$

methyl acrylate   methyl methacrylate   acrylonitrile

bp 80.5°C; very slightly soluble in water   bp 100°C; very slightly soluble in water   bp 78°C; soluble in water

Polymethyl methacrylate is a highly transparent material sold under the name of Plexiglas and used for boat windshields and bubble canopies. Its great transparency and high refractive index make it ideal for piping light beams around corners and into otherwise inaccessible places. Light put into the edge of a sheet of polymethyl methacrylate will emerge wherever the surface is roughened; this is used to make illuminated designs on the plastic.

$$X-(CH_2-\underset{\underset{\underset{CH_3}{|}}{\underset{|}{O}}}{\overset{\overset{\overset{CH_3}{|}}{|}}{C}})-Y$$

polymethyl methacrylate

atactic; becomes rubbery at about 110°C; decomposes at 160°C

Polymethyl methacrylate sheets are made by polymerizing a syrup consisting of partly polymerized monomer between two sheets of heat resistant glass kept apart by a polyvinyl alcohol gasket. Rods are made by filling aluminum tubes with the syrup and heating. Polymethacrylate tubes are made by rotating a partly filled horizontal aluminum tube so that the syrup coats the wall as it polymerizes.

The sheets, rods, and tubes can easily be machined and cut with ordinary tools, but it is important not to let the tool get too hot. Drilling or cutting a piece of acrylic plastic usually generates the characteristic odor of the monomer. The reason is that high temperatures favor a "cracking" or decomposition reaction (Figure 12-6).

Another way of making things from polymethyl methacrylate sheet or rod is to use its property of turning rubbery at temperatures a little above 100°C.

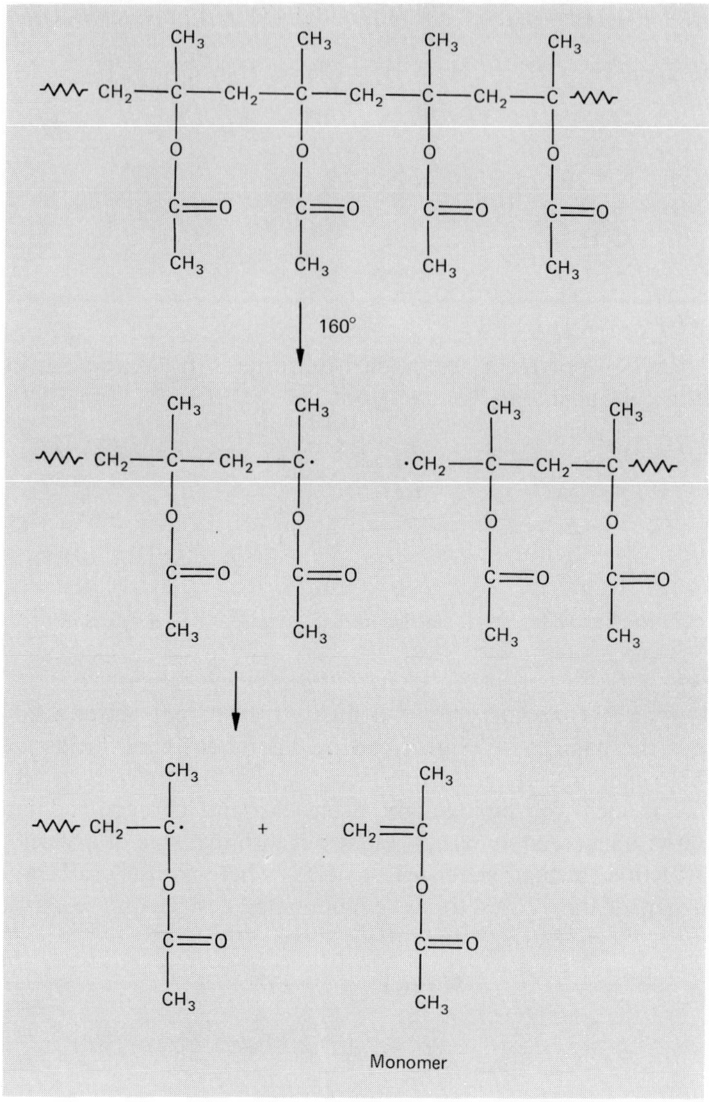

FIGURE 12-6.

The piece is heated over a blow torch (putting out the resulting fire if necessary) until it is rubbery. Then it is bent or twisted to the desired shape and held that way in a bucket of water until it cools.

Polyacrylonitrile is only slightly thermoplastic and difficult to mold, but **copolymers** of acrylonitrile and other vinyl monomers are widely used. To make a copolymer, the monomers are mixed before polymerization. The resulting polymer has the general structure shown on page 257, but with more than one kind of substituent group. Examples are styrene-acrylonitrile copolymers and acrylonitrile-butadiene-styrene (ABS).

POLYTETRAFLUOROETHYLENE

Tetrafluoroethylene polymerizes spontaneously and with violence unless it contains an inhibitor to intercept the growing chains.

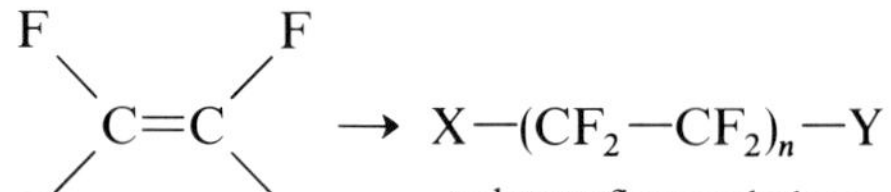

$$\begin{array}{c}F\\ \diagdown\\ \end{array}\!\!C\!=\!C\!\!\begin{array}{c}\\ \diagup\\ F\end{array}\quad\longrightarrow\quad X\!-\!(CF_2\!-\!CF_2)_n\!-\!Y$$

polytetrafluoroethylene
or Teflon
mp 327°C

This polymer has unusual properties because of the stability and unreactivity of the C—F bond. It can withstand high temperatures and is insoluble in almost all solvents except warm fluorocarbons.

As one might expect from the fact that polyfluoroethylene does not dissolve in most substances, it does not stick to them either. The forces that cause different solids to stick to each other are the same as the forces that cause

TABLE 12-1. *Common Names and Trade Names of Some Polymers*

Vinyl Polymers	
ABS	acrylonitrile-butadiene-styrene copolymer
Acrylic	any polymer based on acrylic or methacrylic acid esters or acrylonitrile, most often polymethyl methacrylate
Styrofoam	expanded polystyrene
Formvar	polyvinyl formal
Lucite	polymethyl methacrylate
Perspex	polymethyl methacrylate
Plexiglas	polymethyl methacrylate
PVC	polyvinyl chloride
Saflex	polyvinyl butyral
Teflon	polytetrafluoroethylene or polytetrafluoroethylene-polyhexafluoropropylene copolymer
Orlon	polyacrylonitrile
Marlex	polypropylene
Geon	polyvinyl chloride
Saran	vinylidene chloride ($CH_2\!=\!CCl_2$)-vinyl chloride or acrylonitrile copolymer
Styron	polystyrene
Cycolac	ABS
Polythene	polyethylene
Diene Polymers	
Rubber	*cis*-1,4-polyisoprene
Gutta-percha	*trans*-1,4-polyisoprene
Chicle	mixed *cis*- and *trans*-1,4-polyisoprene
Neoprene	1,4-poly-2-chloro-1,3-butadiene
Polyacetals	
Delrin	polyoxymethylene (acetate terminal)
Polyethers	
Carbowax	polyethylene oxide
Polyesters	
Glyptal	glycerol-phthalic anhydride
Laminac	a laminating resin, unsaturated polyester, crosslinks with vinyl monomer as it cures
Mylar, Dacron, Fortral, Terylene	polyethylene terephthalate

molecules to dissolve in liquids composed of other molecules. The non-stickiness of a polyfluoroethylene surface means that it is also a slippery surface; therefore it can be used to make bearings that need no lubrication. It is also used to make smoothly acting disc drags for fishing reels. Even epoxy glue will not stick to a polyfluoroethylene surface unless the fluorocarbon polymer has been treated with a solution of sodium metal in liquid ammonia, one of the few things with which it will react. Polytetrafluoroethylene, made by some secret art to stick to a frying pan, prevents the frying pan from sticking to a fried egg.

Diene Polymers and Rubber

When La Condamine visited the Amazon basin in 1731, he found that the natives of that region were using rubber-coated cloth and rubber-coated shoes.

Crude rubber is obtained by cutting the bark of the *Hevea brasiliensis* tree, whereupon a milky emulsion of rubber in water, called **latex**, oozes out. The latex is coagulated into a cheeselike material that is either rolled into sheets of crepe rubber or gathered into balls. Crude rubber is called **caoutchouc** (pronounced like a sneeze), the closest European approximation to an Indian word meaning "tears of the wood."

THE STRUCTURE OF RUBBER

The odor of burning rubber is due largely to **isoprene**, which is formed in fairly good yield when rubber is heated. Rubber contains

$$(C_5H_8)_{1500-6000} \xrightarrow{heat} \underset{\text{isoprene}}{CH_2=C(CH_3)-CH=CH_2}$$

rubber

colorless liquid; bp 34°C

double bonds as can be shown by its uptake of bromine, about one C=C for every C_5H_8, or isoprene, unit. X-ray and infrared studies show that the isoprene units are linked together at carbon 1 and carbon 4 and that the double bond is located in the middle.

$$X-\underset{1234}{(CH_2-C(CH_3)=CH-CH_2)_n}-Y$$

The fact that various compounds will add to isoprene in the 1 and 4 positions suggests that a rubber molecule grows by a 1,4 addition of isoprene monomer units. A similar 1,4 addition reaction is given by bromine (equation 12-7). A catalyst such as benzoyl peroxide will

$$Br_2 + CH_2=C(CH_3)-CH=CH_2 \rightarrow Br-CH_2-C(CH_3)=CH-CH_2Br \quad (12\text{-}7)$$

$$\sim\cdot + CH_2=C(CH_3)-CH=CH_2 \rightarrow \sim CH_2-C(CH_3)=CH_2-CH_2\cdot$$

$$\sim CH_2-\underset{\underset{CH_3}{|}}{C}=CH-CH_2\cdot + CH_2=\underset{\underset{CH_3}{|}}{C}-CH=CH_2 \rightarrow$$

$$\sim CH_2-\underset{\underset{CH_3}{|}}{C}=CH-CH_2-CH_2-\underset{\underset{CH_3}{|}}{C}=CH-CH_2\cdot \text{ etc.}$$

initiate the polymerization of isoprene to a polymer, but this substance is quite unlike natural rubber. The difficulty is that all of the double bonds in natural rubber have the cis configuration[5] and ordinary catalysts tend to produce a random mixture of cis and trans double bonds. It was only after the discovery of the Ziegler-Natta catalysts, which, it will be recalled, are also used to make isotactic vinyl polymers, that a material essentially identical to natural rubber could be made synthetically.

$$\ldots CH_2 \diagdown \underset{\underset{}{}}{\overset{\overset{CH_3}{|}}{C}=CH} \diagdown CH_2 \diagup CH_2 \diagdown \underset{\underset{CH_3}{|}}{C}=CH \diagdown CH_2 \diagup CH_2 \diagdown \underset{}{\overset{\overset{CH_3}{|}}{C}= CH} \diagdown CH_2 \ldots$$

cis-1,4-polyisoprene or natural rubber

The elasticity of lightly crosslinked rubber and rubberlike materials is due to the fact that the polymer can exist in two states. *Unstretched* rubber is amorphous, and the polymer molecules are arranged in disorderly coils. This lack of order shows up as an extreme fuzziness in the x-ray diffraction pattern of unstretched rubber. *Stretched* rubber, on the other hand, is crystalline even though it does not look like it. The x-ray diffraction pattern of stretched rubber is sharp, indicating a very orderly or crystalline packing of the atoms. In fact, there is a regular repeating pattern every 9.13×10^{-8} cm corresponding to the length of two isoprene units. Because the isoprene units are all lined up in crystalline rubber, instead of randomly coiled, the polymer molecule is longer in the crystalline or stretched state than in the relaxed or amorphous state (Figure 12-7).

The fact that stretched rubber will shorten and return to the amorphous form spontaneously, and the fact that it takes mechanical work to stretch it again, is an example of the workings of the second law of thermodynamics. Although usually presented in a more inscrutable form, all that the second law really says is that disorder and mess tend to arise spontaneously and that order is restored only by doing work, a fact well known to every housewife. On a molecular level, this phenomenon manifests itself as the force of rubberlike elasticity.

Vulcanization and Crosslinking. Crude rubber is a soft sticky material that becomes even gummier in hot weather. In the vulcanization process, discovered by Charles Goodyear in 1838, the rubber is crosslinked by heating it with sulfur (Figure 12-8). If only a few crosslinks are added (3% sulfur), the polymer is soft and still exhibits rubberlike elasticity. With more crosslinks it becomes harder. Large quantities of sulfur, in the neighborhood of 68%, convert the rubber into a black solid, called ebonite, or hard rubber. Ebonite was one

---

[5] The all-trans material, called gutta percha, also occurs naturally. It is used as the cover of golf balls.

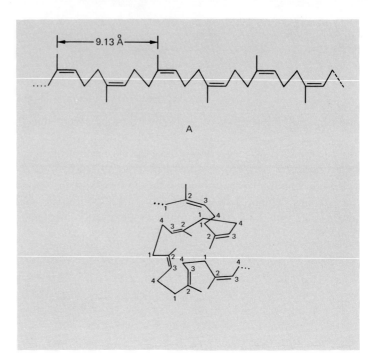

FIGURE 12-7. (**A**) Crystalline rubber (stretched). (**B**) Amorphous rubber (relaxed).

of the first thermosetting plastics; it is still used for making such things as automobile storage batteries, for example.

The latex of *Achras sapota*, a West Indian and Central American plant, provides a thermoplastic material called chicle. Among other things chicle contains mixed *cis*- and *trans*-polyisoprenes; it is used to make chewing gum.

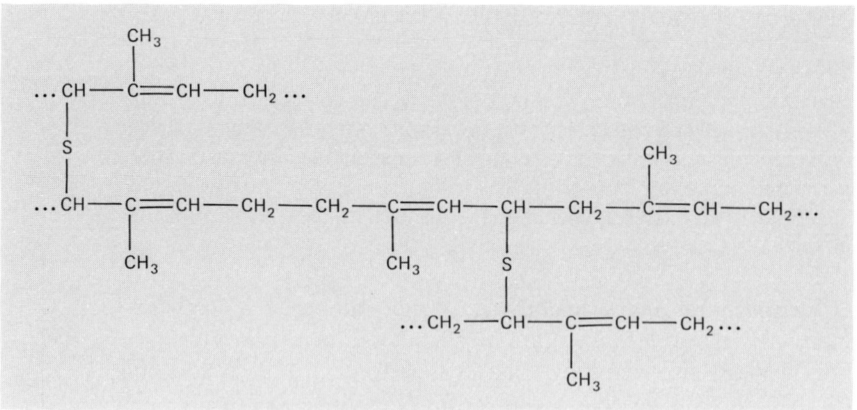

FIGURE 12-8. Vulcanized rubber.

SYNTHETIC ELASTOMERS

Neoprene is a synthetic rubber or **elastomer** made from chloroprene, essentially isoprene with a chlorine atom in place of the methyl group. Neoprene is vulcanized or crosslinked by heating

$$\underset{\substack{\text{2-chloro-1,3-butadiene} \\ \text{or chloroprene}}}{CH_2\!=\!CH\!-\!\underset{\underset{Cl}{|}}{C}\!=\!CH_2} \;\rightarrow\; \underset{\text{neoprene rubber}}{X\!-\!(CH_2\!-\!\underset{\underset{Cl}{|}}{C}\!=\!CH\!-\!CH_2)_n\!-\!Y}$$

with zinc oxide, which converts the chloro substituents to ether linkages.

$$\begin{array}{c}\ldots CH_2\!-\!\underset{\underset{\substack{+\\ZnO\\+\\Cl\\|}}{|}}{\underset{Cl}{C}}\!=\!CH\!-\!CH_2\ldots\\ \ldots CH_2\!-\!C\!=\!CH\!-\!CH_2\ldots\end{array} \;\rightarrow\; ZnCl_2 \;+\; \begin{array}{c}\ldots CH_2\!-\!\underset{|}{C}\!=\!CH\!-\!CH_2\ldots\\ O\\ |\\ \ldots CH_2\!-\!C\!=\!CH\!-\!CH_2\ldots\end{array}$$

The chlorine atoms in place of the methyl groups of rubber make neoprene less hydrocarbonlike and reduce the affinity of the polymer for alkanes. Natural rubber in contact with gasoline swells badly and deteriorates; neoprene does not, in fact neoprene is used for the hoses on gasoline pumps.

Other synthetic elastomers of the vinyl or diene polymer type are copolymers of butadiene with styrene or acrylonitrile, and copolymers of ethyl acrylate with chloroethyl vinyl ether. As we shall see in later sections, polymers of quite different structure can also have rubberlike elasticity.

Oxygen-Containing Chains

The polymers discussed so far in this chapter have all been molecules whose backbones consisted of an unbroken chain of carbon atoms. Many polymers, such as the naturally occurring polypeptides (Chapter 14) and polysaccharides (Chapter 13), contain other atoms such as nitrogen or oxygen in the chain. This structural feature makes the polymer more susceptible to reactions that break it up into smaller molecules, since $C-O$ and $C-N$ bonds are subject to a greater variety of reactions than are $C-C$ bonds. In particular, polypeptides and polysaccharides are easily biodegradable.

POLYACETALS AND POLYETHERS

If compounds with $C=C$ double bonds can be polymerized to form long-chain molecules, it seems logical that compounds with double bonds to oxygen might be able to do the same thing. The **polyacetals** are examples of such polymers.

When a 60% aqueous solution of formaldehyde containing 2% sulfuric acid is distilled, the distillate contains a cyclic trimer, 1,3,5-trioxane (equation 12-8). Aqueous solutions of formaldehyde slowly deposit **paraformaldehyde**, a "low" polymer whose typical formula is $HO-(CH_2-O)_{30}-H$. Paraformaldehyde is a convenient form in which to store formaldehyde since it decomposes to formaldehyde and water above 137°C.

$$\text{(structures of formaldehyde monomers)} \underset{H^+}{\rightleftharpoons} \underset{\substack{\text{1,3,5-trioxane} \\ \text{mp 62°C, bp 115°C}}}{\begin{array}{c}\text{O} \\ \diagup \quad \diagdown \\ CH_2 \quad CH_2 \\ | \quad \quad | \\ O \quad \quad O \\ \diagdown \quad \diagup \\ CH_2\end{array}} \qquad (12\text{-}8)$$

To get a polymer in the molecular weight range useful for resins or fibers, it is necessary to start with extremely dry formaldehyde (0.1% or less water) in a hydrocarbon solvent. This procedure gives linear polyoxymethylenes containing from 600 to 6000 oxymethylene units, depending on how much water was present. Polyoxymethylene, with its HO and H end groups is still not a useful

$$HO-(CH_2-O)_{6000}-H$$
polyoxymethylene

structural material because it falls apart like a zipper every time some base happens to remove the end proton (equation 12-9).

$$\text{base:} + H\frown O\frown CH_2\frown O\frown CH\frown O\frown CH_2\frown O\frown CH_2 \ldots \frown O\frown CH_2\frown OH \rightarrow$$
$$\text{base:} H^+ + O=CH_2 + O=CH_2 + O=CH_2 + O=CH_2 + \ldots O=CH_2 + {}^-OH \quad (12\text{-}9)$$

To block this process, however, it is only necessary to change these sensitive end groups to ester groups. This is done by treating the original polymer with acetic anhydride (equation 12-10). The

$$HO-(CH_2O)_n-CH_2OH + CH_3-\overset{\overset{O}{\|}}{C}-O-\overset{\overset{O}{\|}}{C}-CH_3 \xrightarrow{\overset{\overset{O}{\|}}{NaOCCH_3}}$$

$$CH_3-\overset{\overset{O}{\|}}{C}-O-(CH_2O)_n-CH_2-O-\overset{\overset{O}{\|}}{C}-CH_3$$
$$+ CH_3COOH \qquad (12\text{-}10)$$

product is a stable, tough, and high-melting thermoplastic sold under the name Delrin. This is one of the few examples where the nature of the end group in a large polymer molecule has a decisive effect on the properties of the molecule.

Note that the monomer corresponding to this "polyacetal" is the aldehyde, not an acetal, in spite of the name.

$$\underset{\substack{\text{dimethyl acetal of} \\ \text{formaldehyde}}}{\begin{array}{c} H \quad H \\ \diagdown \diagup \\ C \\ \diagup \diagdown \\ O \quad O \\ | \quad \quad | \\ CH_3 \quad CH_3 \end{array}} \qquad \underset{\substack{\text{part of a polyacetal} \\ \text{chain}}}{\begin{array}{c} H \quad H \\ \diagdown \diagup \\ C \\ \diagup \diagdown \\ O \quad O \\ | \quad \quad | \\ -C- \quad -C- \\ | \quad \quad | \end{array}}$$

The inclusion of an occasional OCH_2CH_2O structure in place of the usual OCH_2 gives a more stable polymer. This is achieved by copolymerizing the formaldehyde with a small amount of ethylene oxide.

$$H-\overset{O}{\underset{\|}{C}}-H + \overset{O}{\overset{/\ \backslash}{CH_2-CH_2}} \rightarrow \ldots O-CH_2-O-CH_2-CH_2\ldots$$
$$\text{ethylene oxide}$$

Ethylene oxide can be polymerized by itself under alkaline conditions to give polyethers (equation 12-11). These polymers look like

$$H_2O + \overset{O}{\overset{/\ \backslash}{CH_2-CH_2}} \xrightarrow{\text{base}} HO-(CH_2CH_2-O)_n-H \quad (12\text{-}11)$$
$$\text{polyethylene oxide}$$
$$\text{(Carbowax, Polyox)}$$

polyethylene or paraffin wax but are completely miscible with water. Like polyvinyl alcohol, they are used to make water-soluble packaging films.

Polysulfide rubber, like polyethylene oxide, has $-CH_2CH_2-$ part structures, but linked with sulfur rather than oxygen. It is used for sealing the seams in boat hulls and decks.

$$Cl-CH_2CH_2-Cl + Na_2S_4 \rightarrow X-(CH_2CH_2-\underset{\underset{S}{\|}}{S}-\underset{\underset{S}{\|}}{S})_n-Y$$

POLYESTERS

Compounds with an ester linkage as a recurring part of the chain are known as polyesters. They are used as fibers, molding compositions, paints, and laminating resins.

To make a polyester it is necessary to start either with a compound that has both an hydroxyl group and a carboxyl group in the same molecule or with a mixture of a dihydroxylic alcohol and a dicarboxylic acid. Polyesters were first used in paint and were

$$n\ HO-R-COOH \rightarrow HO-(R-\overset{O}{\underset{\|}{C}}-O)_n-H + (n-1)\ H_2O$$

$$n\ HO-R-OH + n\ HO-\overset{O}{\underset{\|}{C}}-R-\overset{O}{\underset{\|}{C}}-OH \rightarrow HO-(R-O-\overset{O}{\underset{\|}{C}}-R-\overset{O}{\underset{\|}{C}}-O)_nH + n\ H_2O$$

known as **alkyd** resins, a word allegedly derived from alcohol and acid.

Shellac is a thermosetting polyester of incompletely known structure secreted by the lac insect, a parasite of certain trees in the far east. Each tree yields about 20 lb of the raw material each year. Until about 1950 phonograph records were made from shellac, and it is still used in varnishes.

Laminating resins. A typical polyester laminating resin, such as is used in the construction of glass fiber boats, consists of a viscous yellow liquid which is mixed with a catalyst just before use. The laminated glass fiber structure is constructed by placing layers of glass cloth, mat, or woven roving on a "lay-up" mold and saturating each layer with the resin and catalyst mixture as it is put in place.

When the resin cures or hardens, the result is a composite structure that has the desirable qualities of *both* the plastic and the glass and is stronger than either of them (Figure 12-9).

FIGURE 12-9. *Long John*, a boat made of glass-reinforced polyester.

Composite materials are nothing very new. Steel-reinforced concrete is one example. Other examples are wood, which consists of cellulose fibers in a matrix of a crosslinked polymer called lignin, and bone, which is a mixture of organic macromolecules and calcium compounds.

An early use of a composite material for a boat hull is the "ark of bullrushes daubed with slime and pitch" that launched Moses on his career. According to biblical commentaries the word "slime" should probably have been translated as bitumen, a high molecular weight material left after the evaporation of petroleum. Natural bitumens are still used occasionally as binding materials in cheap molding compositions.

The hardened polyester resin is actually a polymer of *both* the polyester type and the vinyl type. The liquid laminating resin, before adding catalyst, usually consists of a mixture of a low molecular weight polyester and styrene monomer.[6] In a typical example, the polyester might be made from 1,2-dihydroxypropane (1,2-propylene glycol), fumaric acid, and phthalic acid. This is an esterification reaction in which alcohols and acids react to give water and a chain whose parts are derived alternately from the glycol and one of the acids, as shown in Figure 12-10. Note that the low molecular weight polyester contains double bonds. When a catalyst is added to the mixture of this substance with styrene, the styrene polymerizes and incorporates those double bonds into a copolymer as shown in Figure 12-11. If some of the polyester molecules in the liquid resin contain more than one double bond, as is usually the case, the cured polymer will be crosslinked.

The catalyst, or source of free radicals, to initiate the final polymerization and crosslinking is usually a peroxide such as methyl ethyl ketone peroxide.

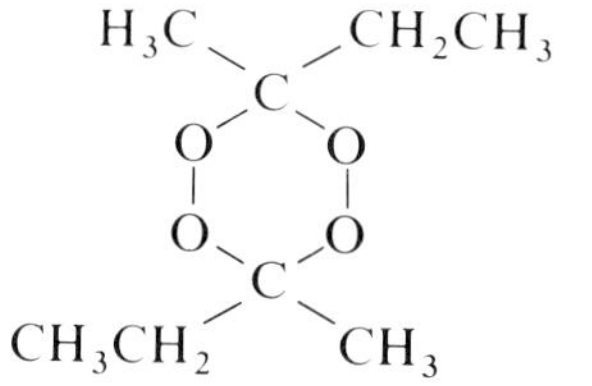

methyl ethyl ketone peroxide

Polyethylene terephthalate. The fiber sold in the United States under the name Dacron, and in Great Britain under the name Terylene, is polyethylene terephthalate. It is made from terephthalic acid and ethylene glycol (equation 12-12). Dacron and Terylene ropes are used as the sheets and halyards on sail boats because they have good resistance to chafing and do not stretch very much.

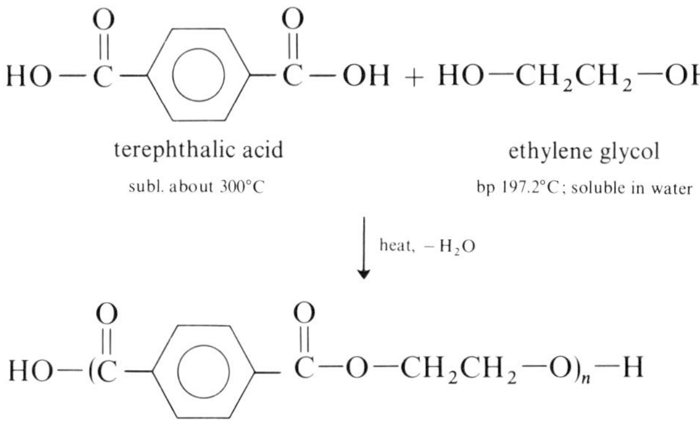

terephthalic acid
subl. about 300°C

ethylene glycol
bp 197.2°C; soluble in water

heat, $-H_2O$

polyethylene terephthalate   (12-12)

A useful property of polyethylene terephthalate, which it shares with Nylon (Chapter 14), is the ability to take a "set" or permanent change of shape when heated. Drip-dry shirts are first put into the desired wrinkle-free shape and then heated to a temperature about 30°C higher than they are likely to encounter in use. So long as this temperature is not exceeded, the fabric will tend to return to the shape in which it was set.

---

[6] The typical odor of the resin and of a newly manufactured glass-plastic object is due to the styrene.

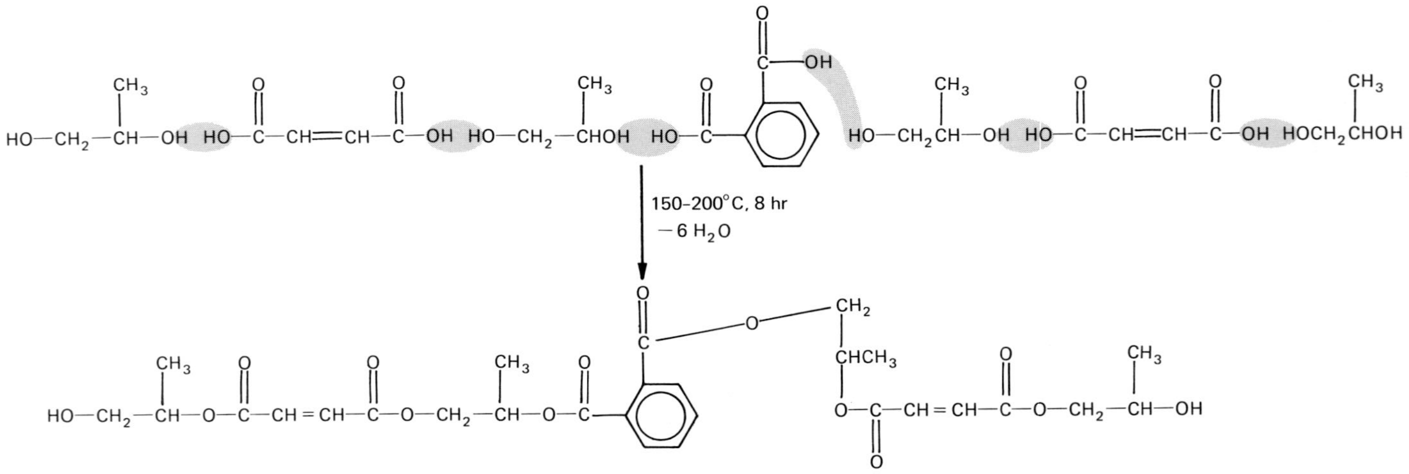

FIGURE 12-10. A typical low-polymer molecule in a polyester laminating resin contains parts derived from a glycol, an unsaturated acid, and a saturated acid.

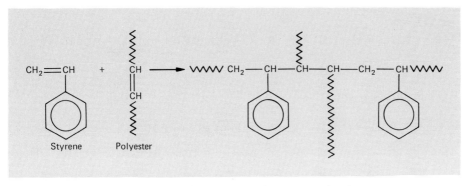

FIGURE 12-11. Reaction between the styrene and unsaturated polyester components of a typical laminating resin.

EPOXY RESINS

An epoxide is a three-membered cyclic ether, usually made by a reaction like equation 12-13 or by oxidizing an alkene (equation 12-14). An "epoxy" resin is a polymer containing epoxide func-

$$\underset{\substack{| \quad \quad | \\ \text{Cl} \quad \text{OH} \\ \text{ethylene} \\ \text{chlorohydrin}}}{CH_2-CH_2} \xrightarrow{Ca(OH)_2} \underset{\substack{\diagdown \diagup \\ O \\ \text{ethylene} \\ \text{oxide}}}{CH_2-CH_2} + CaCl_2 + H_2O \quad (12\text{-}13)$$

$$CH_2=CH_2 \xrightarrow[\text{Ag 250°C}]{O_2} \underset{\substack{\diagdown \diagup \\ O}}{CH_2-CH_2} \quad (12\text{-}14)$$

tional groups along the chain. These react with amine "hardeners" to produce crosslinks during the curing process. Epoxy resins are outstanding for their toughness, low shrinkage on curing, and good adhesive properties.

A typical commercial liquid epoxy resin consists of a mixture of the diglycidyl ether of "bisphenol A" and a polymer (equation 12-15).

$$2\ \underset{\diagdown \diagup \\ O}{CH_2-CH}-CH_2Cl\ +\ HO-\!\!\bigcirc\!\!-\underset{\substack{| \\ CH_3 \\ | \\ CH_3}}{C}-\!\!\bigcirc\!\!-OH$$

bisphenol A

$\downarrow$ NaOH

$$NaCl\ +\ \underset{\diagdown \diagup \\ O}{CH_2-CH}-CH_2-O-\!\!\bigcirc\!\!-\underset{\substack{| \\ CH_3 \\ | \\ CH_3}}{C}-\!\!\bigcirc\!\!-O-CH_2-\underset{\diagdown \diagup \\ O}{CH-CH_2}$$

bisphenol A diglycidyl ether   (12-15)

The polymer that is present in the resin along with the bisphenol A diglycidyl ether is made by further reaction with bisphenol A. This is shown in equation 12-16, where —R— is used as an abbreviation of the bisphenol A structure.

$$HO-R-OH + \overset{O}{\underset{}{CH_2-CH}}-CH_2-O-R-O-CH_2-\overset{O}{\underset{}{CH-CH_2}}$$

$$\rightarrow HO-R-O-CH_2-\underset{OH}{\overset{|}{CH}}-CH_2-O-R-OCH_2-\overset{O}{\underset{}{CH-CH_2}}$$

$$\rightarrow HO-R-O-CH_2-\underset{OH}{\overset{|}{CH}}-CH_2-O-R-OCH_2-\underset{OH}{\overset{|}{CH}}-CH_2-O-R-OH$$

$$\rightarrow HO-(R-O-CH_2-\underset{OH}{\overset{|}{CH}}-CH_2-O-R-O)_n-H \qquad (12\text{-}16)$$

The curing or hardening of the liquid epoxy resin mixture uses reactions of the hydroxy functional group of the polymer and the epoxide functional groups of the glycidyl ether to produce crosslinks. Amines and acids are used as catalysts for the crosslinking reaction and, in addition to catalyzing that reaction, they themselves participate in reactions that extend the chains. Diethylenetriamine will cure or harden bisphenol A diglycidyl ether at room temperature.

$$H_2N-CH_2CH_2-\underset{}{\overset{H}{\underset{|}{N}}}-CH_2CH_2-NH_2$$
diethylenetriamine

Epoxy resins are used to encapsulate or embed delicate electronic components to protect them from handling or from humidity. Encapsulation is also used to make integrated plug-in circuits. If any part of the circuit fails, the entire encapsulated module is replaced.

SILICONES

Dimethyldichlorosilane is made by the reaction of methyl chloride with elemental silicon (equation 12-19). When the carbon analog

$$2\,CH_3Cl + Si \xrightarrow{350°C} CH_3-\underset{Cl}{\overset{Cl}{\underset{|}{\overset{|}{Si}}}}-CH_3 \qquad (12\text{-}19)$$

dimethyldichlorosilane
bp 70°C

of this compound is hydrolyzed, it gives acetone (equation 12-20). However, the silicone corresponding to acetone is not able to persist

$$CH_3-\underset{Cl}{\overset{Cl}{\underset{|}{\overset{|}{C}}}}-CH_3 \xrightarrow{H_2O} [CH_3-\underset{OH}{\overset{OH}{\underset{|}{\overset{|}{C}}}}-CH_3] \rightarrow H_2O + CH_3-\overset{O}{\underset{}{\overset{\|}{C}}}-CH_3 \qquad (12\text{-}20)$$

in the monomeric $>Si=O$ form; instead, it immediately polymerizes (equation 12-21). Silicone polymers are available ranging in

$$CH_3-\underset{\underset{Cl}{|}}{\overset{\overset{Cl}{|}}{Si}}-CH_3 \rightarrow [CH_3-\underset{\underset{OH}{|}}{\overset{\overset{OH}{|}}{Si}}-CH_3] \rightarrow HO-(\underset{\underset{CH_3}{|}}{\overset{\overset{CH_3}{|}}{Si}}-O)_n-H + H_2O$$

dimethylsilicone

(12-21)

physical properties from liquids to Silly Putty to hard solids. Hydrolysis of methyltrichlorosilane gives a highly crosslinked network.

$$n\,CH_3-\underset{\underset{Cl}{|}}{\overset{\overset{Cl}{|}}{Si}}-Cl \xrightarrow[-HCl]{H_2O} n\,[CH_3-\underset{\underset{OH}{|}}{\overset{\overset{OH}{|}}{Si}}-OH] \rightarrow \text{crosslinked network}$$

The amount of crosslinking in a silicone can be adjusted by hydrolyzing mixtures of $(CH_3)_2SiCl_2$ and CH_3SiCl_3. Chains can be kept shorter by adding $(CH_3)_3SiCl$ to the mixture, which causes the polymer to terminate as

$$-\underset{|}{\overset{|}{Si}}-O-\underset{\underset{CH_3}{|}}{\overset{\overset{CH_3}{|}}{Si}}-CH_3$$

The substance known as Silly Putty is a modified silicone rubber in which a few of the silicon atoms have been replaced by boron. In case the reader is not already familiar with its behavior, Silly Putty flows very slowly and acts like a viscous liquid. On the other hand, it also bounces and can be broken by a sharp blow into fragments that have sharp edges and look like pieces of broken pottery. The secret of this material is a matter of timing. Given plenty of time, it is a viscous, slowly flowing liquid. Forced to react to a sharp, fast impulse it is a hard, elastic, but brittle material like glass.

There are a number of resemblances between Silly Putty and glass, particularly Pyrex glass, which is a **boro**silicate. Glass is also a very viscous liquid, but at ordinary temperatures it flows so slowly that we do not notice it. Quartz, or silica, is a crosslinked Si-O polymer something like that made from CH_3SiCl_3, but without the methyl groups.

Silicones have a large number of unusual applications not shared by other polymers. One such use is to make fabrics repel water. This is done either by spraying a solution of the polymer on the fabric or by forming the polymer right on the surface. Barrier creams designed to prevent dirt and grease from sticking to the hands contain silicones. Traces of silicone will prevent foam, which is often a problem in stills and in sewage treatment plants. Vinyltrichlorosilane is used to coat the glass used in glass laminates to make them bond more securely to the resin.

$$\begin{array}{c}\text{Cl}\diagdown\quad\diagup\text{CH}=\text{CH}_2\\ \text{Si}\\ \diagup\quad\diagdown\\ \text{Cl}\qquad\text{Cl}\\ \text{H}\quad\text{H}\\ \text{O}\;\;\text{O}\;\;\text{O}\;\;\text{O}\;\;\text{O}\\ \diagdown\!\diagup\;\diagdown\!\diagup\\ \text{Si}\qquad\text{Si}\\ |\qquad|\\ \text{O}\qquad\text{O}\\ |\qquad|\end{array}\quad\longrightarrow\quad\begin{array}{c}\text{HO}\diagdown\quad\diagup\text{CH}=\text{CH}_2\\ \text{Si}\\ \diagup\quad\diagdown\\ \text{O}\qquad\text{O}\\ |\qquad|\\ \text{Si}\qquad\text{Si}\\ \diagup\;|\;\diagdown\;|\;\diagdown\\ \text{O}\;\;\;\text{O}\;\;\;\text{O}\\ |\qquad|\\ \text{O}\qquad\text{O}\\ |\qquad|\end{array}$$

glass surface

The vinyl groups make the glass surface a part of the polymer molecules as the resin cures.

Paints and Coatings

Paint is usually a suspension of inorganic pigments in an organic liquid which either evaporates, leaving the pigment and dissolved polymer behind, or polymerizes on the surface to be protected. The "vehicle" or film-forming component of the paint can be almost any polymerizable monomer or a polymer solution or some combination thereof. Thus we have epoxy, vinyl, polyester, and silicone paints as well as resins.

One of the oldest paint vehicles is linseed oil. Like some of the modern paints based on vinyl monomers, linseed oil is a vehicle that forms a polymer as it "dries" on the surface. The "drying" of a linseed oil paint is not just the evaporation of the solvent used as a thinner but also includes vinyl polmerization initiated by oxygen from the air. Linseed oil is a mixture of triglycerides of unsaturated fatty acids containing two or three double bonds per acid molecule.

Water-based so-called latex paints are emulsions of polymeric materials. As the water evaporates, the droplets of polymer solution coalesce and form a continuous coating.

SUMMARY

A. Some new terms relating to large molecules are
1. Monomers, dimers, trimers, oligomers, and polymers.
2. End groups.
3. Average molecular weight.
4. Crosslinking.
5. Degree of polymerization, that is, number of monomer units in one polymer molecule.
6. Thermoplastic substances soften when heated.
7. Polymers with asymmetric carbon atoms may be either isotactic (all the same) or atactic (random) in the configurations of their asymmetric carbons.
8. Copolymers have more than one kind of monomer unit in the same molecule.

B. Vinyl polymers
1. Substituted alkenes are made to polymerize into vinyl polymers by means of free radical initiators and by Ziegler-Natta catalysts, $TiCl_4$ plus $Al(CH_2CH_3)_3$. The latter give isotactic polymers.

2. Some well-known vinyl polymers are polyethylene, polypropylene, polyvinyl chloride, polyvinyl acetate, and polyacrylates. They are named after the monomer, or repeating unit, using the prefix poly (Greek: many).
C. Diene polymers and rubber
   1. Natural rubber is *cis*-1,4-polyisoprene.
      (a) Rubberlike elasticity is due to the tendency of stretched, crystalline chain molecules to assume random coiled configurations.
      (b) Rubber is crosslinked (vulcanized) by heating with sulfur.
   2. Neoprene is a synthetic rubber, or elastomer, made from chloroprene or 2-chloro-1,3-butadiene. It has chlorine substituents where rubber has methyl groups.
      (a) Neoprene is crosslinked by heating with ZnO.
      (b) It is resistant to gasoline and hydrocarbons.
D. Polymers with heteroatoms in the chain
   1. Polyacetals have alternating carbon and oxygen atoms. The monomer unit corresponds to an aldehyde.
      (a) Polyethylene oxide (Carbowax) is a related polymer made from ethylene oxide.
   2. Polyesters have ester linkages as recurring parts of the chain. They can be made from
      (a) Molecules containing both a hydroxyl group and a carboxyl group, and
      (b) Mixtures of dihydroxy compounds and dicarboxylic acids, as in Dacron or Terylene.
   3. Polyester laminating resins (for use with glass fibers) are both polyesters and vinyl polymers. They are highly crosslinked.
   4. Epoxy resins are hardened by reaction with amines. The amines convert epoxy functional groups into crosslinks.
   5. Silicones are Si-O polymers resembling the polyacetals in structure.

EXERCISES

1. Define or explain by giving examples:
   (a) polymer
   (b) monomer
   (c) monomer unit
   (d) thermoplastic substance
   (e) thermosetting substance
   (f) end group
   (g) crosslinks
   (h) polymerization initiator
   (i) atactic
   (j) isotactic

2. Give structural formulas for (a) polystyrene, showing four of the monomer units; (b) polystyrene, an abbreviated formula; and (c) polystyrene crosslinked with divinylbenzene, showing one crosslink.

3. The molecules of a "pure" sample of polymer are not exactly alike. Give two ways in which they might differ.

4. What nonpolymeric substance might be expected to result from the vigorous oxidation of polystyrene? Write an equation for the reaction. *Hint:* See Chapter 10, page 197.

5. Write an equation showing how you could convert a sample of polyvinyl acetate into polyvinyl alcohol.

6. Write equations showing how you could convert polymethyl acrylate into (a) polyacrylic acid (*Hint:* See Chapter 10, page 206); (b) polyacrylamide (*Hint:* See Chapter 10, page 208).

7. (Open book) What are
   (a) ABS
   (b) Teflon
   (c) Plexiglas
   (d) Saran

8. Give structures for (a) isoprene and (b) natural rubber, showing four monomer units.

9. Give a complete name for natural rubber, one that tells what its structure is.

10. Explain why neoprene is suitable for gasoline hoses and why natural rubber is not suitable.

11. Define polyester. Why is polyvinyl acetate *not* considered to be a polyester?

12. (a) What reaction might eventually occur in a hot, intimate mixture of polymethyl methacrylate and polyvinyl alcohol? *Hint:* See Chapter 10, page 214.
    (b) Would the resulting product still be soluble in typical organic solvents? *Hint:* See this chapter, page 254.

13. What reaction would you expect to occur on exposure of polyvinyl butyral (page 259) to a dilute water solution of a mineral acid. *Hint:* See Chapter 8, page 145.

14. (Open book) Explain, with structural formulas, how a polyester laminating resin works. Show (a) the liquid resin and (b) what happens when it cures.

15. What is Dacron?

16. Write an equation for a reaction that might occur on boiling Delrin (page 266) with aqueous NaOH. *Hint:* See page 266, this chapter, and page 206, Chapter 10.

17. What is the main purpose of the linseed oil in old-fashioned paint?

18. What substance should be added to a dimethyldichlorosilane reaction mixture to produce a polymer like the one shown below?

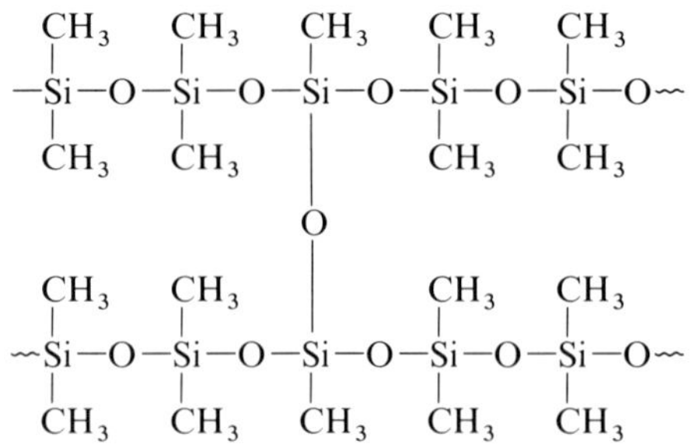

$$\begin{array}{c} CH_3 CH_3 CH_3 CH_3 CH_3 \\ | | | | | \\ -Si-O-Si-O-Si-O-Si-O-Si-O\sim \\ | | | | | \\ CH_3 CH_3 CH_3 CH_3 \\ | \\ O \\ | \\ CH_3 CH_3 CH_3 CH_3 \\ | | | | | \\ \sim Si-O-Si-O-Si-O-Si-O-Si-O\sim \\ | | | | | \\ CH_3 CH_3 CH_3 CH_3 CH_3 \end{array}$$

Chapter 13
Carbohydrates

This chapter discusses a variety of compounds found in plants and animals whose empirical formulas are very often of the form $C_m(H_2O)_n$. To earlier generations of chemists they seemed to be some sort of mysterious hydrates of carbon. The name **carbohydrate** has stuck, although the structures of such substances are now known in detail. They are ordinary organic molecules with hydroxyl, ether, or carbonyl functional groups.

Some of the most important carbohydrates are macromolecules, condensation polymers built up from smaller molecules by eliminating water to form ether linkages (equation 13-1). The small sugar

$$\sim 3000 \text{ D-}C_6H_{12}O_6 \xrightarrow[\text{plants}]{\text{living}} \sim 3000 \text{ H}_2\text{O} + (C_6H_{10}O_5)_{3000} \quad (13\text{-}1)$$

glucose or dextrose — very soluble in water

cellulose and starch

molecules that correspond to the monomer units of polymers such as cellulose and starch are called **monosaccharides** and the polymers are called **polysaccharides**. Monosaccharides are found both as free sugars and as part of the structures of many biologically important molecules, large and small.

Monosaccharides

ALDOHEXOSES

The most wide-spread monosaccharide of all, both as the free sugar and in combined form, is D-glucose, one of a group of sugars

TABLE 13-1. *Some Terms Used in Carbohydrate Chemistry*

Aglycone	The nonsugar part of a glycoside; usually an alcohol, phenol, or amine
Aldohexose	A six-carbon sugar with an aldehyde group or cyclic hemiacetal group
Aldopentose	A five-carbon sugar with an aldehyde or cyclic hemiacetal functional group
Anomers	Sugar isomers differing only in the configuration of the hemiacetal carbon atom; they are designated as α or β
Cyanogenetic glycoside	A glycoside that generates HCN when it is hydrolyzed
Epimers	Sugars that differ only in the configuration at carbon 2
Furanose	Five-membered cyclic sugar, from furan
α- or β-Glucoside	Acetal derived from the cyclic hemiacetal form of α- or β-glucose
α- or β-Glycoside	General term for the acetal derivative of *any* sugar
Ketohexose	A six-carbon sugar with a ketone functional group or cyclic hemiketal group
Monosaccharide	A simple sugar, also the monomer unit of a polysaccharide
Mutarotation	Change in optical rotation as a fresh solution stands
Osazone	Sugar derivative formed from one molecule of the sugar and two molecules of phenylhydrazine
Polysaccharide	A macromolecule composed of many monosaccharides linked by acetal bonds
Pyranose	Six-membered cyclic sugar, from pyran
Reserve carbohydrate	Food sugars stored in the plant or animal in the form of polysaccharides

known as aldohexoses.[1] Although the elucidation and proof of the structure of D-glucose is one of the more interesting tales from the continuing struggle of an inquisitive mankind to wrest truth from an all-too-reticent Nature, we will start our discussion with the structure itself rather than the evidence that led up to it. The structural formula for α-D-glucose is shown in Figure 13-1, both in the conventional form and in a convenient abbreviated form that leaves out five C's but still indicates the shape of the molecule. The reader will notice that α-D-glucose is a six-membered ring compound. Like any cyclic compound with more than one substituent, it has cis and trans isomers. The structural formula in Figure 13-1 has numbers that will be used to refer to the various carbon atoms.

The structure at carbon 1 in α-D-glucose is a hemiacetal (Chapter 8), since that carbon has both an ether linkage and an hydroxyl group

---

[1] See Table 13-1 for definitions of aldohexose and other terms used in carbohydrate chemistry.

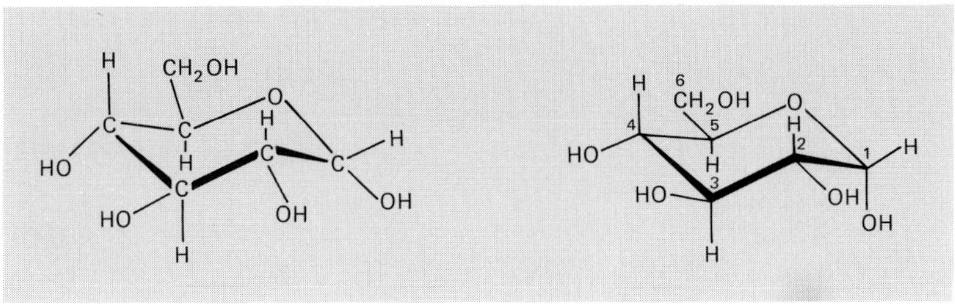

FIGURE 13-1. α-D-Glucose, mp 146°C, $[\alpha]_D = 112.2°$.

attached to it. Like any hemiacetal, glucose in solution is an equilibrium mixture in which some of the molecules are in the alcohol + aldehyde form rather than in the hemiacetal form.

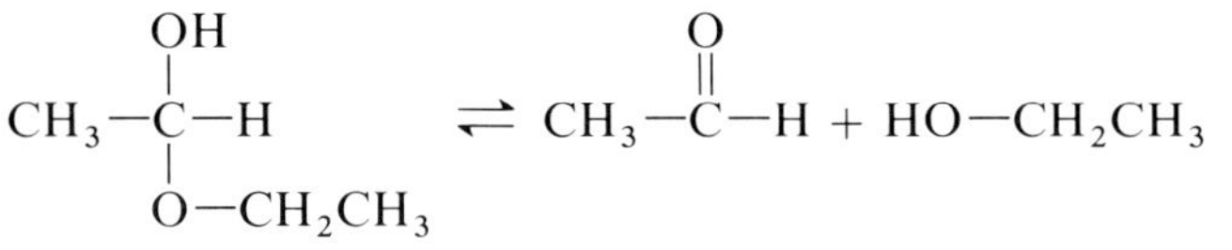

$$\underset{\text{a hemiacetal}}{CH_3-\overset{\overset{\displaystyle OH}{|}}{\underset{\underset{\displaystyle O-CH_2CH_3}{|}}{C}}-H} \rightleftarrows \underset{\text{acetaldehyde}}{CH_3-\overset{\overset{\displaystyle O}{\|}}{C}-H} + \underset{\text{ethanol}}{HO-CH_2CH_3}$$

When α-D-glucose is dissolved in water, the hemiacetal equilibrium converts it into the open-chain aldehyde form and into the isomeric cyclic form, called β-D-glucose (equation 13-2). The α and β cyclic isomers of a sugar are known as **anomers**.

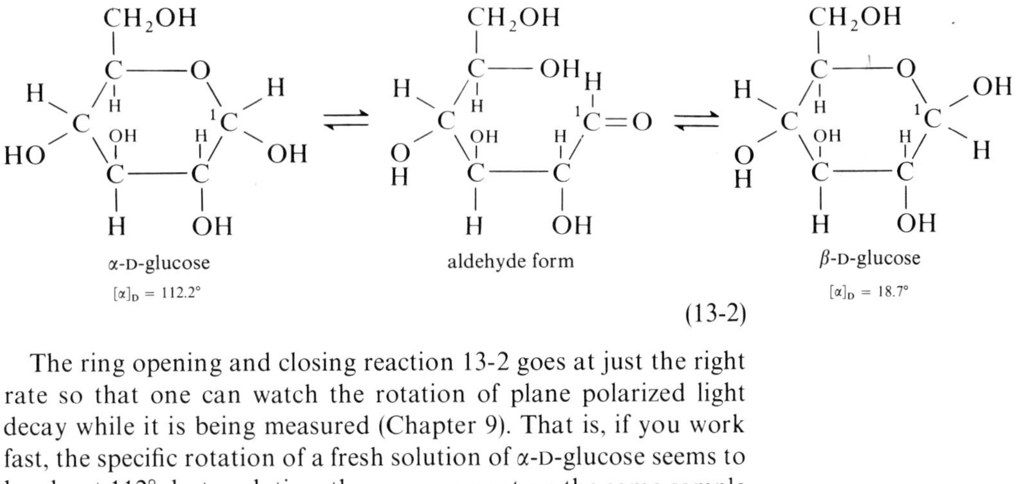

α-D-glucose $[\alpha]_D = 112.2°$ ⇌ aldehyde form ⇌ β-D-glucose $[\alpha]_D = 18.7°$

(13-2)

The ring opening and closing reaction 13-2 goes at just the right rate so that one can watch the rotation of plane polarized light decay while it is being measured (Chapter 9). That is, if you work fast, the specific rotation of a fresh solution of α-D-glucose seems to be about 112°, but each time the measurement on the same sample is repeated the result is less, until finally the value levels off at 52.7°. This final value is an average due to the combined effects of the α and β anomers plus a small amount of the open-chain aldehyde form. The change in rotation on standing is called **mutarotation**.

Because the equilibrium reactions of equation 13-2 are fairly fast, glucose solutions can react as though they consisted of *any* of the three forms. For example, glucose can be made to crystallize out either as the α form or as the β form by choosing the right conditions. It will also react as though it were just an aldehyde; that is, glucose forms typical carbonyl adducts and can be oxidized to the corresponding carboxylic acid.

$$\underset{\text{D-glucose}}{\begin{array}{c}\text{CH}_2\text{OH}\\|\\\text{C—OH}\\\text{(ring structure)}\end{array}} \xrightarrow{[O]} \underset{\substack{\text{D-gluconic acid}\\\text{mp 131°C; }[\alpha]_D = -6.7°\text{; soluble in water}}}{\begin{array}{c}\text{CH}_2\text{OH}\\|\\\text{C—OH}\\\text{(ring structure)}\end{array}} \quad (13\text{-}3)$$

The oxidation reaction (equation 13-3) is used as a test for the presence of glucose or other "reducing" or aldehydic sugars. With Fehling's solution, which consists of cupric ion complexed with tartaric acid, the occurrence of the oxidation reaction is revealed by the appearance of a red precipitate of Cu_2O. Diabetics use this reaction to test for the appearance of sugar in their urine, a sign that their disease is not under proper control.

We can also make full acetal derivatives by reaction of either the α or the β hemiacetal forms of glucose with an alcohol (equation 13-4). Other alcohols will also form acetal derivatives, or glucosides, with glucose.

$$\underset{\alpha\text{-D-glucose}}{\text{(structure)}} + CH_3OH \xrightarrow{H^+} H_2O + \underset{\substack{\text{methyl }\alpha\text{-D-glucoside}\\\text{mp 168°C; }[\alpha]_D = 158.9°}}{\text{(structure)}} \quad (13\text{-}4)$$

The reaction of either α or β-D-glucose with acetic anhydride esterifies all five hydroxyl groups (including the one that is part of the hemiacetal structure) and gives α and β-D-glucose pentacetates.

The reactions discussed so far have rather firmly established that glucose is a pentahydroxy aldehyde that can cyclize to a hemiacetal. Other reactions, such as a sequence that converts glucose into 1-iodohexane, show that the carbon skeleton in glucose is unbranched and uninterrupted by oxygen atoms. This gives structure 13-5 for the aldehyde form of glucose.

$$\underset{\text{glucose}}{\begin{array}{c}\text{H—C=O}\\|\\\text{CHOH}\\|\\\text{CHOH}\\|\\\text{CHOH}\\|\\\text{CHOH}\\|\\\text{CH}_2\text{OH}\end{array}} \quad (13\text{-}5)$$

The next question is, "Which of the five hydroxyl groups is the one that reacts with the aldehyde group to give the α and β cyclic hemiacetals?" This is equivalent to asking how big a ring is formed, since the hydroxyl on carbon 6 would give a seven-membered ring, the hydroxyl on carbon 5 a six-membered ring, and so forth. The answer (carbon 5, six-membered ring) was provided by converting all the hydroxyls to methyl ethers and hydrolyzing the acetal linkage.

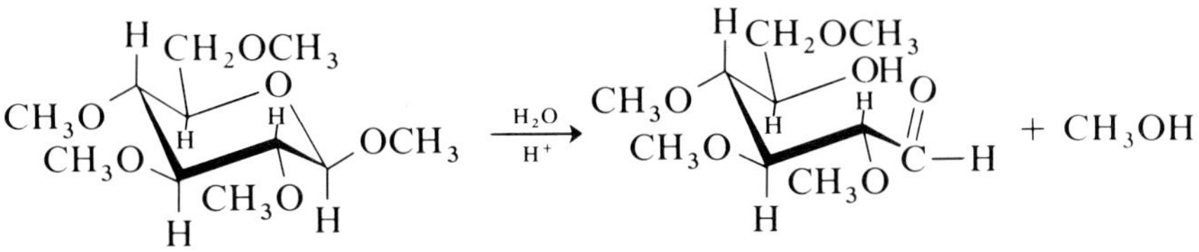

$$\text{(methylated glucopyranose)} \xrightarrow[\text{H}^+]{\text{H}_2\text{O}} \text{(open-chain methylated form)} + \text{CH}_3\text{OH}$$

The hydroxyl group at carbon 5 showed that the cyclic forms of glucose are six-membered rings. Sugars or "-oses" with six-membered rings are called **pyranoses**, after pyran, a six-membered oxygen heterocycle.

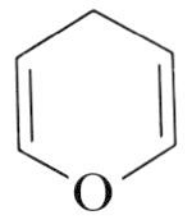

pyran

Thus α-D-glucose can also be called α-D-glucopyranose and β-D-glucose can be called β-D-glucopyranose.

This provides a complete structure for glucose except for one not exactly minor detail: **the aldehyde form of glucose has four asymmetric carbon atoms**. There should be 2^4, or 16, optical isomers all told, or 8 (±) pairs. Thus, corresponding to D-glucose, there should be a mirror image form or enantiomer, L-glucose; further, there should be seven diastereomers of glucose, each with its enantiomer. The names and structures of the eight D-aldohexoses are shown in the top row of Figure 13-2. The structures of the eight L-aldohexoses are of course just the mirror images of the ones given in the figure.

Note that if the hydroxyl on the next-to-the-bottom carbon atom (Figure 13-2) is on the right, the convention is to give the sugar one of the D names. If that hydroxyl is on the left, the sugar belongs to the L series.

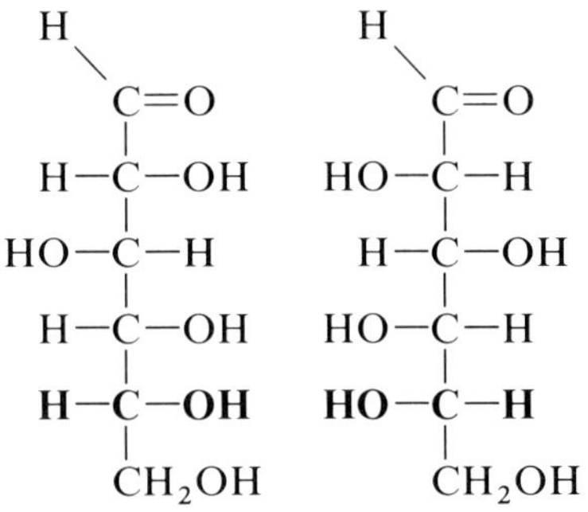

```
     H              H
      \              \
       C=O            C=O
       |              |
    H—C—OH        HO—C—H
       |              |
   HO—C—H         H—C—OH         H              H
       |              |           \              \
    H—C—OH        HO—C—H           C=O            C=O
       |              |            |              |
    H—C—OH        HO—C—H        H—C—OH        HO—C—H
       |              |            |              |
      CH₂OH          CH₂OH        CH₂OH          CH₂OH
```

D- and L-glucose 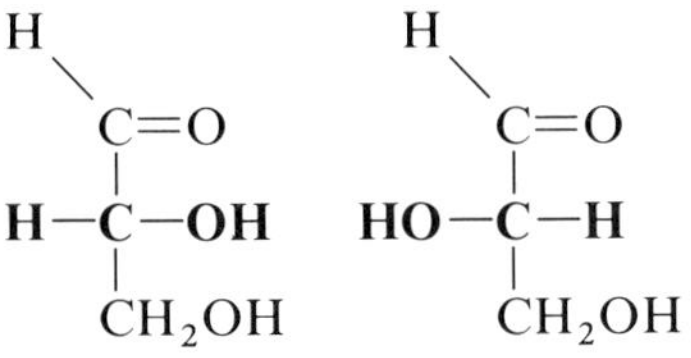   D- and L-glyceraldehyde

Of the eight D-aldohexoses in Figure 13-2, D-glucose is the most important one and its structure should be memorized, preferably in the cyclic form (Figure 13-1).

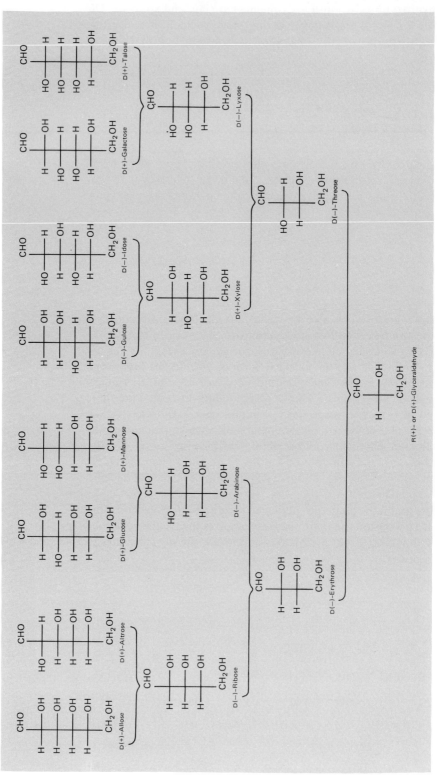

FIGURE 13-2. Fischer projection formulas of the D-aldehyde sugars, or aldoses.

The methods used by the German chemist Emil Fischer to determine which structure went with which sugar are ingenious and interesting, but we will pause only to look at two examples. The first of these reactions is one that relates the sugars in pairs having exactly the same structure for the bottom four carbon atoms.

When any of the aldohexoses reacts with phenylhydrazine, the product is an **osazone** in which not only the aldehyde group but also the adjacent CHOH group have been converted to $C=N-NHC_6H_5$ (equation 13-6). D-Glucose and D-mannose are **epimers**, which means

$$\begin{array}{ccc}
\text{H}\diagdown\!\!\!\!\!\!\!\!\!\!\! & & \text{H}\diagdown\!\!\!\!\!\!\!\!\!\!\! \\
\quad\text{C}=\text{O} & \text{H}-\text{C}=\text{N}-\text{NHC}_6\text{H}_5 & \quad\text{C}=\text{O} \\
\text{H}-\overset{*}{\text{C}}-\text{OH} & \text{C}=\text{N}-\text{NHC}_6\text{H}_5 & \text{HO}-\overset{*}{\text{C}}-\text{H} \\
\text{HO}-\overset{*}{\text{C}}-\text{H} \xrightarrow{C_6H_5NHNH_2} & \text{HO}-\overset{*}{\text{C}}-\text{H} \xleftarrow{C_6H_5NHNH_2} & \text{HO}-\overset{*}{\text{C}}-\text{H} \\
\text{H}-\overset{*}{\text{C}}-\text{OH} & \text{H}-\overset{*}{\text{C}}-\text{OH} & \text{H}-\overset{*}{\text{C}}-\text{OH} \\
\text{H}-\overset{*}{\text{C}}-\text{OH} & \text{H}-\overset{*}{\text{C}}-\text{OH} & \text{H}-\overset{*}{\text{C}}-\text{OH} \\
\text{CH}_2\text{OH} & \text{CH}_2\text{OH} & \text{CH}_2\text{OH} \\
\text{D-glucose} & \text{D-glucosazone} \equiv \text{D-mannosazone} & \text{D-mannose}
\end{array}$$

(13-6)

that they differ only in the configuration of the top asymmetric carbon atom. The formation of an osazone destroys the asymmetry at that point so that both sugars give the same osazone. The other sugars can be sorted into pairs of epimers in the same way. The task of completely assigning the structure is cut in half because, if the structure of any sugar is known, the structure of one other sugar that gives the same osazone is also known.

Another device used to relate the sugars in pairs is to oxidize both the top CHO group and the bottom CH_2OH group to COOH. This makes the top and bottom ends of the molecule identical except for the asymmetric carbon atoms. D-Allose gives an optically inactive meso compound, because its asymmetric carbon atoms have mirror-image configurations at the top and the bottom. Oxidation of L-allose gives the same optically inactive acid as D-allose (equation 13-7). The same procedure reveals that the configuration of the

$$\begin{array}{ccc}
\text{CHO} & & \text{COOH} \\
\text{H}-\overset{*}{\text{C}}-\text{OH} & & \text{H}-\overset{*}{\text{C}}-\text{OH} \\
\text{H}-\overset{*}{\text{C}}-\text{OH} & \xrightarrow{[\text{O}]} & \text{H}-\overset{*}{\text{C}}-\text{OH} \\
\text{H}-\overset{*}{\text{C}}-\text{OH} & & ---\!\!\mid\!\!--- \\
\text{H}-\overset{*}{\text{C}}-\text{OH} & & \text{H}-\overset{*}{\text{C}}-\text{OH} \\
\text{CH}_2\text{OH} & & \text{H}-\overset{*}{\text{C}}-\text{OH} \\
& & \text{COOH} \\
\text{D-allose} & & \text{allaric acid,} \\
& & \text{optically inactive}
\end{array}$$

(13-7)

top half of a D-glucose molecule is like the configuration of the *bottom* half of an L-gulose molecule, and vice-versa (Figure 13-3).

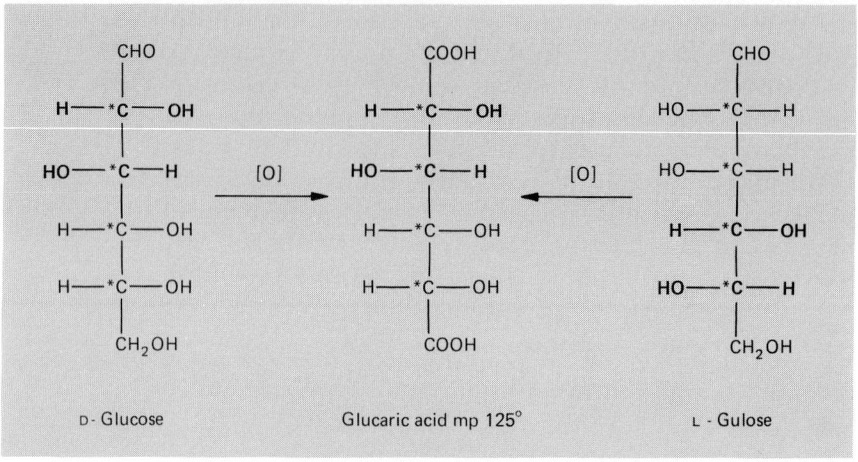

```
    CHO                    COOH                    CHO
H—*C—OH               H—*C—OH                HO—*C—H
HO—*C—H    [O]        HO—*C—H     [O]        HO—*C—H
         ——→                       ←——
H—*C—OH               H—*C—OH                H—*C—OH
H—*C—OH               H—*C—OH                HO—*C—H
    CH₂OH                  COOH                   CH₂OH
  D-Glucose          Glucaric acid mp 125°        L-Gulose
```

FIGURE 13-3. When the ends of the chain are made identical by oxidation, the bottom two asymmetric carbons of L-gulose can be superimposed on the top two of D-glucose, and vice-versa.

D-Glucose occurs widely in both plants and animals, but its enantiomer is represented only by *N*-methyl-L-glucosamine, a hydrolysis product of the antibiotic streptomycin.

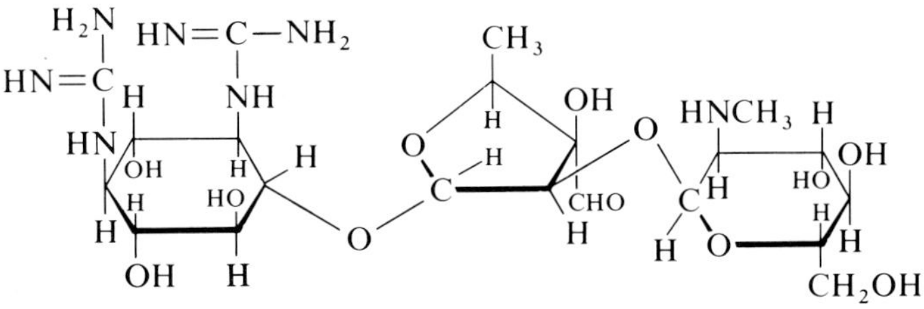

streptomycin

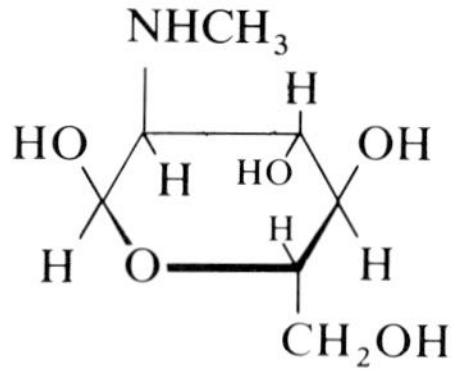

N-methyl-L-glucosamine

Animals usually contain only D- sugars although L-galactose is found in a polysaccharide from the snail *Helix pomatia*. Both D- and L-galactose are found in the polysaccharide from agar.

ALDOPENTOSES AND DEOXYALDOPENTOSES

The most important aldopentoses, or five-carbon aldehyde sugars, are D-ribose and D-2-deoxyribose, which occur in ribonucleic acid (RNA) and deoxyribonucleic acid (DNA). These nucleic acids, which are part of the mechanism by which living cells store and transmit genetic information, will be the subject of Chapter 15.

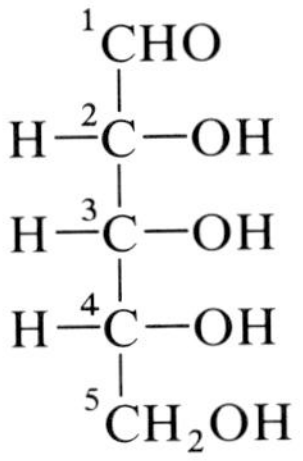

$$^1\text{CHO}$$
$$\text{H}-^2\text{C}-\text{OH}$$
$$\text{H}-^3\text{C}-\text{OH}$$
$$\text{H}-^4\text{C}-\text{OH}$$
$$^5\text{CH}_2\text{OH}$$

D-ribose

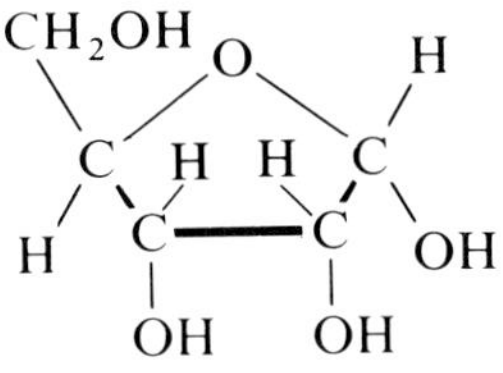

α-D-ribofuranose
mp 87°C

Like glucose, ribose exists both in an open-chain or aldehyde form and in α and β cyclic hemiacetal forms. The latter, at least in nucleic acids, are five-membered rather than six-membered rings. Five-membered sugars are called furanoses after **furan**, a simple oxygen heterocycle.

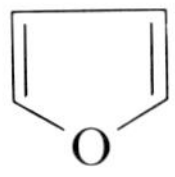

furan
bp 32°C at 758 mm

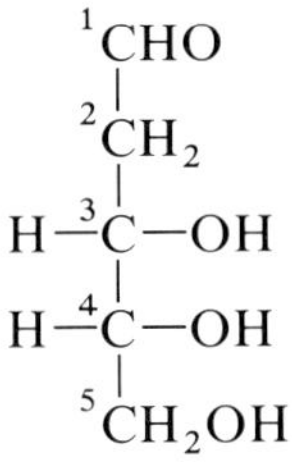

$$^1\text{CHO}$$
$$^2\text{CH}_2$$
$$\text{H}-^3\text{C}-\text{OH}$$
$$\text{H}-^4\text{C}-\text{OH}$$
$$^5\text{CH}_2\text{OH}$$

D-2-deoxyribose
after mutarotating
$[\alpha]_D = -56.2°$

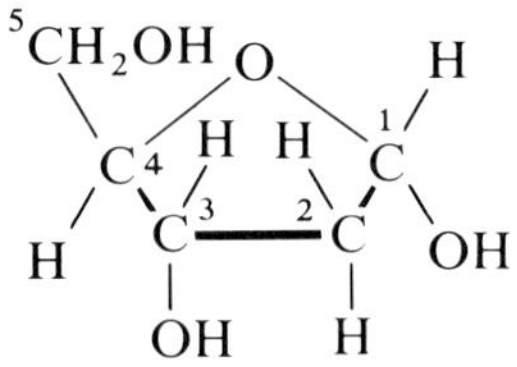

α-D-2-deoxyribofuranose
mp 91°C

The projection formulas for D-ribose and D-2-deoxyribose are easy to remember, because, like allose, these sugars have all the hydroxyl groups on the same side.

There are $2^3 = 8$ isomeric aldopentoses or 4 (±) pairs (Figure 13-2). Besides D-ribose, there are L(+)-arabinose and D(+)-xylose, formed by hydrolyzing various plant polysaccharides. D(−)-Arabinose is found in sponges. Lyxose has not been found in nature. **Apiose**, a branched-chain aldopentose, occurs in parsley as part of the glycoside apiin.

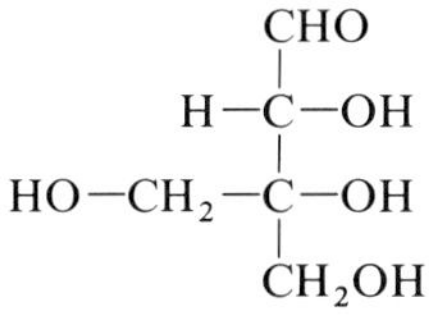

$$\text{CHO}$$
$$\text{H}-\text{C}-\text{OH}$$
$$\text{HO}-\text{CH}_2-\text{C}-\text{OH}$$
$$\text{CH}_2\text{OH}$$

apiose
$[\alpha]_D^{15} = +5.6°$

KETOHEXOSES

The commonly occurring ketoses have the carbonyl group at the second carbon atom. The most abundant is fructose, a ketohexose or six-carbon ketonic sugar. D-Fructose is formed along with glucose

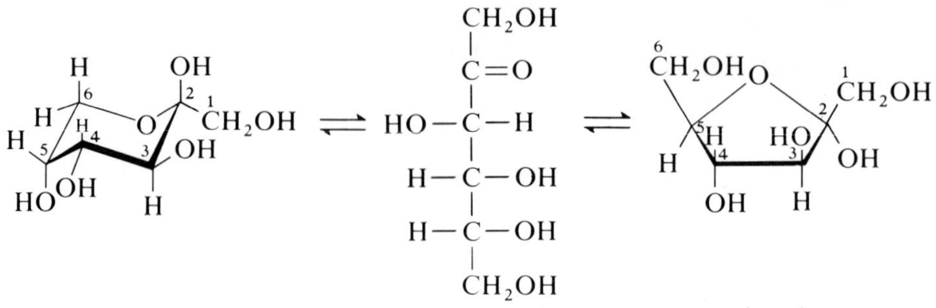

β-D-fructopyranose   D-fructose α-D-fructofuranose
$[\alpha]_D^{20} = -92°$ after mutarotating

in the hydrolysis of cane sugar and is found in the free state in fruit juices and in honey. It is also the monosaccharide formed by the hydrolysis of the polysaccharide inulin, found in dahlia bulbs. Fructose is used to prevent sandiness in ice cream.

Notice that D-fructose and D-glucose, the most common ketohexose and most common aldohexose, have the same configurations about their asymmetric carbon atoms. Fructose reacts with phenylhydrazine to give an osazone identical to that from glucose.

L-Sorbose is used to make ascorbic acid, the antiscorbutic vitamin (vitamin C, equation 13-8). Ascorbic acid owes its acidic

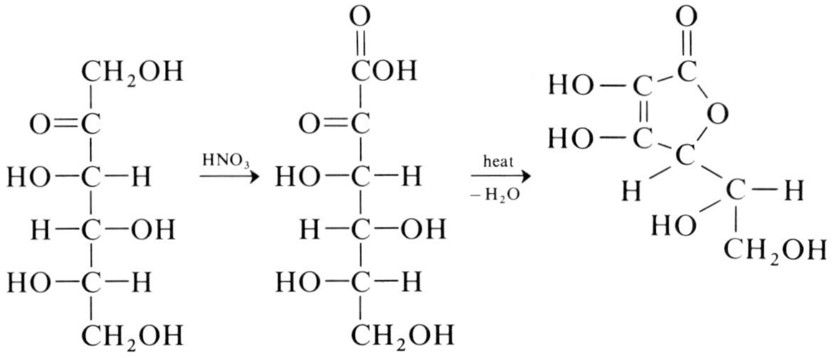

L(−)-sorbose
mp 165°C; $[\alpha]_D^{30} = -42.7°$

ascorbic acid
mp 190–192°C; soluble in water
(see page 126)

(13-8)

properties to the fact that it is an enediol rather than the isomeric ketone. Ascorbic acid is easily oxidized and is used as an antioxidant in foods.

Glycosides

Methyl α-D-glucoside (equation 13-4, page 280) is an example of a class of sugar derivatives called glycosides. Glycosides are acetals derived from the cyclic hemiacetal forms of the sugars and, like the parent sugars, they exist in α and β forms. Glycosides do not mutarotate because the hydrolysis to hemiacetal and alcohol is a slow reaction. Glycosides derive their name from those of the alcohol or phenol and the parent sugar, replacing the **ose** ending with an **oside** ending. For example, the products formed from methyl alcohol and D-mannose (equation 13-9) are known as methyl α- or β-D-mannosides.

CH_3OH + D-mannose → [methyl α-D-mannoside or methyl α-D-mannopyranoside structure] (13-9)

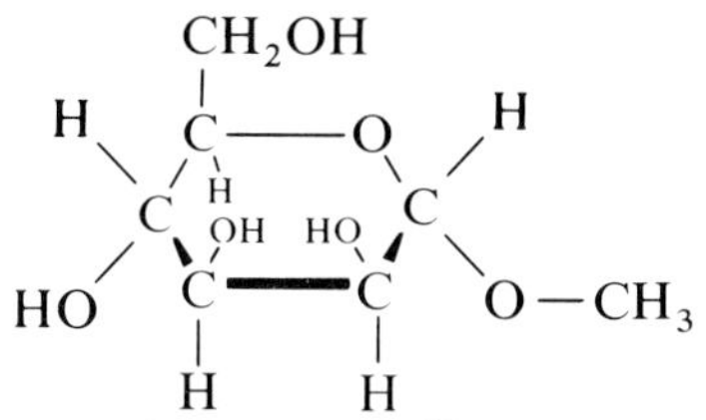

Many naturally occurring alcohols, phenols, and amines are found in plants in the form of glycosides rather than free. An example is the cyanin coloring material in flowers (page 157). Different plants often have essentially the same coloring material except that one may be a glucoside, another a galactoside, and so on.

Hydroquinone is found in the leaves of cranberry and blueberry plants and pear trees in the form of a β-D-glucoside, **arbutin**. Arbutin is easily hydrolyzed by dilute acids or by the enzyme emulsin (equation 13-10). It has been used as a urinary antiseptic.

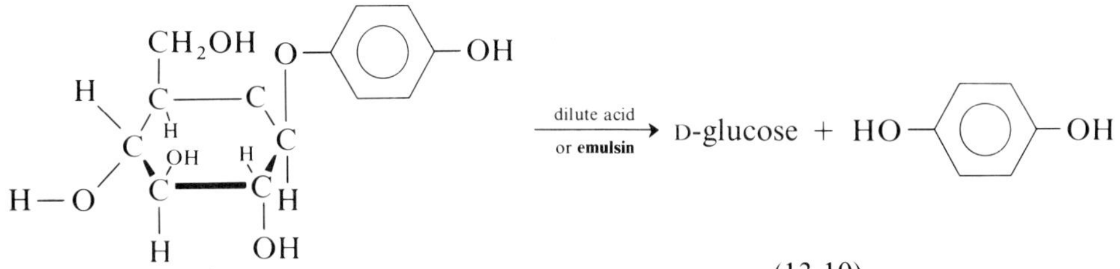

arbutin $\xrightarrow[\text{or emulsin}]{\text{dilute acid}}$ D-glucose + HO—⟨benzene⟩—OH

arbutin
mp 165°C; $[\alpha]_D^{25} = -64°$; soluble in water

(13-10)

The African arrow poison from *Strophanthus* seeds is a mixture of several glycosides in which the nonsugar part, or aglycone, is strophanthidin, a steroid. About 0.07 mg of the glycoside will stop the heart of a 20-g mouse. The glycosides in toad poisons have similar structures and also are cardiac poisons.

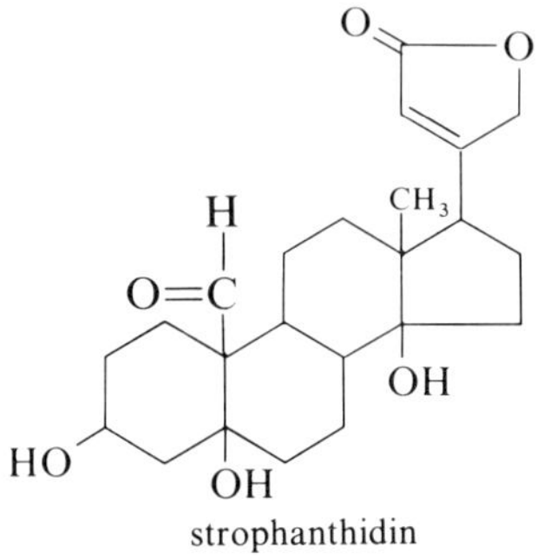

strophanthidin
mp 171°C

Secondary amines also react with hemiacetals, or the cyclic form of sugars, in much the same way that alcohols and phenols do (equation 13-11). The nucleotides of Chapter 15 have this structure.

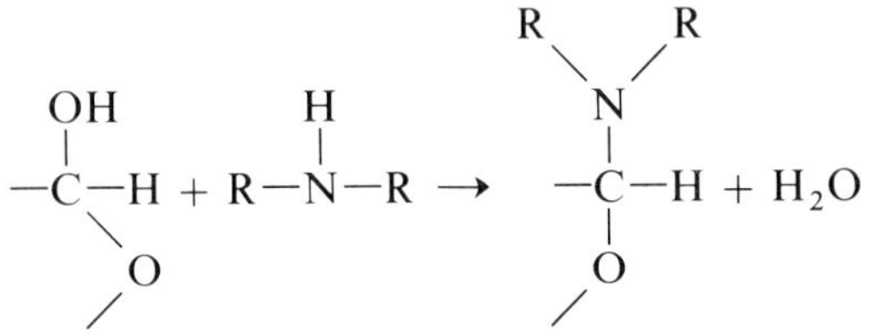

$$\begin{matrix} \text{OH} \\ | \\ -\text{C}-\text{H} \\ | \\ \text{O} \end{matrix} + \text{R}-\overset{H}{\underset{|}{N}}-\text{R} \rightarrow \begin{matrix} \text{R} \diagdown \diagup \text{R} \\ \text{N} \\ | \\ -\text{C}-\text{H} \\ | \\ \text{O} \end{matrix} + \text{H}_2\text{O}$$ (13-11)

Carbohydrates

A cyanogenetic glycoside is one in which the alcohol portion, or aglycone, is a cyanohydrin. An example is amygdalin (page 148).

Disaccharides

When a hydroxyl group of one monosaccharide molecule acts as the alcohol to form a glycoside linkage with the hemiacetal group of a second monosaccharide, the resulting glycoside is called a disaccharide. Further reactions of the same type give tri, tetra, and eventually polysaccharides.

REDUCING AND NONREDUCING DISACCHARIDES

It is convenient to divide disaccharides into two types. In a **reducing** disaccharide, one of the hydroxyl groups used to make the ether or acetal linkage is an ordinary **hydroxyl group not located at the carbonyl-hemiacetal carbon atom**. This leaves the hemiacetal free to react to give the open-chain form of the sugar. Since the open-chain form is a good reducing agent, such sugars will give a positive test with Fehling's solution. An example of a reducing disaccharide is lactose (equation 13-12).

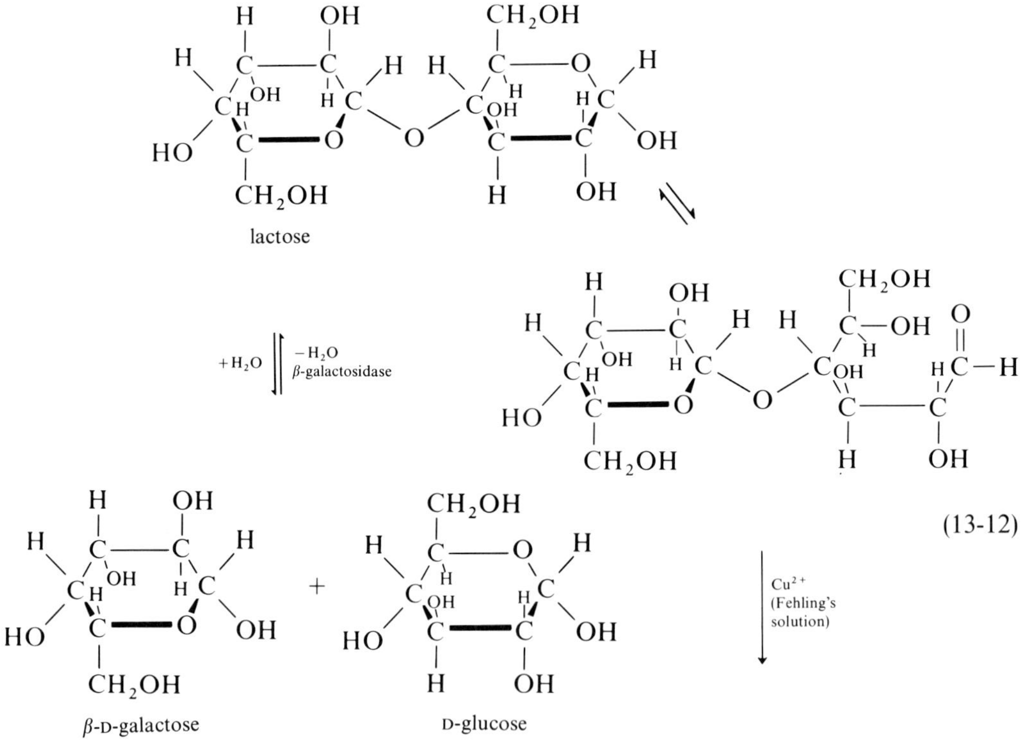

lactose

$+H_2O$ $\Updownarrow$ $\genfrac{}{}{0pt}{}{-H_2O}{\beta\text{-galactosidase}}$

(13-12)

Cu^{2+} (Fehling's solution)

β-D-galactose   +   D-glucose

lactonic acid

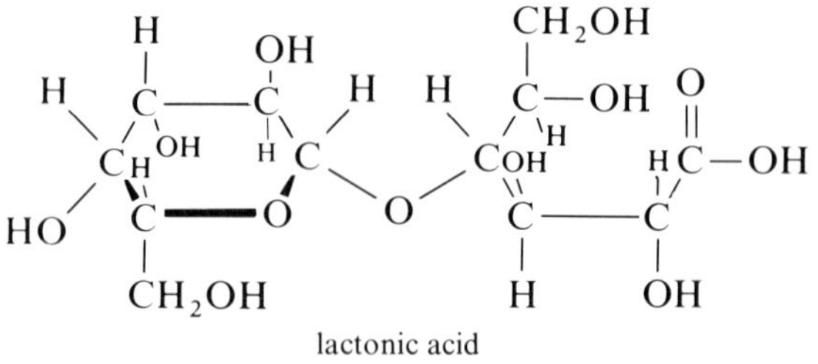

An example of a **nonreducing disaccharide** is **trehalose**. In trehalose there is no hemiacetal structure in either of the monosaccharide units. It is therefore not able to supply an aldehyde group to give the red precipitate of Cu_2O with Fehling's solution (p. 280).

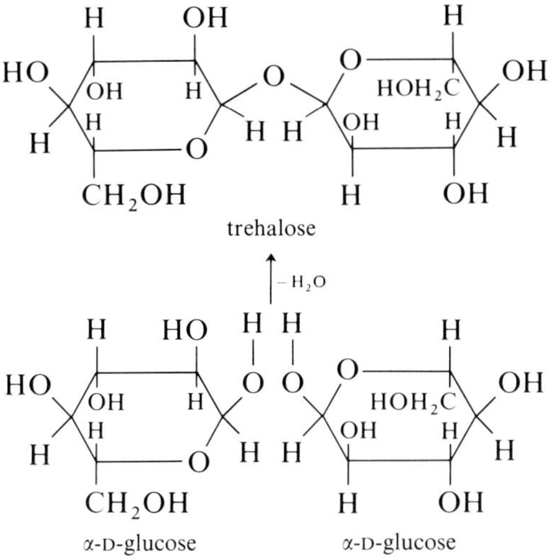

[structure of trehalose] → no ring opening, therefore nonreducing

$\uparrow -H_2O$

α-D-glucose + α-D-glucose

Lactose gets its name (Latin: *lac*, milk) from the fact that it is found in the milk of every mammal with only one known exception, the California sea lion. It is also found in the fruit of the sapodilla tree, from which chicle is obtained.

Most of the important disaccharides have common names, but the systematic name for lactose illustrates how these are put together: 4-*O*-(β-D-galactopyranosyl)-D-glucopyranose. Trehalose is α-D-glucopyranosyl-α-D-glucopyranoside. It is found in a wide variety of algae, lichens, fungi, bacteria and yeasts.

Sucrose is the disaccharide found in sugar bowls. Although it is obtained commercially only from sugar cane and sugar beets, every photosynthetic plant contains some sucrose. It is a nonreducing disaccharide of glucose and fructose (equation 13-13).

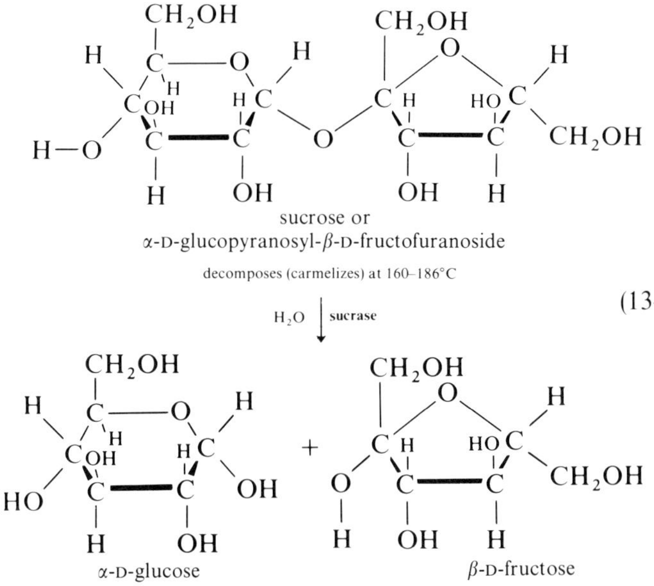

sucrose or
α-D-glucopyranosyl-β-D-fructofuranoside

decomposes (carmelizes) at 160–186°C

H_2O ↓ sucrase

(13-13)

α-D-glucose + β-D-fructose

Carbohydrates

The eight hydroxyl groups of sucrose are available for typical alcohol reactions such as ester formation. Sucrose octaacetate, mp 89°C, is a bitter substance sometimes used to denature alcohol and as an adhesive. It is not very soluble in water but dissolves readily in most organic solvents.

Sucrose esterified with one molecule of a long-chain fatty acid (so that it still has seven hydroxyl groups) is a nonionic detergent. The fatty acid chain tries to dissolve in the grease or oil and the part with the seven hydroxyl groups tries to dissolve in water, much like the ionic end of the more usual detergent molecules. This detergent has the advantage of being destroyed by the bacteria in sewage; that is, it is biodegradable.

α and β Linkages. Like other glycosides, disaccharides are held together either by α or β linkages. In the case of a nonreducing disaccharide, it is necessary to specify the α or β conformation of both monosaccharide units. The hydrolysis of disaccharides is catalyzed by acids, but a much faster method is to use a hydrolytic enzyme. These molecules (proteins, Chapter 14), are specific catalysts. Thus the α-glycosidase of yeast will hydrolyze only α-glycoside linkages and leaves β linkages untouched. Sucrose is hydrolyzed by α-glucosidases, but not by emulsin, a β-glucosidase. Although sucrose contains a β-glycoside linkage, it is a β-fructoside not a β-glucoside. Sucrase, which *does* hydrolyze sucrose, is an enzyme that will hydrolyze β but not α-fructofuranosides. To summarize: enzymes are special catalysts and the enzyme has to be suited not only to the type of glycoside linkage (α or β), but also to the monosaccharide unit.

To recognize α and β linkages in the cyclic structural formulas, stand the ring on end with the glycoside linkage at the top and the oxygen of the ring facing away from the observer. An α-glycoside will have the oxygen of the glycoside linkage on the right, a β-glycoside will have it on the left.

Cellobiose and maltose are reducing disaccharides whose only difference is in the configuration at the glycosidic carbon. Cellobiose is hydrolyzed into two molecules of glucose by emulsin (a β-glucosidase) but not by maltase (an α-glucosidase). The reverse is true of maltose, which has the α linkage.

cellobiose or
4-*O*-(β-D-glucopyranosyl)-D-glucopyranose

maltose or
4-*O*-(α-D-glucopyranosyl)-D-glucopyranose

CELLULOSE

Cellulose is an example of the simpler class of polysaccharide; these are called **homopolysaccharides** because they **contain only one kind of monosaccharide unit**. In the case of cellulose, this unit is glucose linked at positions 1 and 4 and with β-glucoside units exclusively. This is easily proven because, on partial hydrolysis, the only disaccharide that can be identified in the debris is cellobiose, or 4-O-(β-D-glucopyranosyl)-D-glucopyranose. The only monosaccharide formed is D-glucose.

cellulose

The molecular weight of cellulose as determined by means of the ultracentrifuge is about one or two million. Thus n in our structural formula is of the order of 5000. The closest approximation to pure cellulose in nature is cotton fiber, which is about 98% cellulose. It is a major structural fiber in the cell walls of plants of all kinds. Leaves are about 10% cellulose and wood about 50% cellulose.

An animal cellulose, **tunicin**, is the material of the leathery jacket of tunicates, a subphyllum of primitive marine animals (Figure 13-4).

Because cellulose has only β-glucoside linkages, it is not hydrolyzed by the α-glucosidases or amylases found in human saliva and in the intestine. Also, hydrolysis by the hydrochloric acid in the stomach is too slow to be of much use. Cows and termites, on the other hand, can digest cellulose with ease because symbiotic bacteria in their digestive systems make β-glucosidase.

Cellulose has so many hydroxyl groups along the polymer chain that its total insolubility in water deserves some comment. X-ray studies show that cellulose has a rather regular structure that favors hydrogen bonding between the chains rather than hydrogen bonding of the hydroxyl group to water molecules.

Reaction of all or part of the three hydroxyl groups per glucose unit of cellulose gives cellulose esters. Eliminating the hydroxyl groups by esterification makes the cellulose soluble in organic solvents.

The first cellulose esters put to practical use were cellulose nitrates. The conversion of almost all of the hydroxyl groups to $-O-NO_2$, that is, three per glucose unit, gives an acetone-soluble material known as guncotton. Smokeless powder is made from guncotton by adding enough solvent to convert the nitrated fibers into a plastic or gelatinized dough that can be extruded and cut into powder grains, actually perforated cylinders rather than the small lumps of old-fashioned black powder.

A less highly nitrated cellulose, when mixed with alcohol and camphor, gives a plastic material that can be molded or rolled into sheets. After the alcohol has evaporated, the material hardens to a hornlike consistency. This

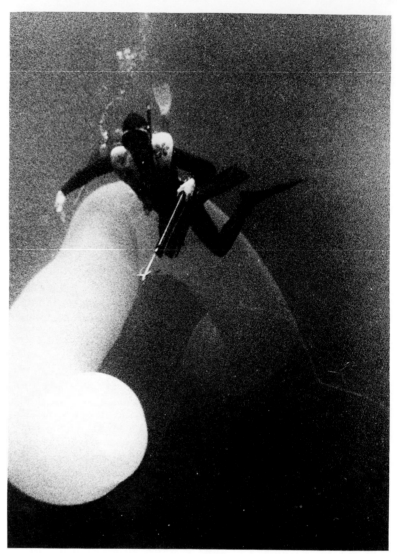

FIGURE 13-4. The giant salp, a compound ascidian or tunicate. [Courtesy of Roger V. Grace, Auckland, New Zealand]

substance, called **celluloid**, was the first plastic made on an industrial scale, and the only one manufactured for about 50 years (1869–1920). It was used for such things as billiard balls (otherwise made of ivory) and celluloid collars. Besides the distinctive smell of the camphor used to keep it from becoming too brittle, celluloid also had the disadvantage of being highly flammable. Persons wearing celluloid collars were well advised not to smoke. Other uses for cellulose nitrate are in gelatin dynamite, in which the cellulose nitrate is plasticized with nitroglycerine, and in cellulose nitrate lacquers.

Cellulose acetate is still an important commercial plastic used for making both films and fibers for cloth (Figure 13-5).

Rayon is a fiber prepared from cellulose that has been made temporarily soluble in water by conversion of some of the hydroxyl groups to xanthate groups, $-OCSS^-Na^+$ (equation 13-14). The viscous xanthate solution is

$$R-OH + CS_2 + NaOH \rightarrow R-O-\overset{\overset{S}{\|}}{C}-S^-Na^+ \qquad (13\text{-}14)$$

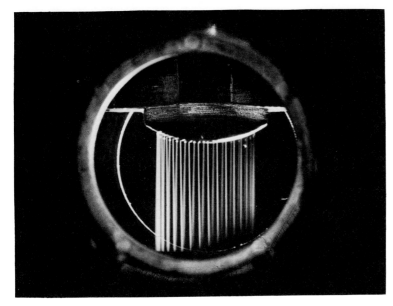

FIGURE 13-5. Spinnerets making filaments from cellulose acetate. [Courtesy of E. I. du Pont de Nemours and Company]

forced through spinnerets to form fibers which are then coagulated and converted back to cellulose by passage through a bath of sodium bisulfate. **Cellophane** is the same material extruded through a slot rather than through a spinneret, and kept soft by the addition of a little glycerol. Cellulose "sponges" are made by coagulating blocks of the viscous solution in which crystals of $Na_2SO_4 \cdot 10H_2O$ have been suspended. Washing with water leaches out the crystals and leaves the holes that make the product a sponge. Natural sponges, which are the skeletons of a marine animal related to coral, are still superior, however.

Lignin, the other main structural material of wood, is a polymeric resin in which the cells, consisting mostly of cellulose, are embedded. Wood is thus a composite material, like glass-reinforced plastic. Although the detailed structures of the lignins from various trees are not known, the lignin of conifers appears to be a polymer in which coniferyl alcohol has been oxidized as well as polymerized. Alkaline oxidation of this lignin gives vanillin in good yield. Unfortunately the demand for vanillin is not great enough to prevent the waste lignin from paper mills from being dumped into streams.

coniferyl alcohol
mp 73°C

vanillin
mp 81–82°C

STARCH

Starch is found mainly as microscopic granules in seeds and in tubers such as the potato where it acts as a reserve supply of food and energy for the growing plant. Although it is also a polyglucoside, starch is a more complicated substance than cellulose. By utilizing differences in solubility, most starches can be separated into two components, **amylose** and **amylopectin** in a ratio of about 1 to 3 or 4, although some grains produce only amylopectin.

Amylose. Amylose is only sparingly soluble even in hot water, but it is the component of starch that gives the blue color of the starch-iodine test. The blue color is due to an **inclusion compound** in which the amylose polymer is arranged in a helical structure with the iodine molecules inside the helix (Figure 13-6). Inclusion compounds are unusual in that they do not depend on ordinary chemical bonds to hold them together; they are essentially molecular boxes with another molecule inside.

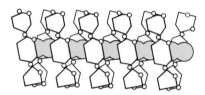

FIGURE 13-6. Model of helical starch chain with iodine molecules in the center of the helix. [After R. E. Rundle and R. R. Baldwin, *J. Amer. Chem. Soc.*, **65**:554 (1943)]

Complete hydrolysis of amylose gives D-glucose. Partial hydrolysis by enzymes gives different products depending on the enzyme used. One enzyme breaks glucoside linkages randomly along the chain. Another attacks only the second link from the "nonreducing" end of the chain, splitting off one molecule of the disaccharide maltose at a time. Since the yield of maltose is quite good, amylose is essentially a poly-(1,4-α-glucoside); it is like cellulose except that amylose has α rather than β linkages. The part of the molecule that is *not* hydrolyzed readily to maltose by this enzyme is believed to have side chains or branches in which the linkages are β (Figure 13-7). The molecular weight of amylose can range from a few thousand to a million and, even within a given sample, some of the molecules are ten times as big as others.

Amylopectin. Amylopectin is also a polyglucoside, but it is a more highly branched molecule than amylose. Apparently it consists of chains of poly-(1,4-α-glucoside) containing about 20 or 25 glucose units; these chains in turn are connected to each other by 1,6-α linkages. The molecular weight is of the order of several million.

Bacillus macerans grown on starch converts it into a number of polyglucosides in which the ends of the polyglucoside chain are joined to form a ring. Known as Schardinger dextrins, these compounds have from six to twelve 1,4-α glucose units in the ring, and are named according to the ring size. For example, the compound with six glucose units is called cyclohexaglucose. It forms inclusion compounds like the starch-iodine complex, but only with molecules that are just the right size—with benzene but not bromobenzene, for example.

GLYCOGEN

Glycogen is a reserve carbohydrate found in animals, some of the molecules being bound to proteins (Chapter 14). Hydrolysis of a

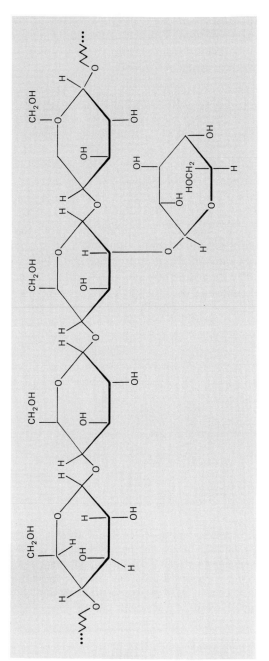

FIGURE 13-7. Part of an amylose chain, showing one β-glucoside branch.

glycogen-containing tissue with alkali breaks the bond to protein and gives the polysaccharide. Glycogen is like amylopectin but more highly branched.

CHITIN

Chitin is the structural material of lobster shells, insects, and fungi. It is essentially identical to cellulose except for an acetylamino group, CH_3CONH-, instead of hydroxyl at carbon 2. Chitin is soluble in concentrated hydrochloric acid. A deacylated chitin obtained commercially from shrimp shells has been used to coat wool and make it shrink-resistant.

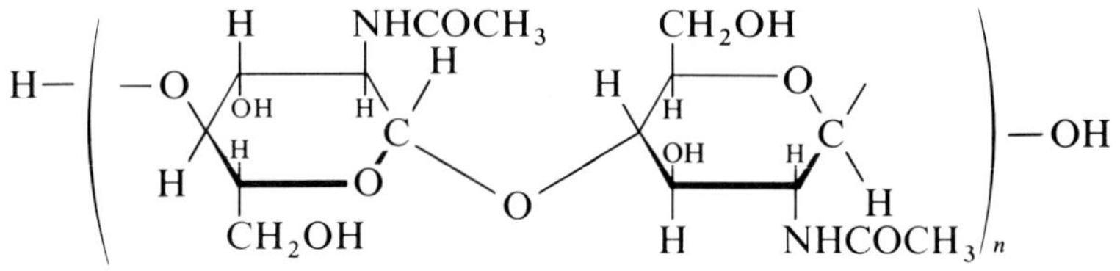

chitin

INULIN

Inulin is a carbohydrate somewhat like starch found in many plants and easily isolated from dahlia bulbs. It is a polyfructoside, consisting of chains of about 30 fructose units with a glucose unit at one end. The fructoside linkages are β, and the disaccharide unit that makes up the end with the glucose is probably like sucrose.

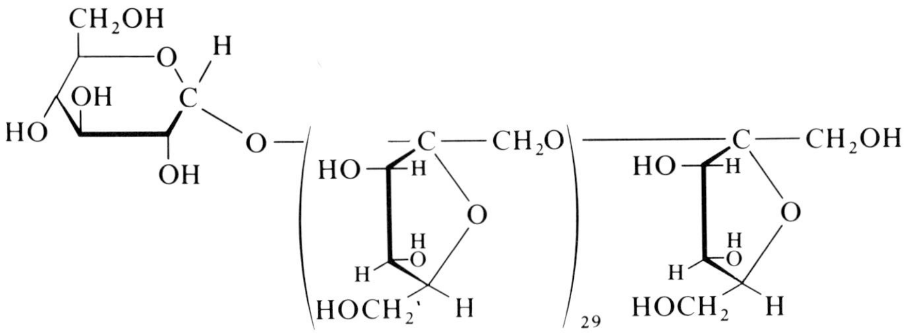

inulin

AGAR

The gel agar, used in biology laboratories to grow colonies of bacteria, comes from seaweed. Like starch, it has more than one component, and these probably differ in the amount of branching. The main component, agarose, is a polysaccharide in which D-galactose units alternate with 3,6-anhydro-L-galactose units. The latter have an ether bond across the ring in place of the normal hydroxyl groups at the 3 and 6 positions.

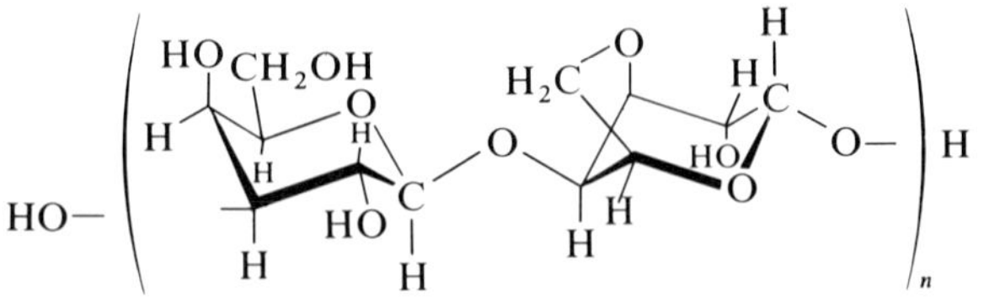

agarose

OTHER POLYSACCHARIDES

Hemicelluloses are polysaccharides found with cellulose in the cell walls of plants. They are smaller molecules than cellulose and have from 50 to 100 monosaccharide units. A typical example is **xylan**, a poly-(1,4-β-xylopyranoside), essentially cellulose with the CH_2OH groups replaced by hydrogen.

Gum arabic, used as mucilage or glue, comes from acacia and is mainly the calcium salt of a polysaccharide called **arabic acid**. Hydrolysis of arabic acid gives L-arabinose, L-rhamnose, D-galactose, and D-glucuronic acid.

The outer covering of type III pneumococcus is an important substance because it is the part of the organism that is "recognized" by our specific immunological defence mechanisms. It is a polysaccharide with a molecular weight of about one million, consisting of β-D-glucose and β-D-glucuronic acid units.

glucuronic acid

Hyaluronic acid is found in the gel-like vitreous humor of the eye. It is probably a polysaccharide in which glucuronic acid and N-acetylglucosamine alternate along the chain. N-Acetylglucosamine is the monosaccharide unit of chitin.

Energy and Carbohydrates

The synthesis of carbohydrates from carbon dioxide and water, with oxygen as a byproduct, is possible in green plants only because the chlorophyll in the chloroplasts absorbs the necessary energy from sunlight and uses it to make this energetically uphill reaction go. The amount of heat required to make 1 g of glucose according to equation 13-15 is about 3750 cal. This same amount of energy is

$$6\,CO_2 + 6\,H_2O \rightarrow C_6H_{12}O_6 + 6\,O_2 \qquad (13\text{-}15)$$

liberated either as heat or partly as heat and partly as work whenever 1 g of glucose is burned to carbon dioxide and water, whether in an engine or by being metabolized in a human body.

The calorie used here is the unit used by chemists, defined as the heat required to raise the temperature of 1 g of water 1°C, from 3.5°C to 4.5°C. The dietician's Calorie, which they treat very reverently and always spell with a capital C, is a unit about 1000 times as large. Thus the heat produced by burning 1 g of glucose is 3750 cal or 3.75 Cal.

Some of the energy made available by metabolizing or burning food is always wasted because it appears as heat rather than as work. If none were wasted as heat, the energy from the combustion of 1 g of glucose would be enough to lift a 1 kg object (a large bird, for example) a distance of about 1.6 km. Therefore a bird would have to eat about the equivalent of 1 g of glucose to obtain the energy to climb (without the help from the thermals that get sailplane enthusiasts up there) to a height of 1 km or so. This seems like a very reasonable amount of food. However, if the same calculation is done without noticing that the calories in nutritional tables are really calories × 1000, the bird has to eat several pounds of food before each flight and the logistics begin to resemble those of a rocket trip to the moon.

The combustion of 1 g of a typical fat (triglyceride) gives about 9500 cal, so fat is a much more energy-rich food than glucose. This is not surprising, since glucose already has a large number of C-O and H-O bonds and might be considered well on the way to carbon dioxide and water to begin with. The combustion of one gram of glycogen, the reserve carbohydrate or "fuel tank" of animals, gives about 4300 cal.

SUMMARY

1. Typical monosaccharides include
    (a) Aldohexoses such as glucose.
    (b) Aldopentoses such as ribose and deoxyribose.
    (c) Ketohexoses such as fructose.
2. Structure of monosaccharides
    (a) The aldohexoses and aldopentoses are aldehydes in equilibrium with a cyclic hemiacetal.
        (i) Opening of the hemiacetal ring causes mutarotation.
        (ii) The aldoketoses are ketones in equilibrium with a cyclic hemiketal.
3. Glycosides are acetal derivatives formed from the hemiacetal form of a sugar and an alcohol or secondary amine functional group.
    (a) They are either α or β depending on the configuration at the acetal carbon atom.
4. Disaccharides are glycosides in which the alcohol functional group is part of a second monosaccharide molecule.
    (a) The linkage may be either α or β.
    (b) Disaccharides are either reducing or nonreducing towards Fehling's solution.
5. Cellulose is a polysaccharide made up of glucose units linked at the 1 and 4 positions with β-glucoside linkages.
    (a) Its molecular weight is about 5000.
    (b) Esterification of the hydroxyl functional groups converts cellulose into materials soluble in organic solvents.
6. Starches are 1,4-polyglucosides with α linkages.
    (a) Amylose has a few side chains with β linkages.
    (b) In amylopectin several 1,4-polyglucoside chains are connected to each other by 1,6-α-glucoside linkages.
7. Other polysaccharides found in nature include glycogen, chitin, inulin, and agar.
8. The combustion of glucose liberates about 3750 cal/g.

EXERCISES

1. Explain what is meant by the terms monosaccharide and disaccharide.

2. Explain what is meant by the term (a) aldohexose; (b) aldopentose; (c) ketohexose.

3. What functional group is characteristic of the cyclic form of a monosaccharide?

4. Draw structural formulas for D-glucose in both the open and cyclic forms. Number the carbon atoms in both formulas.

5. Will the 2,3,4,5,6-pentamethyl ether of glucose mutarotate? Explain your answer.

6. Define (a) α and β anomers and (b) mutarotation.

7. What are the products of the reaction of D-glucose with Fehling's solution? What is the reaction used for?

8. Define
   (a) glucoside
   (b) glycoside
   (c) pyranose
   (d) furanose
   (e) furanoside

9. Write a structural formula for the open chain form of an aldotetrose. How many optical isomers are there?

10. Do the same for a ketotetrose.

11. How can one tell (by looking at a model or at a projection formula) whether a sugar is a D sugar or an L sugar?

12. What is the direction of rotation of the plane of polarized light by a D sugar?

13. Write the open-chain Fischer projection formulas for all the pairs of aldohexoses that would give the same osazone. See equation 13-6 for an example.

14. Write open-chain Fischer projection formulas for all the aldohexoses whose oxidation would give *optically inactive* glycaric acids (1,6-dicarboxylic acids). See equation 13-7 for the oxidation reaction.

15. Write the open-chain Fischer projection formulas for (a) all possible D-aldopentoses and (b) all possible D-2-ketohexoses.

16. (a) How can we prove that the sugar D-allose does not have the structure shown below?
    (b) Look this sugar up (in Figure 13-2, p. 282) and name it

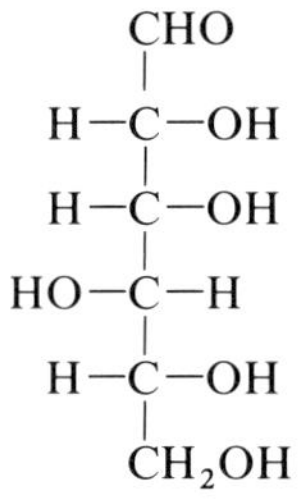

$$\begin{array}{c} \text{CHO} \\ | \\ \text{H} - \text{C} - \text{OH} \\ | \\ \text{H} - \text{C} - \text{OH} \\ | \\ \text{HO} - \text{C} - \text{H} \\ | \\ \text{H} - \text{C} - \text{OH} \\ | \\ \text{CH}_2\text{OH} \end{array}$$

17. Assuming that we know the structure of glucose already, what is an easy way to prove that gulose is

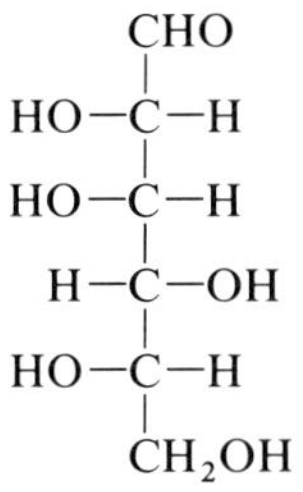

$$\begin{array}{c} \text{CHO} \\ | \\ \text{HO} - \text{C} - \text{H} \\ | \\ \text{HO} - \text{C} - \text{H} \\ | \\ \text{H} - \text{C} - \text{OH} \\ | \\ \text{HO} - \text{C} - \text{H} \\ | \\ \text{CH}_2\text{OH} \end{array}$$

18. Give the structures of α-D-ribofuranose and α-D-2-deoxyribofuranose. What is a good way to remember them?

19. Give the structures of D-fructose and sucrose.

20. Give examples of two naturally occurring glucosides.

21. Explain why maltose and lactose are "reducing" and sucrose is non-reducing. Use structural formulas.

22. Give two examples of naturally occurring disaccharides.

23. Explain why starch, but not cellulose, is digestible by humans.

24. Would it be possible to make cellobiose from the wooden parts of a musical instrument?

25. (a) What is the structure of cellulose?
    (b) What is the main difference between cellulose and amylose?

26. Give part structures for two esters of cellulose and explain why they are soluble in organic solvents whereas cellulose is not.

27. Give a brief description of the structure and function in nature of any two polysaccharides *other* than starch or cellulose.

Chapter 14
Polyamides, Polypeptides, and Proteins

The subject of this chapter is natural and synthetic polyamides in which the monomer units are linked by the amide functional group, $-\overset{H}{\underset{|}{N}}-\overset{O}{\underset{}{\overset{\|}{C}}}-$. The natural polymers are called **polypeptides** and are built up from amino acids, monomer units which contain the necessary amino and carboxyl groups in the same molecule. We will also discuss conjugated proteins, which are polypeptides that also have nonpeptide parts called prosthetic groups.

Proteins play many important roles in living systems. They are found in structural materials like cartilage, finger nails, or insect cocoons, in muscle and contractile fibers, and in the almost incredibly efficient and selective catalysts which we call enzymes.

In the field of protein chemistry the amide linkage was given the special name **peptide bond** because when meat or other protein is digested these bonds are hydrolyzed and broken (equation 14-1). The word **peptide** comes from the Greek πεπτειν (peptein), which means to digest. The same root is found in words like dyspeptic and eupeptic.

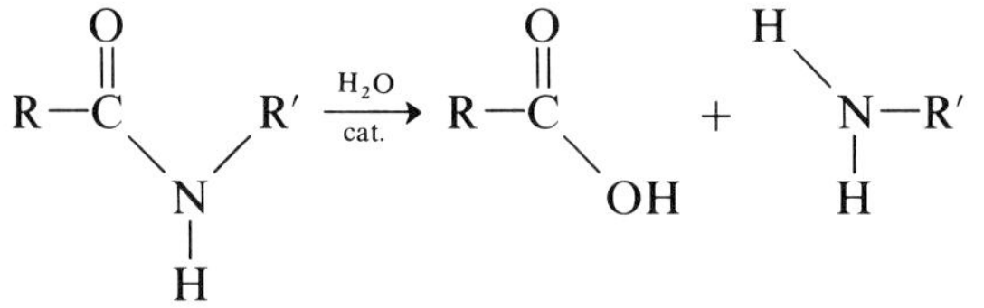

$$R-\overset{O}{\overset{\|}{C}}\diagdown_{\underset{H}{\overset{|}{N}}}\diagup R' \xrightarrow[\text{cat.}]{H_2O} R-\overset{O}{\overset{\|}{C}}-OH + \overset{H}{\underset{H}{\overset{|}{N}}}-R' \qquad (14\text{-}1)$$

the peptide bond

The acid and amino groups formed by the hydrolysis of an amide or peptide bond of course tend to react to form salts (equation 14-2).

$$R-\overset{O}{\underset{\|}{C}}-OH + H_2NR' \leftrightarrows RCOO^- \; H_3\overset{+}{N}R' \qquad (14\text{-}2)$$

If an amide is hydrolyzed with a strong acid such as HCl as the catalyst, the products will be the free carboxylic acid RCOOH and a **salt**, $R'NH_3^+Cl^-$, of the amine. Hydrolysis of an amide with a basic catalyst such as NaOH gives the free amine RNH_2 and the salt $RCOO^-Na^+$ of the carboxylic acid.

Amides are sometimes made from the corresponding acid and amine as in equation 14-3.

$$RCOOH + H_2NR' \longrightarrow RCOO^- H_3\overset{+}{N}R' \xrightarrow{heat} R-\overset{O}{\underset{\|}{C}}-\overset{H}{\underset{|}{N}}-R' + H_2O \qquad (14\text{-}3)$$

Nylon

The most common kind of nylon is known as nylon 66. Introduced into commercial production in 1939, it is made from a six-carbon dicarboxylic acid and a six-carbon diamine.

$$HOOCCH_2CH_2CH_2CH_2COOH + H_2NCH_2CH_2CH_2CH_2CH_2CH_2NH_2$$

adipic acid   hexamethylene diamine

↓

"nylon salt"

↓ heat, pressure 220–200°C

$$HOOC(CH_2)_4-\overset{O}{\underset{\|}{C}}-\left(\overset{H}{\underset{|}{N}}-CH_2CH_2CH_2CH_2CH_2CH_2-\overset{H}{\underset{|}{N}}-\overset{O}{\underset{\|}{C}}-CH_2CH_2CH_2CH_2-\overset{O}{\underset{\|}{C}}\right)_{\sim 60}$$

$$-\overset{H}{\underset{|}{N}}-(CH_2)_6-NH_2 + 120\,H_2O$$

nylon 66

mp 260°C; soluble in HCOOH and phenol

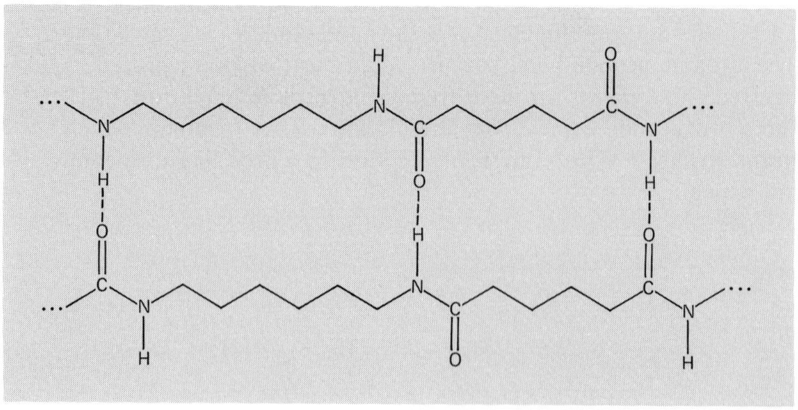

FIGURE 14-1. Hydrogen bonding between adjacent molecules in cold-drawn nylon 66.

To make fibers from a nylon, the molten material is extruded as filaments. After being cooled, the filaments are stretched or cold-drawn to about four times their original length, a process that orients the molecules parallel to the axis of the fiber. The oriented or drawn fibers obtain additional strength because of hydrogen bonds between adjacent parallel molecules (Figure 14-1).

FIGURE 14-2. Ahah!

Silk Fibroin

Silk fibers, spun by the Japanese silkworm, are composed of **fibroin** (Figure 14-3) one of the simplest naturally occurring polypeptides. Like nylon, fibroin consists of long-chain molecules with amide or peptide linkages between the monomer units. It resembles cold-drawn nylon in that the polymer molecules are arranged in a regular way that encourages hydrogen bonding between adjacent chains. Because of the greater frequency of peptide linkages and hydrogen bonds between the chains, silk is completely insoluble and decomposes or chars before it melts.

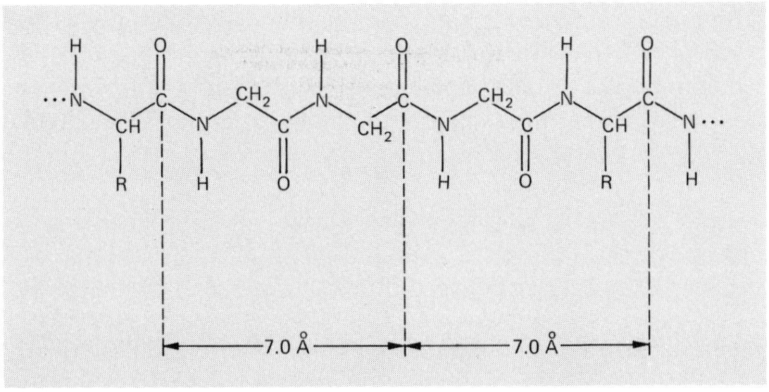

FIGURE 14-3. Silk fibroin. The groups R at intervals along the chain are variously CH_3-, $HO-\langle\bigcirc\rangle-CH_2-$, and $HO-CH_2-$.

Polyamides, Polypeptides, and Proteins

Both nylon and silk or wool fibers are subject to degradation by hydrolysis of the amide or peptide bonds. The hydrolysis is catalyzed by strong acids or bases, which is one reason for wearing old clothes in the laboratory. In addition, the hydrolysis of natural polymers such as silk and wool is catalyzed by proteolytic enzymes.

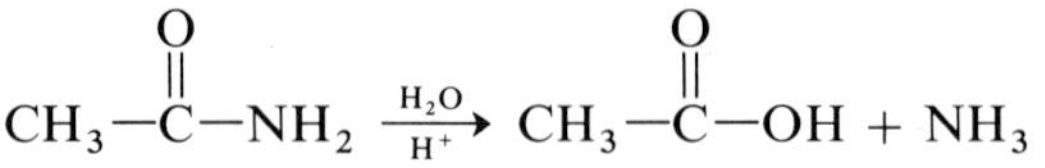

$$CH_3-\underset{acetamide}{\overset{O}{\underset{\|}{C}}-NH_2} \xrightarrow[H^+]{H_2O} CH_3-\underset{acetic\ acid}{\overset{O}{\underset{\|}{C}}-OH} + \underset{ammonia}{NH_3}$$

...C—N—CH—C—N—CH$_2$—C—N—CH$_2$—C... part of a fibroin molecule and some water molecules

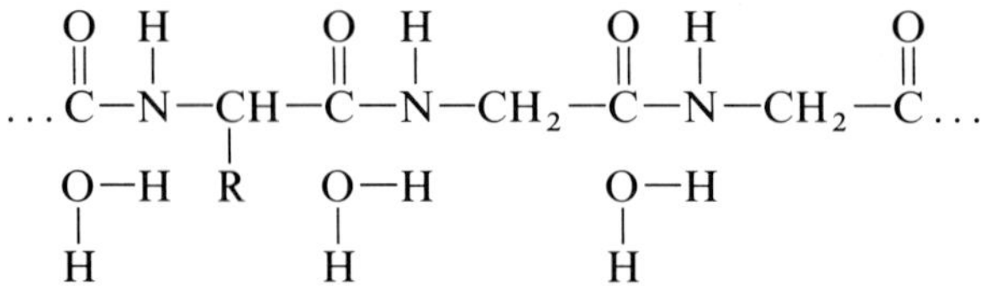

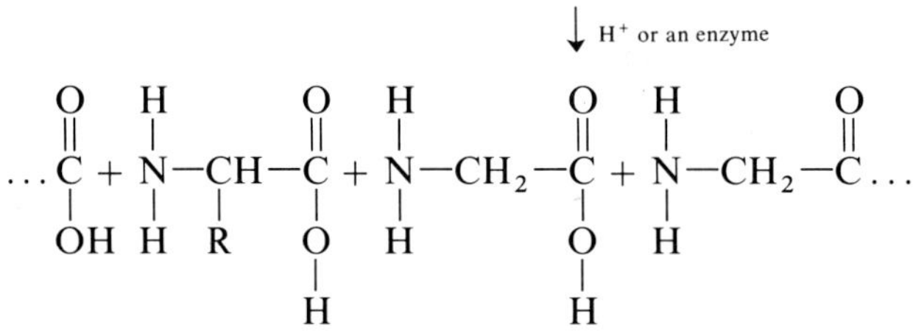

(with O-H, R, O-H, O-H groups and H's shown)

↓ H$^+$ or an enzyme

...C + N—CH—C + N—CH$_2$—C + N—CH$_2$—C... amino acids

$$H_2N-\underset{CH_3}{\underset{|}{CH}}-\overset{O}{\underset{\|}{C}}-OH \qquad H_2N-\underset{\underset{\underset{OH}{|}}{\underset{|}{\bigcirc}}}{\underset{|}{\underset{CH_2}{|}}}CH-\overset{O}{\underset{\|}{C}}-OH \qquad H_2N-\underset{\underset{OH}{|}}{\underset{CH_2}{\underset{|}{CH}}}-\overset{O}{\underset{\|}{C}}-OH \qquad H_2N-CH_2-\overset{O}{\underset{\|}{C}}-OH$$

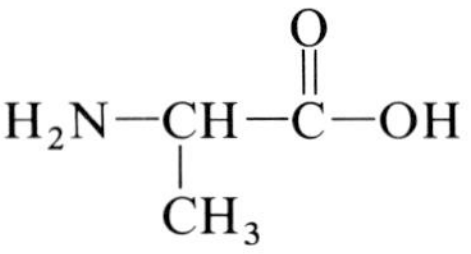

alanine 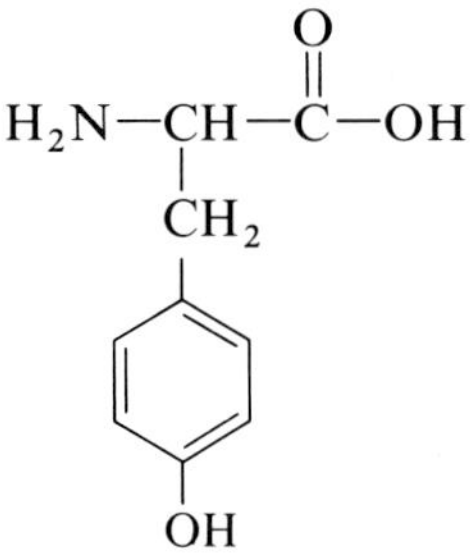   tyrosine 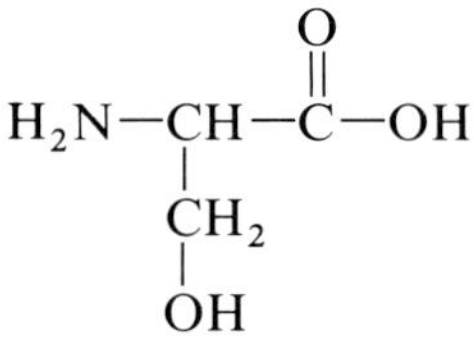   serine 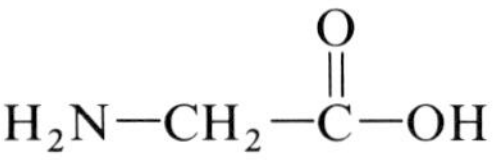   glycine

Amino Acids

STRUCTURE

By convention, the formulas for amino acids are usually written as we have done, with amino and carboxyl groups. However, these compounds are in fact **dipolar ions** containing ammonium and carboxylate rather than uncharged functional groups. This dipolar

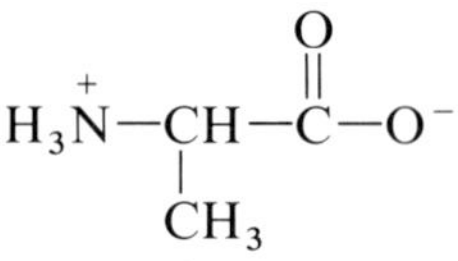

$$H_3\overset{+}{N}-\underset{CH_3}{\underset{|}{CH}}-\overset{O}{\underset{\|}{C}}-O^-$$
alanine

ionic structure means that most amino acids are readily soluble in water, not very soluble in organic solvents, and that they decompose on heating to the melting point.

The reaction of an amino acid with a proton from a strong acid converts the dipolar ion into a cation. Similarly, the loss of a proton by reaction with a strong base converts the dipolar ion into an anion (Figure 14-4). Just what concentration of H^+ or HO^- it takes to convert the amino acid dipolar ion into the cation or the anion depends on the particular amino acid. That is, at a given concentration of H^+, one amino acid will be in the cationic form, whereas another is still mostly dipolar ion.

This is the basis for **electrophoresis**, a method of separating amino acids (and also proteins). The pH ($-\log[H^+]$) at which a given substance has no net positive or negative charge is called the **isoelectric point**. At its isoelectric point, the amino acid or protein will not migrate in an electric field. In solutions more acidic than the isoelectric point, the substance migrates to the cathode; in solutions more basic than the isoelectric point, the substance migrates to the anode (Figure 14-4).

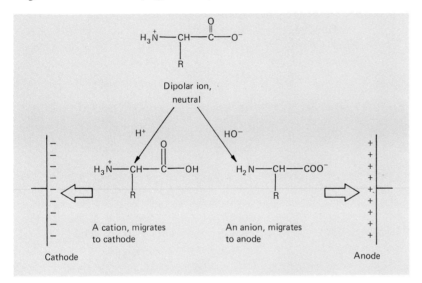

FIGURE 14-4. Reaction of an amino acid with acid or base and the behavior of the product in an electric field.

Most amino acids have just one amino and one carboxyl group, and their isoelectric points are near pH 7. There are a few, however, that have either an extra amino group or an extra carboxyl group. Lysine is an example of a basic amino acid, and aspartic acid is an example of an acidic amino acid. The isoelectric point of lysine is 9.74, a very alkaline solution. The isoelectric point of aspartic acid is 2.77, a very acidic solution.

$$\begin{array}{cc}
\text{H} & \text{H} \\
| & | \\
H_2N-C-COOH & H_2N-C-COOH \\
| & | \\
CH_2 & CH_2 \\
| & | \\
CH_2 & COOH \\
| & \\
CH_2 & \\
| & \\
CH_2-NH_2 & \\
\text{L-lysine} & \text{L-aspartic acid} \\
\text{mp } 224°\text{C (dec.)}; [\alpha]_D^{20} = +14.6° & \text{mp } 270°\text{C}; [\alpha]_D^{20} = +25.0° \text{ in } 6\ N\ \text{HCl}
\end{array}$$

Polyamides, Polypeptides, and Proteins

Degradation of lysine by the action of the enzymes of the comma bacillus on meat removes the carboxyl group and gives cadaverine. Cadaverine is also found in the discharge from the bowels of cholera victims. Highly poisonous,

$$H_2NCH_2CH_2CH_2CH_2CH_2NH_2$$
cadaverine
mp 9°C; bp 178–180°C

cadaverine and the closely related putrescine (1,4-diaminobutane) are called ptomaines. They may be partly responsible for the toxic effects of spoiled meat.

TABLE 14-1. *Amino Acids from Protein*

Name	Abbreviation	Formula
A. Neutral		
1. Glycine	Gly	$CH_2(NH_2)-COOH$
2. Alanine	Ala	$CH_3-CH(NH_2)-COOH$
3. Phenylalanine*	Phe	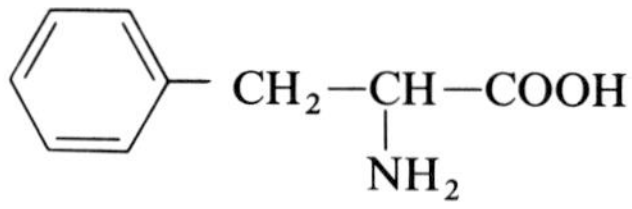 $C_6H_5-CH_2-CH(NH_2)-COOH$
4. Tyrosine	Tyr	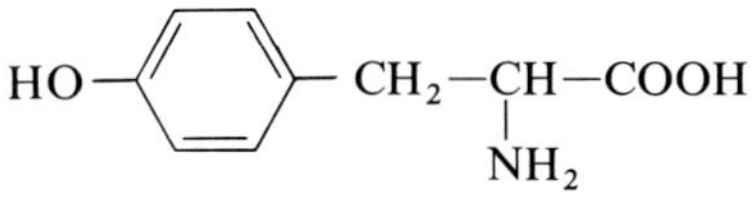 $HO-C_6H_4-CH_2-CH(NH_2)-COOH$
5. Tryptophan*	Try	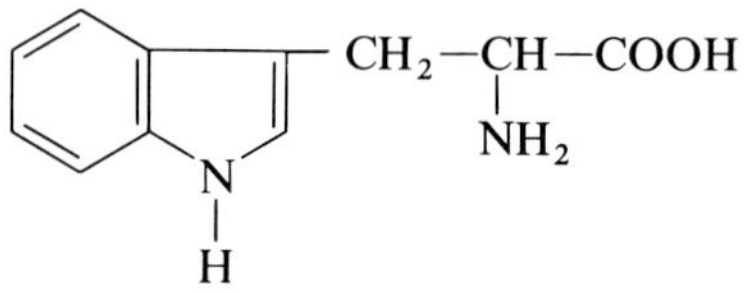 indolyl-$CH_2-CH(NH_2)-COOH$
6. Thyroxine	Thy	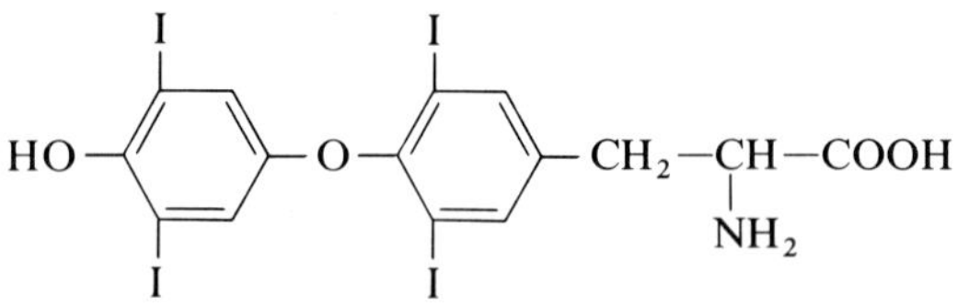 $HO-C_6H_2I_2-O-C_6H_2I_2-CH_2-CH(NH_2)-COOH$
7. Serine	Ser	$CH_2(OH)-CH(NH_2)-COOH$
8. Threonine*	Thr	$CH_3-CH(OH)-CH(NH_2)-COOH$
9. Valine*	Val	$CH_3-CH(CH_3)-CH(NH_2)-COOH$
10. Leucine*	Leu	$CH_3-CH(CH_3)-CH_2-CH(NH_2)-COOH$
11. Isoleucine*	Ile	$CH_3-CH_2-CH(CH_3)-CH(NH_2)-COOH$

TABLE 14-1. *Amino Acids from Protein (cont.)*

Name	Abbreviation	Formula
12. Proline	Pro	$\begin{array}{c} H_2C - CH_2 \\ | \quad\quad | \\ H_2C \quad CH-COOH \\ \backslash N / \\ H \end{array}$
13. Hydroxyproline	HPro	$\begin{array}{c} HO-HC - CH_2 \\ | \quad\quad | \\ H_2C \quad CH-COOH \\ \backslash N / \\ H \end{array}$
14. Cysteine	CyS	$HS-CH_2-CH(NH_2)-COOH$
15. Cystine	CyS-SCy	$\begin{array}{c} S-CH_2-CH(NH_2)-COOH \\ | \\ S-CH_2-CH(NH_2)-COOH \end{array}$
16. Methionine*	Met	$CH_3-S-CH_2-CH_2-CH(NH_2)-COOH$

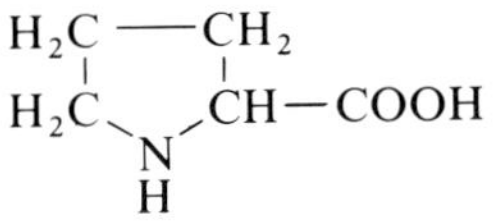

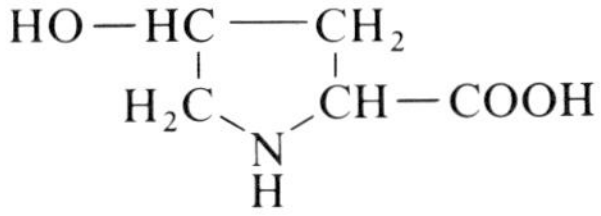

B. Basic amino acids

Name	Abbreviation	Formula
17. Arginine	Arg	$HN=C(NH_2)-N(H)-CH_2-CH_2-CH_2-CH(NH_2)-COOH$
18. Lysine*	Lys	$CH_2(NH_2)-CH_2-CH_2-CH_2-CH(NH_2)-COOH$
19. Histidine	His	$\begin{array}{c} HC=C-CH_2-CH(NH_2)-COOH \\ | \quad | \\ N \quad NH \\ \backslash CH / \end{array}$

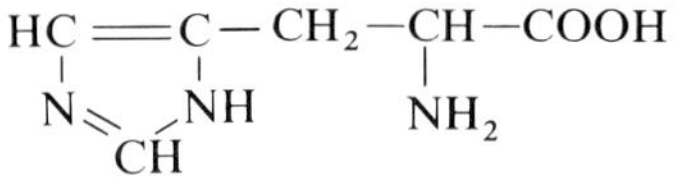

C. Acidic amino acids

Name	Abbreviation	Formula
20. Aspartic acid	Asp	$HOOC-CH_2-CH(NH_2)-COOH$
21. Glutamic acid	Glu	$HOOC-CH_2-CH_2-CH(NH_2)-COOH$

D. Amides

Name	Abbreviation	Formula
22. Asparagine	Asn or AspNH$_2$	$H_2N-CO-CH_2-CH(NH_2)-COOH$
23. Glutamine	Gln or GluNH$_2$	$H_2N-CO-CH_2-CH_2-CH(NH_2)-COOH$

* Amino acids essential to the rat and probably also essential to man.

The sodium salt of glutamic acid (Table 14-1) is used to enhance the flavor of meat and is likely to be an ingredient of soy sauce. A few people are sensitive to glutamic acid, and their symptoms have come to be known as the Chinese restaurant syndrome. It is only the naturally occurring L-enantiomer that has the flavor-enhancing property.

FIGURE 14-5. There were no footprints here and no amino acids. Samples of soil from the moon have, in fact, been examined for amino acids and polypeptides. If these substances are present at all, the amounts are extremely small. [Courtesy National Aeronautics and Space Administration, Houston, Tex.]

OPTICAL ACTIVITY

With a few exceptions, the amino acids obtained by hydrolyzing naturally occurring proteins or polypeptides such as silk fibroin have the general structure shown in Figure 14-6. Except for glycine, in which R is just a second hydrogen atom, the amino acids from protein (Table 14-1) are all optically active and all belong to the L series. There are good reasons for using amino acids all from the same series in the construction of a protein. For example, silk fibroin is an isotactic polymer and forms stronger fibers than it would if it were an atactic polymer made from a random sequence of D- and L-amino acid units.

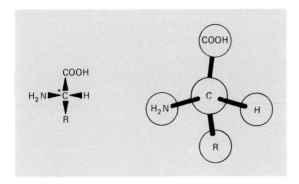

FIGURE 14-6. Projection diagram and model of an L-amino acid.

ESSENTIAL AMINO ACIDS

The amino acids marked with a star in Table 14-1 are those essential in the diet of the rat and probably also of man. Some amino acids, but not others, are necessary in the diet because our bodies have lost the ability to synthesize these particular substances during the course of evolution. This evolutionary path has been possible only because of our symbiotic relationship with other animals and plants that have kept the ability to synthesize the essential amino acids.

When a certain essential amino acid is present in insufficient quantity, the tissues simply make less of any protein containing that acid; the result is a deficiency in various important enzymes. Having just the right amino acid in just the right place in a protein is so important that if one amino acid is unavailable it is not just left out. Instead the synthesis of that protein ceases completely.

The Primary Structure of Polypeptides

ANALYSIS OF THE HYDROLYSATE

The first step in determining the structure of a polypeptide or a protein is to hydrolyze it and find out what kind of amino acid units it contains and their relative proportions. This is usually done by chromatographing the mixture on an **ion-exchange resin**. The ion-exchange resin is typically a crosslinked polystyrene with sulfonic acid groups (SO_3H). These are converted to the sodium salt by treatment with an acidic buffer containing Na^+. When the mixture of amino acids is introduced at the top of the chromatographic column, amino acid cations take the place of Na^+ as in equation 14-4. As the column is washed with a buffered salt solution,

$$\text{\textasciitilde\textasciitilde} - H_2C-CH-\text{\textasciitilde\textasciitilde} \begin{array}{c} \\ \\ \text{C}_6H_4 \\ | \\ SO_3^-Na^+ \end{array} + H_3\overset{+}{N}-\underset{R}{\overset{H}{\underset{|}{C}}}-COOH \longrightarrow \text{\textasciitilde\textasciitilde} - H_2C-CH-\text{\textasciitilde\textasciitilde} \begin{array}{c} \\ \\ \text{C}_6H_4 \\ | \\ SO_3^-H_3\overset{+}{N}-\underset{R}{\overset{H}{\underset{|}{C}}}-COOH \end{array} + Na^+$$

(14-4)

the amino acid cations are in turn displaced from the ion-exchange resin and tend to move down the column. Because they are displaced and move at different rates, different acids arrive at the bottom of the column at different times. Because the job of analyzing amino acid mixtures is one that needs to be done very often, the whole procedure has been made automatic (Figure 14-7).

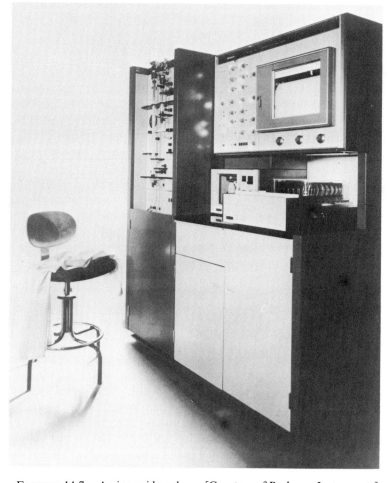

FIGURE 14-7. Amino acid analyzer. [Courtesy of Beckman Instruments]

END-GROUP ANALYSIS

A simple polypeptide chain should have only one α-amino terminal group and only one α-carboxyl terminal group.[1] That is, its structure can be represented by

$$H_2N-\underset{R}{\underset{|}{C}}H-\underset{}{\overset{O}{\overset{\|}{C}}}-\underset{}{\overset{H}{\overset{|}{N}}}\ldots\ldots\ldots\overset{O}{\overset{\|}{C}}-\underset{}{\overset{H}{\overset{|}{N}}}-\underset{R}{\underset{|}{C}}H-\overset{O}{\overset{\|}{C}}OH$$

Some proteins contain more than one polypeptide chain, and this can be detected by the presence of more than one α-amino terminal group and more than one α-carboxyl terminal group (Figure 14-8).

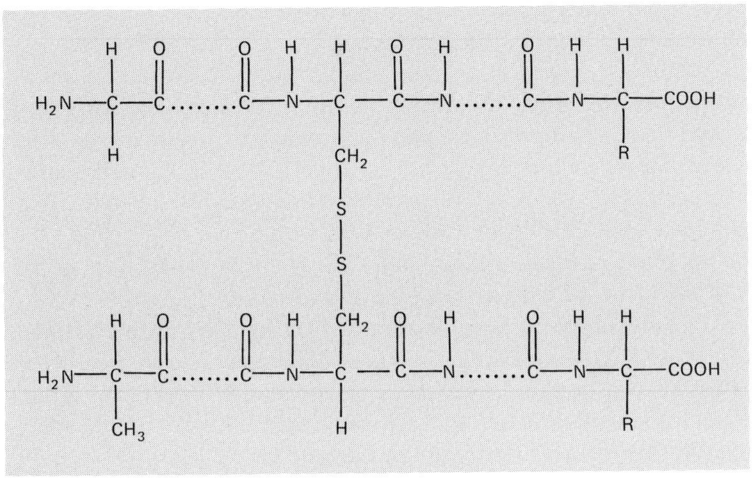

FIGURE 14-8. Two polypeptide chains combined into a single molecule by a cystine crosslink.

α-Amino end groups are determined by converting them to dinitrophenyl derivatives, hydrolyzing the polypeptide, and noting how many molecules of the N-(dinitrophenyl)amino acid are produced. First the polypeptide is treated with 2,4-dinitrofluorobenzene as in equation 14-5. Next, the polypeptide is hydrolyzed

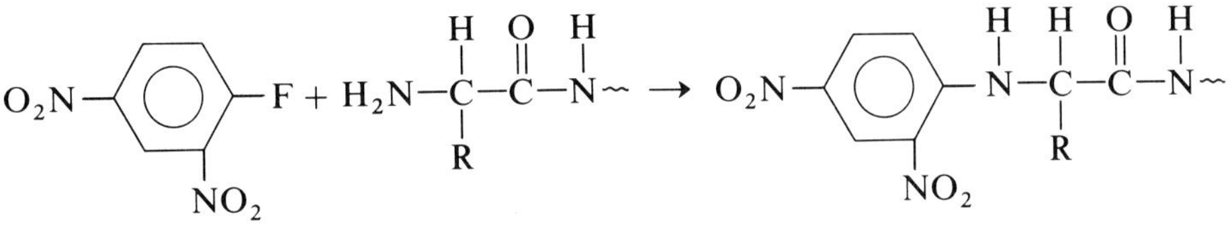

$$O_2N-\underset{NO_2}{\underset{}{\bigcirc}}-F + H_2N-\underset{R}{\underset{|}{C}}H-\overset{O}{\overset{\|}{C}}-\overset{H}{\overset{|}{N}}\sim \rightarrow O_2N-\underset{NO_2}{\underset{}{\bigcirc}}-\overset{H}{\overset{|}{N}}-\underset{R}{\underset{|}{C}}H-\overset{O}{\overset{\|}{C}}-\overset{H}{\overset{|}{N}}\sim$$

(14-5)

---

[1] If the polypeptide contains basic or acidic amino acids, it will have additional amino and carboxyl groups, but they will not be α groups. For example, if it contains lysine, the α-amino group of that amino acid will be part of the peptide bond and only the epsilon(ε)-amino group will be left free.

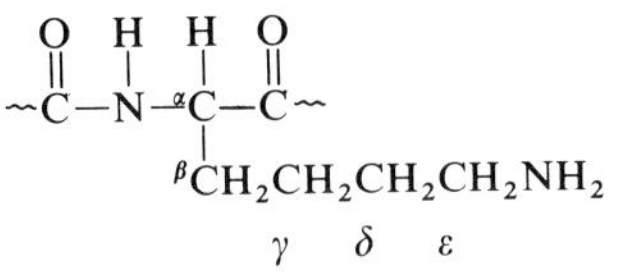

$$\sim\overset{O}{\overset{\|}{C}}-\overset{H}{\overset{|}{N}}-\overset{H}{\underset{\beta CH_2CH_2CH_2CH_2NH_2}{\overset{|}{\underset{}{C}}}}-\overset{O}{\overset{\|}{C}}\sim$$

γ δ ε

as in equation 14-6.

$$O_2N-\underset{NO_2}{\underset{|}{\bigcirc}}-\overset{H}{\underset{|}{N}}-\overset{H}{\underset{|}{C}}-\overset{O}{\overset{\|}{C}}-\overset{H}{\underset{|}{N}}-\overset{H}{\underset{|}{C}}-\overset{O}{\overset{\|}{C}}-\overset{H}{\underset{|}{N}}\sim \xrightarrow{H_2O}$$

$$O_2N-\underset{NO_2}{\underset{|}{\bigcirc}}-\overset{H}{\underset{|}{N}}-\overset{H}{\underset{|}{C}}-COOH + H_2N-\overset{H}{\underset{R}{\underset{|}{C}}}-COOH + \ldots \quad (14\text{-}6)$$

It is very easy to separate the 2,4-dinitrophenyl-substituted amino acids from the ordinary ones and determine their amounts. An additional bonus is the identification of the amino-terminal (N-terminal) amino acid from the known N-(2,4-dinitrophenyl)-amino acid obtained. For example, the substance in Figure 14-8 would have two molecules of N-(2,4-dinitrophenyl)amino acid for each molecule of the polypeptide. One of these would be a derivative of glycine, H_2NCH_2COOH, and the other would be a derivative of alanine, $H_2NCH(CH_3)COOH$. We would conclude that this polypeptide had two chains, one with a glycyl N-terminal group and the other with an alanyl N-terminal group.

The polypeptide insulin, a hormone that regulates the metabolism of sugar, has two different polypeptide chains, one with a glycyl N-terminal group and one with a phenylalanyl N-terminal group. Human hemoglobin has four polypeptide chains, all four of which have valine as the N-terminal amino acid.

Carboxy-terminal groups can be counted and the carboxy-terminal amino acid identified by treating the polypeptide with hydrazine. This reagent converts all the amino acid units to **hydrazides**, except for the carboxy-terminal unit. For example, the tripeptide Gly-Ala-Leu would react as shown in equation 14-7. This mixture is easily separated, and only leucine would be isolated as the free amino acid rather than as the hydrazide. Therefore the carboxy-terminal unit would be known to be leucine.

$$H_2N-\overset{H}{\underset{H}{\overset{|}{C}}}-\overset{O}{\overset{\|}{C}}-\overset{H}{\underset{|}{N}}-\overset{H}{\underset{CH_3}{\overset{|}{C}}}-\overset{O}{\overset{\|}{C}}-\overset{H}{\underset{|}{N}}-\overset{H}{\underset{CH_2-C(CH_3)_2H}{\overset{|}{C}}}-\overset{O}{\overset{\|}{C}}-OH + 2\,H_2NNH_2 \rightarrow$$

glycylalanylleucine
(Gly-Ala-Leu)   hydrazine

$$H_2N-CH_2-\overset{O}{\overset{\|}{C}}-NHNH_2 + H_2N-\overset{H}{\underset{CH_3}{\overset{|}{C}}}-\overset{O}{\overset{\|}{C}}-NHNH_2 + H_2N-\overset{H}{\underset{CH_2-C(CH_3)_2H}{\overset{|}{C}}}-\overset{O}{\overset{\|}{C}}-OH \quad (14\text{-}7)$$

glycyl hydrazide   alanyl hydrazide   leucine

THE AMINO ACID SEQUENCE

After the number of end groups and polypeptide chains has been determined, the next step in establishing the structure of the polypeptide is to determine the sequence in which the amino acids occur along the chain. A dipeptide composed of glycine and alanine has two isomers, one with glycine as the *N*-terminal unit and the other with alanine in that position.

$$\underset{\text{Gly-Ala}}{H_2N-CH_2-\underset{\underset{}{\|}}{\overset{O}{C}}-\underset{\underset{}{H}}{\overset{H}{N}}-\underset{\underset{CH_3}{|}}{\overset{H}{C}}-COOH} \qquad \underset{\text{Ala-Gly}}{H_2N-\underset{\underset{CH_3}{|}}{\overset{H}{C}}-\underset{\underset{}{\|}}{\overset{O}{C}}-\underset{\underset{}{H}}{\overset{H}{N}}-CH_2-COOH}$$

A tripeptide composed of glycine, alanine, and leucine has six isomers. Using the convention that the *N*-terminal unit is put on the left end, these isomers can be represented by Gly-Ala-Leu, Gly-Leu-Ala, Ala-Gly-Leu, Ala-Leu-Gly, Leu-Ala-Gly, and Leu-Gly-Ala. With increasing size of the molecule, the number of

$$H_2N-CH_2-\overset{O}{\underset{\|}{C}}-\overset{H}{\underset{|}{N}}-\underset{\underset{CH_3}{|}}{CH}-\overset{O}{\underset{\|}{C}}-\overset{H}{\underset{|}{N}}-\underset{\underset{\underset{H}{|}}{\underset{CH_3-C-CH_3}{|}}{\underset{CH_2}{|}}}{\overset{H}{C}}-COOH$$

Gly-Ala-Leu
(glycylalanylleucine)

$$H_2N-CH_2-\overset{O}{\underset{\|}{C}}-\overset{H}{\underset{|}{N}}-\underset{\underset{\underset{H}{|}}{\underset{CH_3-C-CH_3}{|}}{\underset{CH_2}{|}}}{\overset{H}{C}}-\overset{O}{\underset{\|}{C}}-\overset{H}{\underset{|}{N}}-\underset{\underset{CH_3}{|}}{\overset{H}{C}}-COOH$$

Gly-Leu-Ala
(glycylleucylalanine)

isomers goes up very fast. A polypeptide with 20 amino acids each used just once would have 2×10^{18} isomers. This would be of no great consequence if all 2×10^{18} isomers were equally effective (as enzymes or hormones, for example), but biochemical specificity is so great that a change of position of even one amino acid unit is likely to have a large effect. This means that a mere analysis of amino acid content is not enough; we need ways of determining the structures of particular polypeptides.

The amino acid sequence of a *small* peptide molecule, say a tetrapeptide, can be determined easily by the method of partial hydrolysis with aqueous HCl. Thus Gly-Arg-Leu-Val on partial hydrolysis will give the following:

1. All four amino acids, because some of the molecules will be completely hydrolyzed.
2. Two tripeptides, Gly-Arg-Leu and Arg-Leu-Val.
3. Three dipeptides, Gly-Arg, Arg-Leu, and Leu-Val.

This mixture of nine compounds can still be separated without too much trouble, and the known structures of the di- and tripeptides makes the structure of the original tetrapeptide obvious.

With large polypeptides, the very large number of hydrolysis products makes hydrolysis with aqueous HCl impractical; **proteolytic enzymes** are used to catalyze the reaction instead. Enzymatic hydrolysis is usually quite specific for a particular type of bond. For example, **crystalline trypsin** will hydrolyze only peptide bonds in which the carbonyl end is part of a lysine or arginine unit. Used on our hypothetical Gly-Arg-Leu-Val, tripsin-catalyzed hydrolysis would give only the two peptides Gly-Arg and Leu-Val instead of a mixture of nine compounds. Other proteolytic enzymes are specific catalysts for hydrolyzing other types of peptide bond.

Insulin. A classic example of the use of these methods was the elucidation of the structure of bovine insulin, shown in Figure 14-9.

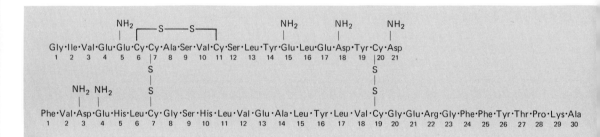

FIGURE 14-9. The structure of bovine insulin, established in 1955 by Sanger and his colleagues.

The *N*-terminal groups are at the left ends of the chains. Bovine insulin contains 51 amino acid units, one *N*-terminal glycine and one *N*-terminal phenylalanine. This tells us that there are two peptide chains, Gly . . . and Phe . . . , linked together in some way. The two chains were separated by a reaction that cleaves the S-S bonds of Cy-S-S-Cy, converting the crosslink into two $CySO_3H$, or cysteic acid units, one in each chain. Next, the two resulting polypeptides, called A and B, were separated and studied one at a time.

$$\text{Bovine insulin} \xrightarrow{\text{H–C(=O)–OOH}} \begin{array}{c} \boxed{\text{chain A}} \\ || \\ SO_3H SO_3H \\ SO_3H SO_3H \\ || \\ \boxed{\text{chain B}} \end{array} \quad (14\text{-}8)$$

Chain B turned out to be a basic peptide containing 30 amino acid units and an *N*-terminal phenylalanine unit. Chain A turned out to be an acidic polypeptide of 21 units with a glycine *N*-terminal group.

$$\text{H}_2\text{N}-\text{CH}_2-\overset{\text{O}}{\underset{\|}{\text{C}}}-\left(\underset{\underset{\text{R}}{|}}{\overset{\text{H}}{\underset{|}{\text{N}}}}-\overset{\text{H}}{\underset{|}{\text{C}}}-\overset{\text{O}}{\underset{\|}{\text{C}}}\right)_{19}-\overset{\text{H}}{\underset{|}{\text{N}}}-\underset{\underset{\text{R}}{|}}{\overset{\text{H}}{\underset{|}{\text{C}}}}-\text{COOH}$$

chain A

$$\text{H}_2\text{N}-\overset{\text{H}}{\underset{\underset{\text{C}_6\text{H}_5}{|}}{\text{C}}}-\overset{\text{O}}{\underset{\|}{\text{C}}}-\left(\underset{\underset{\text{R}}{|}}{\overset{\text{H}}{\underset{|}{\text{N}}}}-\overset{\text{H}}{\underset{|}{\text{C}}}-\overset{\text{O}}{\underset{\|}{\text{C}}}\right)_{28}-\overset{\text{H}}{\underset{|}{\text{N}}}-\underset{\underset{\text{R}}{|}}{\overset{\text{H}}{\underset{|}{\text{C}}}}-\text{COOH}$$

chain B

Partial hydrolysis of the 2,4-dinitrophenyl derivative of A gave not only N-(2,4-dinitrophenyl)glycine, but also a number of known peptides beginning with an N-(dinitrophenyl)glycyl unit. This established the sequence of the first four units. The initial sequence for B was determined in the same way.

Gly-Ile-Val-Glu . . .   Phe-Val-Asp-Glu . . .
    chain A   chain B

The complete sequence of each chain was pieced together by an overlap study (Figure 14-10) of various small peptides obtained in the hydrolysis. These left only one possibility for the structure of the entire chain. The positions of the crosslinks between chains A and B were determined by isolating fragments such as that shown in Figure 14-11.

```
Ser·Leu          Glu·Leu              Asp·Tyr
   Leu·Tyr          Leu·Glu              Tyr·CySO₃H
      Tyr·Glu            Glu·Asp
Ser·Leu·Tyr          Leu·Glu·Asp
   Leu·Tyr·Glu             Glu·Asp·Tyr
         Glu·Leu·Glu
Ser·Leu·Tyr·Glu          Glu·Asp·Tyr·CySO₃H

Ser·Leu·Tyr·Glu·Leu·Glu·Asp·Tyr·CySO₃H
_____/
              Complete sequence
```

FIGURE 14-10. "Overlap" study of the bovine insulin A-chain sequence.

```
              NH₂
               |
Tyr——Cy——Asp          (A chain A sequence)
 19   |  20  21
      S
      |
      S
      |
Val——Cy——Gly——Glu     (A chain B sequence)
 18   19   20   21
```

FIGURE 14-11. The structure of this fragment shows that one of the crosslinks between chains A and B of bovine insulin must be between position 20 of chain A and position 19 of chain B.

The S—S bond, as illustrated by the insulin molecule, is the most common form of crosslinking between polypeptide chains. By reaction with **ammonium thioglycolate** it is possible to convert S—S bonds to SH (sulfhydryl) groups (equation 14-9).

$$\begin{array}{c}\text{NH}_2\\|\\\text{HOOCCHCH}_2\text{S}\\|\\\text{HOOCCHCH}_2\text{S}\\|\\\text{NH}_2\end{array} + 2\,\text{NH}_4^+\,{}^-\text{OOCCH}_2\text{SH} \rightarrow \begin{array}{c}\text{NH}_4^+\,{}^-\text{OOCCH}_2\text{S}\\|\\|\\\text{NH}_4^+\,{}^-\text{OOCCH}_2\text{S}\end{array} + 2\,\begin{array}{c}\text{NH}_2\\|\\\text{HOOCCHCH}_2\text{SH}\end{array} \qquad (14\text{-}9)$$

cystine   ammonium thioglycolate

A chemical method of putting permanent waves in hair is based on this reaction. First some of the crosslinks that hold the hair fibers in their natural shape are converted to SH groups. Then the hair is put into the desired shape and the S—S crosslinks are reestablished in new positions. This is done by oxidizing the SH groups with potassium bromate. A crosslink appears wherever two SH groups are sufficiently close together.

$$2\,\text{RSH} \xrightarrow{[O]} \text{RS—SR} + \text{H}_2\text{O} \qquad (14\text{-}10)$$

A number of other proteins beside insulin have now been investigated in enough detail to give their amino acid sequences.

Cytochrome c. Different proteins have entirely different sequences unless they are proteins that serve the same function in related organisms. Then the degree of resemblance between the amino acid sequences depends on how closely the species are related.

Cytochrome c, a pigment found in the cells of a large number of higher plants and animals, is the best-worked-out example of the relationship between amino acid sequence and heredity. The various cytochromes c are basic proteins consisting of a single polypeptide chain of 104 or more amino acid units attached to an iron-porphyrin prosthetic group hemin (page 228) that is responsible for the color. The sequence of amino acid units in the cytochrome c from human heart cells is shown in Figure 14-12. The iron-porphyrin, which is not shown, is attached to sulfur atoms of two of the cysteine units.

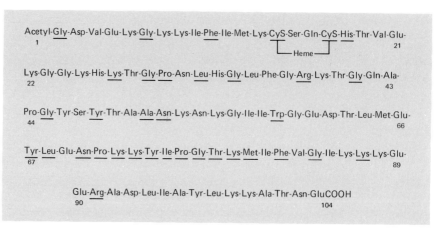

FIGURE 14-12. The polypeptide from human heart cytochrome c.

The function of a cytochrome c in the cell is connected with respiration. Thus, cells such as heart cells that work hard and consume lots of oxygen are exceptionally rich in this pigment. Cytochrome c acts as an oxidizing agent in energy-liberating metabolic reactions. In order for it to do that job, certain parts of the cytochrome c molecule are necessary; other parts are merely an accident of the evolution of the particular organism in which the pigment is found. The necessary parts are the iron-porphyrin prosthetic group and *the amino acids units underlined* in Figure 14-12. These portions of cytochrome c are identified as necessary parts by the fact that they are *always* present, not only in human heart cells but also in such widely divergent species as fungi and plants. The other amino acid units can apparently be interchanged without spoiling the biochemical workings of the cytochrome c molecule.

It is by the use of these interchangeable amino acids that we can trace relationships between different organisms. Thus, human cytochrome c differs from the variant found in rhesus monkeys only in the amino acid unit at position 58 (Figure 14-13); the amino

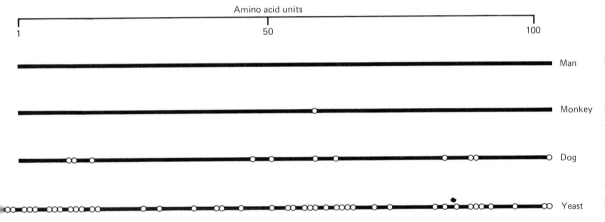

FIGURE 14-13. Cytochrome c from man, monkey, dog, and yeast cells. Circles indicate locations of amino acid units that are different from the corresponding ones in human heart cells.

acid at this position is isoleucine in man and threonine in the monkey. It may be that the common ancestor of man and the other primates had the monkey amino acid sequence and that, during the course of evolution, a mutation changed the amino acid at position 58 in man. The relationship between man and dog is more distant, and so we find that the amino acid sequence in canine cytochrome c differs from that in the human polypeptide at 11 out of 104 positions. A *really* remote relative of ours, the baker's yeast fungus, has four extra amino acid units at the beginning of the chain and lacks one at the end. Within the chain it differs from human cytochrome c at 40 positions.

Secondary Structure of Polypeptides

In protein chemistry the term **primary structure** refers to the ordinary structural formula in contrast to secondary and tertiary structures, which have to do with the shape of the molecule. As will be recalled from earlier chapters most molecules are quite flexible, and a given structural formula can correspond to a large number of conformations or molecular shapes. It will also be recalled that in some molecules one of the possible shapes is more stable than the others. For example, *t*-butylcyclohexane will exist in a chair conformation with the bulky *t*-butyl group in an equatorial position. Proteins, too, tend to prefer certain conformations. However, in the case of proteins the reason that a particular conformation is preferred is that it gives rise to a greater amount of hydrogen bonding, like that shown in Figure 14-1.

The most common and most important secondary structure is the **α-helix**, shown in Figure 14-14. An α-helix can be either a right-handed spiral or a left-handed spiral. By definition, the right-handed spiral curves to the right as the chain is traversed from the *N*-terminus to the carboxy-terminus. The right-handed helix is favored by the naturally occurring L-amino acids. The size and bond angles of an individual peptide unit are such that it takes about 3.7 amino acid units to complete one turn of the helix. Each carbonyl group is connected by hydrogen bonding to an NH group of an amino acid unit in the next turn of the spiral.

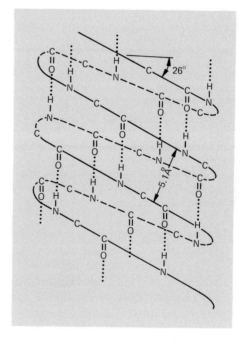

FIGURE 14-14. A few turns of an α-helix.

The details of the α-helix structure are deduced from x-ray scattering experiments, the same technique used to determine the structures of molecules packed in a crystal. Indeed, a regular structure like an α-helix can be considered to *be* a sort of crystal, even though the regular array is achieved by arranging segments of a large chain molecule instead of arranging separate molecules.

The simplest polypeptides in which the α-helix structure has been observed are **homopeptides** in which all the peptide units correspond to the same amino acid. These polypeptides do not occur in nature, but they can be made by heating the L-amino acid. The α-helix structure is also found in certain **keratins**, fibrous proteins of hair, wool, fingernails, and the like.

Tertiary Structures

In more complex proteins it is not unusual to find several α-helixes. These may be made from a single polypeptide chain but with amorphous regions of the chain separating the parts that have the α-helix conformation. It is the spatial arrangement of these α-helixes and the prosthetic groups relative to one another that is called the tertiary structure of the protein.

DENATURATION

A complicated protein will retain its biological activity (for example its ability to act as an enzyme) only so long as its tertiary structure is not disrupted. When a protein loses its biological activity without being hydrolyzed, it is said to be **denatured**. Since denaturation usually involves a mere change in molecular conformation, it is extraordinarily easy to bring about. Most enzymes are denatured by heating to a temperature that could be regarded as a sort of "internal melting point" and also by treatment with strong acids or alkalis or even salts or organic solvents. Sometimes denaturation is reversible, as might be expected if the native conformation is the most stable one.

Conjugated Proteins

A conjugated protein is a polypeptide to which some nonpeptide structure is attached. The nonpeptide part is called a **prosthetic group** and is often the most important part of the molecule for its biological function, as in the case of many enzymes.

Examples of conjugated proteins already encountered are cytochrome c (page 316) and hemoglobin, whose prosthetic group, an iron-porphyrin (hemin) was shown on page 228 of Chapter 11.

Enzymes

An enzyme is a naturally occurring catalyst. A catalyst is a substance that makes a reaction go faster without itself being permanently used up. A catalyst does its job by combining with one of the reagents, called a **substrate**, and converting it into a more reactive species. After the reaction, the catalyst is formed again to cause the reaction of another molecule of substrate, and so on. Enzymes differ from ordinary catalysts in three important ways:

1. They can usually make the reaction go thousands or billions of times faster than an ordinary catalyst would.

2. They will work under mild conditions (room temperature, 1 atm pressure).
3. They are extremely specific or choosy about which reaction they will catalyze.

HOW CATALYSTS WORK

Before any further discussion of enzymes it is a good idea to review what we already know about the action of an ordinary catalyst such as H^+. The hydrolysis of an acetal will do as an example. *Without* the catalyst the reaction is so slow that, for practical purposes, we can say that reaction 14-11 does not go at all. In the

$$R-C\begin{matrix}H\\ \diagup\\ \diagdown\end{matrix}\begin{matrix}OCH_3\\ \\ OCH_3\end{matrix} + H_2O \xrightarrow{\text{very slow}} R-C\begin{matrix}H\\ \diagup\\ \diagdown\end{matrix}\begin{matrix}OH\\ \\ OCH_3\end{matrix} + CH_3OH \quad (14\text{-}11)$$

an acetal

presence of the H^+ catalyst, a catalyst-substrate complex is formed. Actually, in this example (equation 14-12) the catalyst-substrate complex is merely the acetal with a proton bonded to one of the unshared electron pairs on oxygen. The "catalyst-substrate com-

$$R-C\begin{matrix}H\\ \diagup\\ \diagdown\end{matrix}\begin{matrix}OCH_3\\ \\ OCH_3\end{matrix} + H^+ \rightleftharpoons R-C\begin{matrix}H\\ \diagup\\ \diagdown\end{matrix}\begin{matrix}{}^+OCH_3H\\ \\ OCH_3\end{matrix} \quad (14\text{-}12)$$

substrate   catalyst   catalyst-substrate complex

plex" reacts rapidly to give the reaction product and regenerate the catalyst (equation 14-13).

$$R-C\begin{matrix}H\\ \diagup\\ \diagdown\end{matrix}\begin{matrix}{}^+OCH_3H\\ \\ OCH_3\end{matrix} \xrightarrow{\text{fast}} CH_3OH + [R-\overset{H}{\underset{}{C^+}}-OCH_3]$$

$$[R-\overset{H}{\underset{}{C^+}}-OCH_3] + H_2O \xrightarrow{\text{fast}} R-\overset{H}{\underset{OH}{\overset{|}{C}}}-OCH_3 + H^+ \quad (14\text{-}13)$$

product   catalyst

Note that the catalyst is not used up in the reaction. One of the most important features of catalysis is that it takes only a small amount of catalyst to bring about the reaction of very large amounts of substrate.

This is the reason for the dramatic effect of very small amounts of certain vitamins in deficiency diseases. The vitamin is used as a prosthetic group in an enzyme that catalyzes a reaction important to the health of the organism.

Hydroxide ion (OH$^-$) acts as an **inhibitor** for reactions catalyzed by H$^+$ because OH$^-$ ties up the catalyst, H$^+$, as an ineffective catalyst-inhibitor complex, which happens in this case to be a water molecule.

$$H^+ + HO^- \rightleftharpoons H-O-H \quad (14\text{-}14)$$

catalyst   inhibitor   "catalyst-inhibitor complex"

ENZYME-SUBSTRATE COMPLEXES

Enzymes act in much the same way as ordinary catalysts such as H$^+$, except that they are much more specific and give much faster reactions. Thus when the enzyme **α-glucosidase** catalyzes the hydrolysis of an α-glucoside, it first forms an enzyme-substrate complex; this complex then reacts to give glucose and the alcohol; finally the α-glucosidase is turned loose to seek another α-glucoside molecule (equations 14-15 and 14-16).

substrate + enzyme ⇌ enzyme-substrate complex (14-15)

H$_2$O + enzyme-substrate complex $\xrightarrow{\text{fast}}$ glucose + ROH + enzyme (14-16)

ENZYME INHIBITORS AND ANTIMETABOLITES

Many physiologically active compounds owe their activity to the fact that they either destroy or tie up some important enzyme. In many cases these **enzyme inhibitors** are compounds almost, but not quite, the same as the compound whose reaction the enzyme is supposed to catalyze. The enzyme is lured into forming a complex with the unreactive inhibitor instead of with its proper substrate.

An example is found in the action of sulfanilamide on the growth of bacteria. Sulfanilamide, a typical "sulfa" drug, was routinely sprinkled on wounds in World War II. The inhibition of the growth of bacteria by sulfanilamide is reversed by giving the bacteria lots of *p*-aminobenzoic acid. *p*-Aminobenzoic acid is a metabolite needed by the bacteria in order to grow; the **antimetabolite**, sulfanilamide is a molecule of closely the same size and shape as *p*-aminobenzoic acid. The two compounds are also quite a bit alike chemically, since the amide hydrogens of a sulfonamide are slightly acidic. It is believed that an enzyme in sulfanilamide-sensitive bacteria is inhibited by binding with sulfanilamide in place of its normal substrate, *p*-aminobenzoic acid. Adding an excess of *p*-aminobenzoic acid allows it to compete with the sulfanilamide and restores the normal metabolic reaction.

sulfanilamide
mp 164.5–166.5°C; K_A 3.7 × 10^{-11}

p-aminobenzoic acid
mp 187.0–187.5°C; K_A about 2 × 10^{-5}

Other enzyme inhibitors are not at all like the normal substrate, but happen to have a special reactivity towards the enzyme. For example, if the enzyme has sulfhydryl groups, SH, it can be destroyed by forming an insoluble precipitate with a heavy metal ion such as Pb^{2+}, a well-known poison. Cyanide ion, another well-known poison, combines with the Fe^{3+} or Cu^{2+} ions that are essential parts of certain enzymes that catalyze oxidation-reduction reactions.

Inhibition of enzyme activity is part of the normal mechanism by which a cell keeps its myriad reactions under proper control. For example, in some cases the product of an enzyme-catalyzed series of reactions acts as an inhibitor for one of the enzymes. The presence of too much product slows down the reaction, too little speeds it up.

ENZYME SPECIFICITY

In a conjugated protein the polypeptide chains and α-helixes are usually folded around the prosthetic group to give a very complicated tertiary structure. An enzyme-substrate complex is like a conjugated protein with the substrate as an additional prosthetic group. The enzyme and its substrate must be related in such a way that parts of the enzyme that are able to interact with parts of the substrate—for example by making hydrogen bonds—are in the right position for this to happen. Although the enzyme may change its tertiary structure, or shape, somewhat in order to accommodate the substrate, there are limits to what it can do. The relationship has been compared to that between a lock and a key, except that the lock is slightly rubbery and makes modest changes in shape to fit the key. In spite of this flexibility, the lock (enzyme) will only fit a limited variety of keys (substrates).

This specificity also extends to discrimination between D and L configurations of the substrate. Just as a D monomer unit will not fit comfortably into an all L isotactic polymer, a D substrate will not fit an enzyme designed for an L substrate. And, of course, an enzyme designed for a D substrate will have nothing to do with an L substrate. An example is D-amino acid oxidase, which catalyzes the oxidation of D-amino acids to keto acids but has no effect on L-amino acids.

In addition to their substrate specificity, or reluctance to work with anything but a particular kind of starting material, enzymes are specific about the structure of the product. For example, where an ordinary reaction would produce a (±) mixture of a pair of enantiomorphs, an enzyme-catalyzed reaction will make only *one* of the two mirror-image products. Thus when *cis*-aconitic acid adds water across its double bond in a reaction catalyzed by the enzyme aconitase, the product is optically active and has the particular configuration shown in Figure 14-15, rather than one of the other three possible optically isomeric configurations. In solution the two sides of the plane containing the double bond are equivalent, and isocitric acid would be accompanied by an equal amount of its mirror-image isomer.

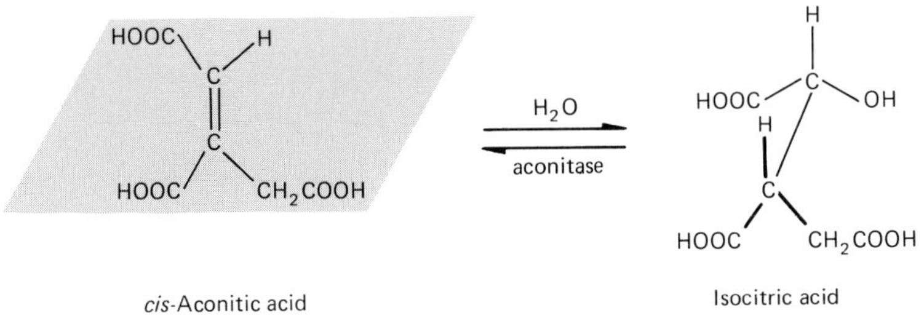

FIGURE 14-15.

Although we do not actually know the structure of the aconitic acid–aconitase enzyme-substrate complex, we can easily imagine a structure that would account for the stereospecific product. Suppose that the *cis*-aconitic acid is held to the surface of the enzyme molecule by bonds (probably hydrogen bonds) at the three points a, b, and c on the enzyme surface as in Figure 14-16A. Then we imagine that a hydroxyl group is put on one end of the double bond from below the plane and a hydrogen put on the other end of the double bond from above the plane, as in Figure 14-16B. Next, the product, isocitric acid, is released from its three points of attachment to the enzyme —and we see that it is indeed the correct stereoisomer.

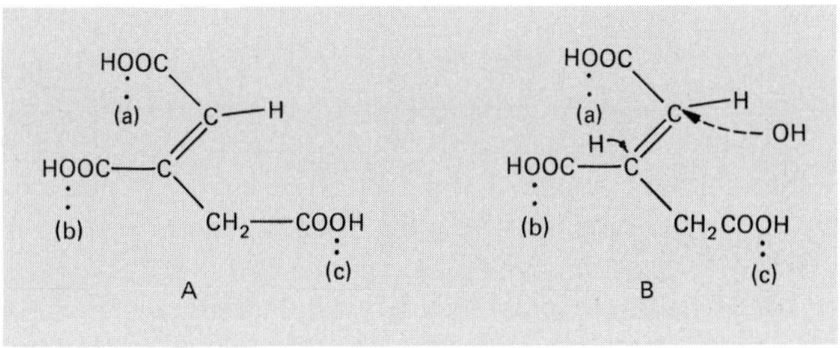

FIGURE 14-16. (A) The enzyme is represented by three points of attachment (a), (b), (c). (B) The hydroxyl group is attached from the enzyme side of the plane containing the double bond.

But are we sure that we couldn't also make the mirror image of isocitric acid this way? Let's try it and see. First we place the *cis*-aconitic acid molecule on the enzyme in a position that is the *mirror image* of its position in Figure 14-16A, hoping that this will cause the HO group to be attached to the other side of the molecule. In Figure 14-17, the CH$_2$COOH group has been allowed to hydrogen bond at position (b) of the enzyme, and we let one of the COOH groups be hydrogen bonded at position (c). But this leaves the second

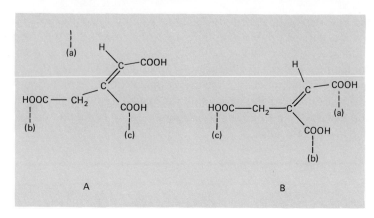

FIGURE 14-17. **(A)** Substrate and enzyme. **(B)** Substrate and *mirror image* of the enzyme.

carboxyl group not attached to anything, because position (a) is in the wrong place. To make the enantiomer of isocitric acid, we would need *the enantiomer, or mirror image of the enzyme*. Since the enzyme is asymmetric, it is different from its mirror image and cannot play that part.

THE NATURE OF THE ACTIVE SITE

In the case of enzymes that are conjugated proteins with highly specialized prosthetic groups, it is easy to conjecture, and probably correct, that the prosthetic group is the **active site**, that is, the actual location at which the enzyme-catalyzed reaction takes place. This does not mean, however, that adjacent polypeptide parts of the molecule do not take part in the catalysis. More remote parts of the polypeptide chain probably serve to establish the tertiary structure that is responsible for the "flexible lock" and key relationship of enzyme and substrate.

However, some enzymes are just polypeptides, and here the location of the active site is less obvious. Chymotrypsin is an example of an enzyme that has no prosthetic group. Found in the gastrointestinal tract, it catalyzes the hydrolysis of the peptide links of amino acid units having aromatic or large aliphatic side chains and also the hydrolysis of ester linkages.

On the basis of chemical inactivation experiments, it is believed that the active site of chymotrypsin involves just one of the enzyme's 28 serine units, located at position 198, the histidine unit located at position 57, and an unidentified acidic group. Although the reactive serine and histidine units are far apart in the primary structure (that is, not near each other on the chain) the folding of the tertiary structure is such as to put them close together in space.

A possible mechanism for the hydrolysis of an ester by chymotrypsin is outlined in Figure 14-18. The substrate, an ester **R'OOCR**, is shown in bold face. According to this mechanism, an acid group A—H somewhere in the enzyme helps to pull off the alkoxy group, **R'O**, as an alcohol **R'OH**. Simultaneously, an oxygen atom of a serine unit pulls on the carbonyl end of the ester linkage as shown in equation 14-17. The serine CH_2OH group is helped in forming

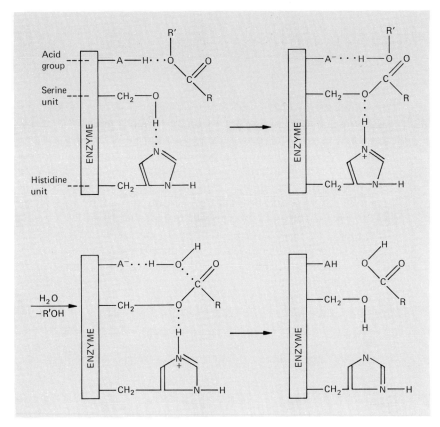

FIGURE 14-18. A mechanism that has been suggested for the chymotrypsin-catalyzed hydrolysis of an ester. It involves a cooperative effort by an unknown acid group, A—H, the hydroxyl group of a serine unit, and the imidazole ring of a histidine unit.

its bond to the carbonyl by the adjacent imidazole group, which temporarily takes its hydrogen atom. In short, the picture is of a conspiracy in which several different reactive groups gang up on the ester substrate and cooperate in its destruction. It is this cooperative effect that most probably accounts for the high rate of reaction brought about by enzymes as compared with ordinary catalysts.

Polypeptide Synthesis

A polypeptide, or rather a horrible mixture of isomeric polypeptides, can be made just by heating a mixture of amino acids (equation 14-18). The trouble is that the sequence of amino acid units in such a material is almost random and the product has none

$$H_2N-\underset{R}{\underset{|}{\overset{H}{\overset{|}{C}}}}-COOH \rightarrow H_2N-\underset{R}{\underset{|}{\overset{H}{\overset{|}{C}}}}-\overset{O}{\overset{\|}{C}}-(N-\underset{R}{\underset{|}{\overset{H}{\overset{|}{C}}}}-\overset{O}{\overset{\|}{C}})_n-\underset{}{\underset{}{\overset{H}{\overset{|}{N}}}}-\underset{R}{\underset{|}{\overset{H}{\overset{|}{C}}}}-COOH \quad (14\text{-}18)$$

(various R)

of the interesting enzymatic or other biological activity of naturally occurring proteins. In order to make a polypeptide of a given, known structure it is necessary to add the amino acid units one step at a time, first making a dipeptide, then a tripeptide, and so on.

A problem encountered, even at the dipeptide stage, is how to ensure that the two amino acids are in the proper sequence, for example, Gly-Ala and not Ala-Gly. If a mixture of the two amino acids is merely heated, the product includes not only both of the dipeptides, but also Gly-Gly, Ala-Ala, and a number of tripeptides and higher peptides. The yield of the desired product would be very low. To ensure reaction in the desired sequence it is necessary to block one of the amino groups temporarily. This is done by means of a reaction of benzyloxycarbonyl chloride (equation 14-19), with an ester of the amino acid (equation (14-20).

$$C_6H_5\text{—}CH_2OH + Cl\text{—}\overset{O}{\overset{\|}{C}}\text{—}Cl \rightarrow C_6H_5\text{—}CH_2\text{—}O\text{—}\overset{O}{\overset{\|}{C}}\text{—}Cl + HCl \quad (14\text{-}19)$$

benzyl alcohol   phosgene   benzyloxycarbonyl chloride

$$C_6H_5\text{—}CH_2\text{—}O\text{—}\overset{O}{\overset{\|}{C}}\text{—}Cl + H_2N\text{—}\overset{H}{\overset{|}{C}}\text{—}\overset{O}{\overset{\|}{C}}\text{—}OCH_3 \rightarrow$$

methyl glycinate

$$C_6H_5\text{—}CH_2\text{—}O\text{—}\overset{O}{\overset{\|}{C}}\text{—}\underset{H}{\underset{|}{\overset{H}{\overset{|}{N}}}}\text{—}\overset{H}{\overset{|}{C}}\text{—}\overset{O}{\overset{\|}{C}}\text{—}OCH_3 + HCl \quad (14\text{-}20)$$

Next, the methyl ester is converted to an **acid azide**, $CO-N_3$, in a two step process (equation 14-21). Acid azides react like acid

$$C_6H_5CH_2\text{—}O\text{—}\overset{O}{\overset{\|}{C}}\text{—}\overset{H}{\overset{|}{N}}\text{—}\underset{H}{\underset{|}{\overset{H}{\overset{|}{C}}}}\text{—}\overset{O}{\overset{\|}{C}}\text{—}OCH_3 + H_2NNH_2 \longrightarrow$$

$$C_6H_5CH_2\text{—}O\text{—}\overset{O}{\overset{\|}{C}}\text{—}\overset{H}{\overset{|}{N}}\text{—}\underset{H}{\underset{|}{\overset{H}{\overset{|}{C}}}}\text{—}\overset{O}{\overset{\|}{C}}\text{—}\overset{H}{\overset{|}{N}}NH_2 + CH_3OH$$

$$\downarrow HNO_2$$

$$C_6H_5CH_2\text{—}O\text{—}\overset{O}{\overset{\|}{C}}\text{—}\overset{H}{\overset{|}{N}}\text{—}\underset{H}{\underset{|}{\overset{H}{\overset{|}{C}}}}\text{—}\overset{O}{\overset{\|}{C}}\text{—}N_3 \quad (14\text{-}21)$$

chlorides, so this product can be used to react with another molecule of an amino acid methyl ester, making a peptide linkage (equation 14-22).

$$C_6H_5CH_2-O-\overset{O}{\underset{\|}{C}}-\overset{H}{\underset{|}{N}}-\overset{H}{\underset{|}{C}}-\overset{O}{\underset{\|}{C}}-N_3 + H_2N-\overset{H}{\underset{|}{C}}-\overset{O}{\underset{\|}{C}}-OCH_3 \rightarrow$$
$$ H CH_3$$
$$\text{methyl alaninate}$$

$$C_6H_5CH_2-O-\overset{O}{\underset{\|}{C}}-\overset{H}{\underset{|}{N}}-\overset{H}{\underset{|}{C}}-\overset{O}{\underset{\|}{C}}-\overset{H}{\underset{|}{N}}-\overset{H}{\underset{|}{C}}-\overset{O}{\underset{\|}{C}}-OCH_3 + HN_3 \quad (14\text{-}22)$$
$$ H CH_3$$

We now have our dipeptide, but need to get rid of the unwanted benzyloxycarbonyl and methyl ester groups. This can be accomplished by the reactions indicated in equation 14-23.

$$C_6H_5CH_2-O-\overset{O}{\underset{\|}{C}}-\overset{H}{\underset{|}{N}}-CH_2-\overset{O}{\underset{\|}{C}}-\overset{H}{\underset{|}{N}}-\overset{H}{\underset{|}{C}}-\overset{O}{\underset{\|}{C}}-OCH_3 \xrightarrow[\text{then H}^+]{\text{HO}^-}$$
$$ CH_3$$

$$C_6H_5CH_2-O-\overset{O}{\underset{\|}{C}}-\overset{H}{\underset{|}{N}}-CH_2-\overset{O}{\underset{\|}{C}}-\overset{H}{\underset{|}{N}}-\overset{H}{\underset{|}{C}}-COOH \xrightarrow{H_2/Pd}$$
$$ CH_3$$

$$C_6H_5CH_3 + CO_2 + H_2N-\overset{H}{\underset{|}{C}}-\overset{O}{\underset{\|}{C}}-\overset{H}{\underset{|}{N}}-\overset{H}{\underset{|}{C}}-COOH \quad (14\text{-}23)$$
$$ H CH_3$$
$$\text{Gly-Ala}$$
$$\text{(glycylalanine)}$$

This procedure gives the desired dipeptide, Gly-Ala, in six steps. To make a tripeptide, say Gly-Ala-Gly, the sequence shown in equation 14-24 would have been used, giving the tripeptide in a total of nine steps.

$$C_6H_5CH_2-O-\overset{O}{\underset{\|}{C}}-\overset{H}{\underset{|}{N}}-CH_2-\overset{O}{\underset{\|}{C}}-\overset{H}{\underset{|}{N}}-\overset{H}{\underset{|}{C}}-\overset{O}{\underset{\|}{C}}-OCH_3 \xrightarrow{H_2NNH_2} [\;] \xrightarrow{HNO_2}$$
$$ CH_3$$

$$C_6H_5CH_2O-\overset{O}{\underset{\|}{C}}-\overset{H}{\underset{|}{N}}-CH_2-\overset{O}{\underset{\|}{C}}-\overset{H}{\underset{|}{N}}-\overset{H}{\underset{|}{C}}-\overset{O}{\underset{\|}{C}}-N_3 \xrightarrow{H_2N-CH_2-\overset{O}{\underset{\|}{C}}-OCH_3}$$
$$ CH_3$$

$$C_6H_5CH_2O-\overset{O}{\underset{\|}{C}}-\overset{H}{\underset{|}{N}}-CH_2-\overset{O}{\underset{\|}{C}}-\overset{H}{\underset{|}{N}}-\overset{H}{\underset{|}{C}}-\overset{O}{\underset{\|}{C}}-\overset{H}{\underset{|}{N}}-CH_2-\overset{O}{\underset{\|}{C}}-OCH_3 \xrightarrow{\text{HO}^-, \text{then H}^+} [\;] \xrightarrow{H_2/Pd}$$
$$ CH_3$$

$$C_6H_5CH_3 + CO_2 + H_2N-CH_2-\overset{O}{\underset{\|}{C}}-\overset{H}{\underset{|}{N}}-\overset{H}{\underset{|}{C}}-\overset{O}{\underset{\|}{C}}-\overset{H}{\underset{|}{N}}-CH_2-\overset{O}{\underset{\|}{C}}-OH$$
$$ CH_3$$
$$\text{Gly-Ala-Gly} (14\text{-}24)$$

Of course, every time one of the intermediate products is isolated and purified, some is lost, so by the time Gly-Ala-Gly has been reached there is not very much product. A nonapeptide could be made in much the same way by combining three tripeptides. However, each of the tripeptides is, as we have seen, very hard to get, and the further manipulations needed to make the nonapeptide would reduce the yield still more.

The Protein-Making Machine

The problem of eliminating the wasteful separation procedures at each step in a polypeptide synthesis was solved by making use of some developments in polymer chemistry. It is an example of the way in which a discovery in one field can have large and unanticipated effects on another field.

In Chapter 12 it was noted that crosslinking can convert a soluble polymer into an insoluble network. The functional groups on such a network molecule can still undergo their usual reactions and, unless the reaction is one that breaks crosslinks, the reaction product is also insoluble. *The important thing about such reactions is that excess reagent and small by-product molecules can be removed simply by washing the polymer.* The polymer, which might be in the form of small beads, is shaken with a solution of the other reagent until the reaction is over, then the excess reagent and small-molecule by-products are drained off, and the beads are washed with fresh solvent.

The product of this reaction, still in the form of functional groups on the surface of polymer beads, can be treated in the same way with a solution of a second reagent, to bring about the second step in a multistep synthesis. Again, impurities and by-products are washed off and the desired product, *all* of it, remains as part of the insoluble polymer molecule. None of the product is lost.

The polymer beads used in the protein-making machine start out as polystyrene crosslinked with divinyl benzene to make them insoluble (structure 14-25). The first reaction is an electrophilic

(14-25)

aromatic substitution that introduces chloromethyl groups, $ClCH_2$, into some of the phenyl substituents along the polymer chains (equation 14-26). This product, of course, is still a crosslinked polymer and the excess CH_3OCH_2Cl, catalyst, and CH_3OH can be removed just by washing the beads with solvent.

The chloromethyl groups on the surface of the polymer beads are reactive and provide anchor points at which polypeptide chains can be built up (equation 14-27). The last step in the synthesis is a reaction that cuts the polypeptide chains loose so that they can also be washed from the surface of the beads.

$$\underset{\displaystyle}{\sim\!\!\sim\!\!\sim\text{CH}_2-\text{CH}\sim\!\!\sim\!\!\sim} + \text{CH}_3\text{OCH}_2\text{Cl} \xrightarrow{\text{SnCl}_4 \text{ catalyst}}$$

(with phenyl ring attached)

$$\sim\!\!\sim\!\!\sim\text{CH}_2-\text{CH}\sim\!\!\sim\!\!\sim + \text{CH}_3\text{OH} \quad (14\text{-}26)$$

(with phenyl ring bearing CH_2Cl)

$$\underset{\text{chloromethylated polymer}}{\sim\!\!\sim\!\!\sim\text{CH}_2-\text{CH}\sim\!\!\sim\!\!\sim\;\text{(Ar-CH}_2\text{Cl)}} + \underset{\substack{\text{amino acid salt with} \\ \text{protected amino group}}}{\text{C}_6\text{H}_5\text{CH}_2\text{O}-\overset{\overset{\text{O}}{\|}}{\text{C}}-\overset{\overset{\text{H}}{|}}{\text{N}}-\underset{\underset{R_1}{|}}{\text{CH}}-\overset{\overset{\text{O}}{\|}}{\text{C}}-\text{O}^-} \rightarrow$$

$$\underset{\text{polymer with first unit of polypeptide chain attached}}{\sim\!\!\sim\!\!\sim\text{CH}_2-\text{CH}\sim\!\!\sim\!\!\sim \;\text{Ar-CH}_2-\overset{\overset{\text{O}}{\|}}{\text{O}-\text{C}}-\underset{\underset{R_1}{|}}{\text{CH}}-\overset{\overset{\text{H}}{|}}{\text{N}}-\overset{\overset{\text{O}}{\|}}{\text{C}}-\text{O}-\text{CH}_2\text{C}_6\text{H}_5} + \text{Cl}^- \quad (14\text{-}27)$$

After the chloromethyl groups of the polymer have been converted to the amino acid derivative as in equation 14-27, the next step is to unblock the amino groups. This is done by shaking the beads with dilute HBr (equation 14-28). Next the beads are immersed in a

$$\sim\!\!\sim\!\!\sim\text{CH}_2-\text{CH}\sim\!\!\sim\!\!\sim\;\text{Ar-CH}_2-\overset{\overset{\text{O}}{\|}}{\text{O}-\text{C}}-\underset{\underset{R_1}{|}}{\text{CH}}-\overset{\overset{\text{H}}{|}}{\text{N}}-\overset{\overset{\text{O}}{\|}}{\text{C}}-\text{O}-\text{CH}_2\text{C}_6\text{H}_5 \xrightarrow[\text{H}_2\text{O}]{\text{HBr}}$$

$$\sim\!\!\sim\!\!\sim\text{CH}_2-\text{CH}\sim\!\!\sim\!\!\sim\;\text{Ar-CH}_2-\overset{\overset{\text{O}}{\|}}{\text{O}-\text{C}}-\underset{\underset{R_1}{|}}{\text{CH}}-\text{NH}_2 + \text{CO}_2 + \text{C}_6\text{H}_5\text{CH}_2\text{OH} \quad (14\text{-}28)$$

reagent consisting of a second amino acid unit with its amino group blocked by the usual carbobenzoxy ($C_6H_5CH_2OCO$) substituent (equation 14-29). The sequence of reactions is repeated with various

$$\text{bead-}CH_2\text{-}O\text{-}\underset{\text{O}}{\underset{\|}{C}}\text{-}\underset{R_1}{\underset{|}{CH}}\text{-}NH_2 + HO\text{-}\underset{\text{O}}{\underset{\|}{C}}\text{-}\underset{R_2}{\underset{|}{\underset{|}{C}H}}\text{-}N\text{-}\underset{\text{O}}{\underset{\|}{C}}\text{-}O\text{-}CH_2C_6H_5 \rightarrow$$

$$\text{bead-}CH_2\text{-}O\text{-}\underset{\text{O}}{\underset{\|}{C}}\text{-}\underset{R_1}{\underset{|}{CH}}\text{-}N\text{-}\underset{\text{O}}{\underset{\|}{C}}\text{-}\underset{R_2}{\underset{|}{\underset{|}{C}H}}\text{-}N\text{-}\underset{\text{O}}{\underset{\|}{C}}\text{-}OCH_2C_6H_5 + H_2O \quad (14\text{-}29)$$

groups R until the desired polypeptide chain has been built up. In the final step the last $C_6H_5CH_2OCO$ protecting group is detached from the end of the chain and the chain itself is detached from the polymer so that the polypeptide can be washed off of the beads (equation 14-30).

$$\text{bead-}CH_2\text{-}O\text{-}\underset{\text{O}}{\underset{\|}{C}}\text{-}\underset{R_1}{\underset{|}{CH}}\text{-}N\text{-}\underset{\text{O}}{\underset{\|}{C}}\text{-}\underset{R_2}{\underset{|}{\underset{|}{C}H}}\text{-}N\cdots \underset{\text{O}}{\underset{\|}{C}}\text{-}\underset{R}{\underset{|}{CH}}\text{-}N\text{-}\underset{\text{O}}{\underset{\|}{C}}\text{-}O\text{-}CH_2\text{-}C_6H_5$$

$$\downarrow \text{HBr} / F_3\text{CCOOH}$$

beads-CH_2Br + $HO\text{-}\underset{\text{O}}{\underset{\|}{C}}\text{-}\underset{R_1}{\underset{|}{CH}}\text{-}N\text{-}\underset{\text{O}}{\underset{\|}{C}}\text{-}\underset{R_2}{\underset{|}{\underset{|}{C}H}}\text{-}N\cdots \underset{\text{O}}{\underset{\|}{C}}\text{-}\underset{R}{\underset{|}{CH}}\text{-}NH_2 + CO_2 + C_6H_5CH_2Br$

(beads)   polypeptide   (14-30)

As far as the physical operations are concerned, only two things are ever done to the polymer beads. They are shaken with a liquid at some controlled temperature for a predetermined length of time, and then that liquid is drained off. Automation is fairly straightforward because all that is required is to open and close valves and turn heating or cooling devices on or off at predetermined times. Of course the process is not really as simple as it sounds because the time allowed for each reaction and the best concentration of reagent must be determined experimentally, and the final product will still need some purification.

In Dr. Merrifield's protein machine (Figure 14-19), the valves are opened and closed electrically at the command of switches tripped by pegs mounted on a rotating drum.

FIGURE 14-19. Dr. R. B. Merrifield and his protein machine. [From *Chem. Eng. News*, August 2, 1972]

SUMMARY

A. Polyamides and the peptide bond
   1. Nylons are macromolecules in which six- to ten-carbon monomer units are held together by amide bonds, CONH.
   2. Silk fibroin, a polypeptide, is similar to nylon except that the monomer units are α-aminoacids, RCH(NH$_2$)COOH.
      (a) Silk is an isotactic polymer.

3. Amide or peptide bonds are hydrolyzed to form amino and carboxyl functional groups. This degrades the polymer.
B. Amino acids
  1. Amino acids at neutral pH exist largely as dipolar ions.
  2. Basic amino acids have an extra amino group; acidic amino acids have an extra carboxyl group.
  3. Most of the naturally occurring amino acids (in polypeptides) are of the L configuration.
C. Polypeptides
  1. The complete characterization of a polypeptide involves determining
     (a) The primary structure (number of chains, sequence of amino acids in each chain, position of crosslinks).
     (b) The secondary structure, or conformation.
        (i) This is stabilized by hydrogen bonding.
        (ii) The α-helix is an example.
     (c) The tertiary structure, or mode of arrangement of several secondary structures (such as α-helixes) in space relative to one another and to any prosthetic group.
  2. A protein is denatured (loses its biological activity) if its tertiary structure is disrupted.
D. Conjugated proteins and enzymes
  1. A nonpeptide part of a conjugated protein is called a prosthetic group.
     (a) Examples are the catalytically active sites of enzymes.
  2. Enzymes are naturally occurring catalysts.
     (a) They change the reactivity of their substrates by forming enzyme-substrate complexes.
     (b) Enzymes are highly specific, even discriminating between enantiomers.
     (c) Many enzymes have prosthetic groups. Vitamins may function by providing the prosthetic groups for enzymes.
E. Polypeptide synthesis
  1. Methods of ensuring the desired amino acid sequence involve blocking one of the amino groups.
  2. The process has been automated, using reactions on the surface of a polymer.

EXERCISES

1. Define (a) dipeptide; (b) polypeptide; (c) prosthetic group of a protein.

2. Write equations showing how to make nylon 66.

3. Explain what happens when a newly formed nylon filament is stretched.

4. Draw a structure showing a typical section of a molecule of silk fibroin.

5. Write equations to show what happens when (a) nylon 66 and (b) silk are exposed to a strong acid.

6. Give the dipolar ion structures for any three amino acids. Explain why most amino acids have high melting points or decompose on heating.

7. Write equations for the reaction of an amino acid in its dipolar form with (a) a strong acid and (b) a strong base.

8. (a) Explain what the isoelectric point of an amino acid is.
   (b) Describe the behavior of an amino acid in an electric field at various H^+ concentrations.

9. Give structures for (a) any two naturally occurring neutral amino acids; (b) any two naturally occurring basic amino acids; (c) any two naturally occurring acidic amino acids.

10. (a) What is the predominant configuration of naturally occurring amino acids, D or L?
    (b) Why should one form predominate over the other?

11. (a) What is an essential amino acid?
    (b) Give three examples.
    (c) How can a mutation occur that makes an animal unable to synthesize a needed amino acid without causing the animal to become extinct?

12. (Open book) Describe the ion-exchange method of analyzing mixtures of amino acids.

13. (a) What is end-group analysis and what does it tell about the structure of a protein?
    (b) Write equations showing the reaction of 2,4-dinitrofluorobenzene with Ala-Gly-Phe, followed by hydrolysis.
    (c) A dipeptide on hydrolysis gives alanine and glycine, and is therefore either Ala-Gly or Gly-Ala. Show with equations how you could tell which of these is the correct structure.

14. A tripeptide on reaction with 2,4-dinitrofluorobenzene, followed by hydrolysis, gave the following compounds.
    N-(2,4-dinitrophenyl)glycine
    N-(2,4-dinitrophenyl)glycylalanine
    alanylleucine
    alanine
    leucine
    What is the structure of the original tripeptide?

15. Write structural formulas and names for all the possible di-, tri-, and tetrapeptides that can be constructed using only glycyl, alanyl, and leucyl units.

16. How many possible tetrapeptides can be constructed from one molecule of glycine and three of alanine?

17. Write a part structure showing a disulfide crosslink between two polypeptide chains.

18. (Open book) Explain how carboxy-terminal groups can be counted, using an equation.

19. Describe two methods used to determine the sequence of amino acids in a polypeptide chain.

20. Explain what is meant by the primary, secondary, and tertiary structures of a protein.

21. Can a stable α-helix be formed from a polypeptide chain that contains both L and D amino acid units?

22. Define denaturation and explain its connection with the tertiary structure of a protein.

23. Define (a) conjugated protein and (b) substrate.

24. Give three ways in which enzymes differ from most ordinary catalysts.

25. Explain (a) how a substance resembling the substrate sometimes interferes with an enzyme-catalyzed reaction and (b) how a cell can automatically keep from making too much of a given compound.

26. (Open book) Give two examples of enzyme specificity.

27. Explain how even a small amount of a vitamin can have a large effect on the health of an animal, whereas essential amino acids are needed in larger quantities.

28. Explain briefly the working principle of the polypeptide-synthesizing machine.

29. Make an intelligent guess about the way in which a particular polypeptide might be made in a living cell.

Chapter 15
Nucleic Acids

Introduction

Despite the apparent wide diversity of different species, genera, and kingdoms, different organisms are animated by essentially the same kinds of biochemical machinery. To understand one organism completely would be to have an almost complete understanding of all organisms. They all use similar reactions to accomplish similar things, whether the synthesis of protein or carbohydrate or the conversion of food or light into chemically useful forms of energy.

This basic similarity of all earthly life forms extends also to the genetic apparatus that determines what a developing organism is to be. Thus, the chromosomes and other components of the cells of different organisms work in essentially the same way, even though each cell grows into an organism of its own particular species and not just any organism.

An easy way to understand how the same machinery can produce such drastically different results is to use the analogy of a modern factory controlled by an on-line digital computer. Such a factory can be caused to make a different product just by changing the tape on which the computer program is encoded. It is not necessary to change either the computer or the lathes, milling machines, or lasers that do the actual work. A living cell is that kind of a factory. **Its product is another cell complete with a duplicate of the computer program**. Changes in the computer program are called **mutations**. They may produce a slightly different organism, in which case they are called viable mutations or, as is most often the case, they may produce only junk.

The thing in the cell that corresponds to a computer program is a set of instructions called **genes**. The "tape" on which these instructions are encoded is called a **chromosome**. The chromosome is a macromolecule, deoxyribonucleic acid, and the genes or instructions on the "tape" are different sequences of monomer units in the nucleic acid chain.

Nucleic Acids Defined

The backbone of the nucleic acid polymer is a phosphate-sugar polyester. Phosphoric acid has three acidic hydroxyl groups. Two of these are used to establish ester linkages with the hydroxyl groups of adjacent sugar molecules (equation 15-1). The polymer is acidic

$$\text{HO}-\overset{\overset{O}{\|}}{\underset{\underset{H}{O}}{P}}-\text{OH} + 2\,\text{ROH} \rightarrow 2\,\text{H}_2\text{O} + \text{R}-\text{O}-\overset{\overset{O}{\|}}{\underset{\text{OH}}{P}}-\text{O}-\text{R} \quad (15\text{-}1)$$

and forms salts because of the third hydroxyl group of the phosphoric acid part structures. It is called a nucleic acid because it is found mainly in the cell nucleus, although there is some in the protoplasm as well. A **nucleoprotein**, the form in which nucleic acids often occur, is a saltlike complex of nucleic acid and protein.

Figure 15-1 shows a section of the chain of a typical nucleic acid. Attached to each sugar part structure is a nitrogen base. The nitrogen base is attached at carbon 1 of the sugar and these compounds are β-glycosides. The structure shown in Figure 15-1 is **deoxyribonucleic acid**, or **DNA**, because the sugar is **deoxyribose**. The sugar in the other commonly occurring type of nucleic acid, **ribonucleic acid**, **RNA**, is D-ribose.

Stepwise hydrolysis of a nucleoprotein first divides it into nucleic acid and protein, then cleaves the nucleic acid into units called **nucleotides**. These represent the fundamental repeating structure of the polymer and contain one fragment each of phosphoric acid, sugar, and nitrogen base. Hydrolysis of the phosphoric acid ester linkage of a nucleotide gives phosphoric acid and a **nucleoside**, a β-glycoside consisting of the sugar and the nitrogen base. This is summarized in Figure 15-2. The good news in Figure 15-2 is that a given nucleic acid contains only *one* kind of sugar, either D-ribose or deoxyribose. Additional good news is that there are only a limited number of different kinds of nitrogen base. These are all either purines (Chapter 11, page 233) or pyrimidines. Further, most of the purines are adenine and guanine, and most of the pyrimidines are uracil, thymine, and cytosine.

The predominant purines found in nucleic acids are adenine and guanine. The arrow in Figure 15-3 points to the 9 position at which the purines are attached to the sugar moiety as β-glycosides. Guanine is shown in the isomeric **6-keto** form, because this is the structure that is most important in the nucleic acid. The predominant pyrimidines of nucleic acids are uracil, thymine, and cytosine.

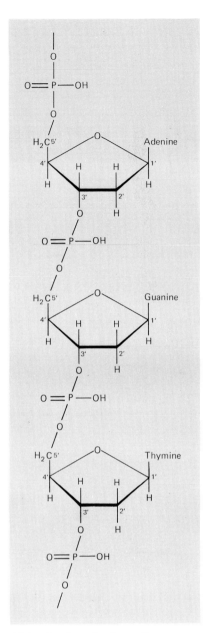

FIGURE 15-1. Portion of a deoxyribonucleic acid chain.

These compounds are also present in the nucleic acids in the keto form (Figure 15-4) rather than the isomeric hydroxy form on which the name is based.

Nucleotides and Nucleosides

Before returning to the main topic of nucleic acids it is desirable to discuss nucleotides (phosphoric acid-sugar-base compounds) and nucleosides (sugar-base compounds) (see Table 15-1). These compounds occur free as well as in nucleic acids, and some of them have important biochemical roles of their own.

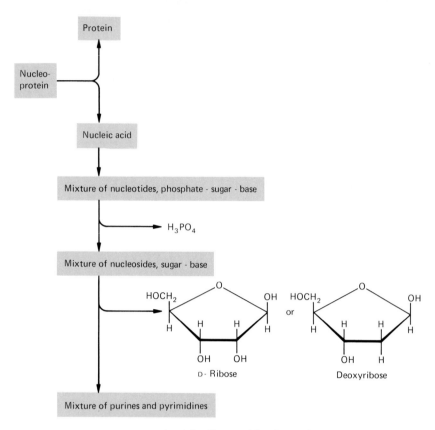

FIGURE 15-2. The disassembly of a nucleoprotein.

TABLE 15-1. *Nucleotides and Nucleosides*

Base	Nucleotide*		Nucleoside	
	Ribonucleotide	Deoxyribonucleotide	Ribonucleoside	Deoxyribonucleoside
Purines				
Adenine	5'-adenylic acid or adenosine 5'-phosphate	deoxy-5'-adenylic acid	adenosine	deoxyadenosine
Guanine	5'-guanylic acid or guanosine 5'-phosphate	deoxy-5'-guanylic acid	guanosine	deoxyguanosine
Pyrimidines				
Uracyl	5'-uridylic acid or uridine 5'-phosphate	deoxy-5'-uridylic acid	uridine	deoxyuridine
Thymine		5'-thymidylic acid		thymidine
Cytosine	5'-cytidylic acid or cytidine 5'-phosphate	deoxy-5'-cytidylic acid	cytidine	deoxycytidine

* Other isomers besides the 5' ones illustrated are possible.

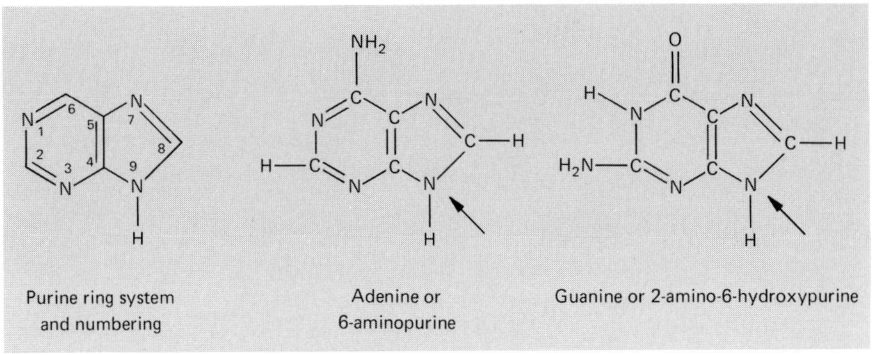

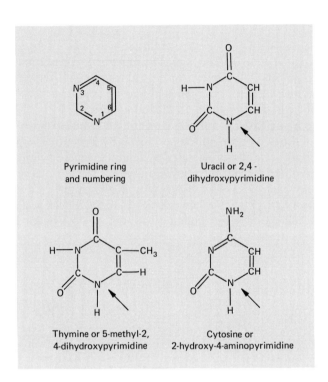

FIGURE 15-3.

FIGURE 15-4.

NOMENCLATURE OF NUCLEOSIDES AND NUCLEOTIDES

The nucleosides formed from D-ribose and a nitrogen base have common names ending in **ine** and suggestive of the name of the nitrogen base. Thus the base guanine is found in the nucleoside **guanosine**, adenine in **adenosine**, cytosine in **cytidine**, and uracil in **uridine**. The sugar is not mentioned if it is ribose, but if it is deoxyribose the name of the nucleoside indicates this by a **deoxy** prefix. Thus we have deoxyadenosine, deoxycytidine, and so forth. Thymidine, infuriatingly enough, is a deoxyribonucleoside. The deoxy prefix is left out because thymine is found mainly in deoxyribonucleic acid in any case.

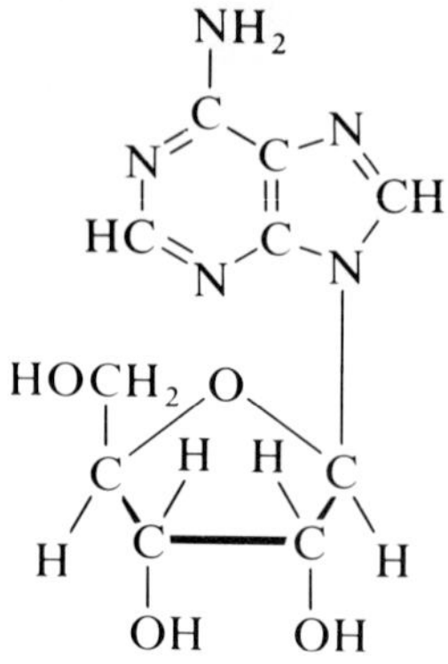

adenosine
(9-β-D-ribofuranosyladenine)

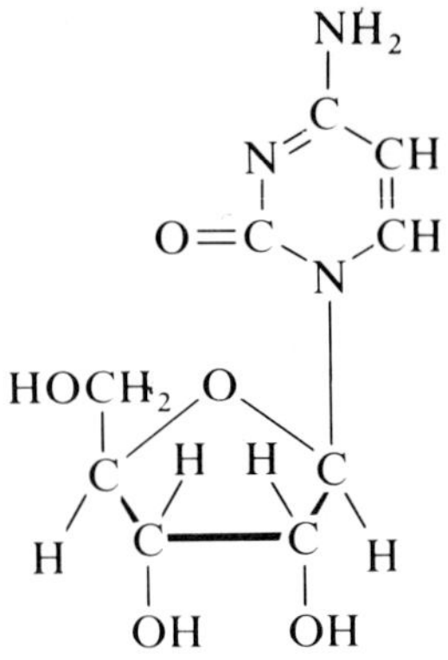

cytidine
(1-β-D-ribofuranosylcytosine)

Nucleotides are given names ending in **-ylic acid** and have a root indicating the nature of the base. The sugar is not mentioned unless it is deoxyribose. Thus we have the names adenylic acid, guanylic acid, thymidylic acid, cytidylic acid and uridylic acid; also deoxyadenylic acid, deoxyguanylic acid, and so forth.

A given nucleic acid can give isomeric nucleotides depending on which of the two phosphate-sugar links is hydrolyzed. The position of the phosphoric acid is indicated by a number marked with a prime to differentiate it from the numbers used for the various positions in the nitrogen base. Thus hydrolysis of the sample nucleic acid shown in Figure 15-1 could give deoxy-3'-adenylic acid or deoxy-5'-adenylic acid and corresponding isomers of deoxyguanylic and thymidylic acids.

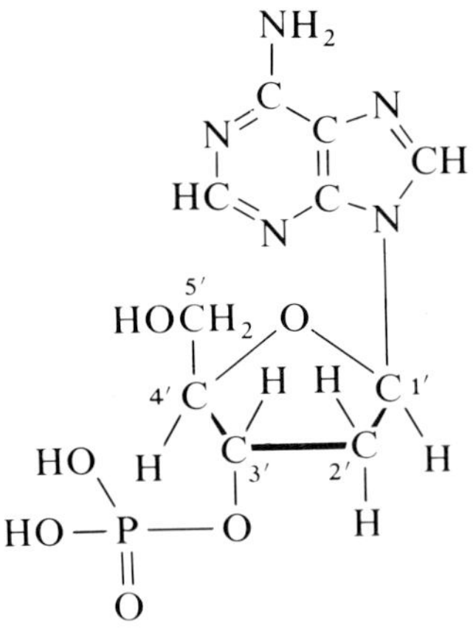

deoxy-3'-adenylic acid
(deoxyadenosine 3'-phosphate)

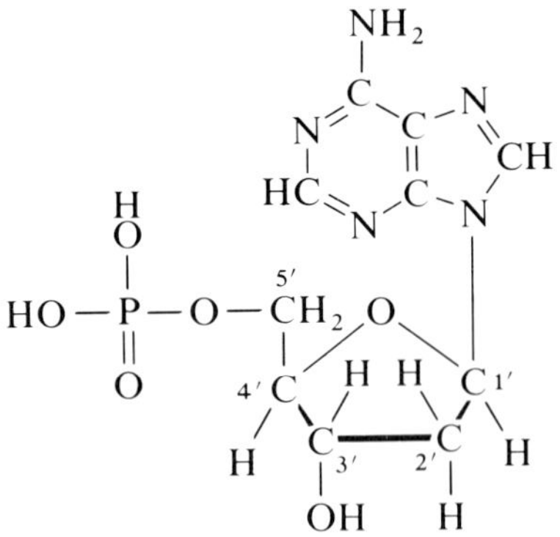

deoxy-5'-adenylic acid
(deoxyadenosine 5'-phosphate)

ADENOSINE TRIPHOSPHATE (ATP)

Every living organism is a consumer of energy which it gets from sources such as sunlight or the oxidation of foods. The energy is used in many ways but mostly to make biochemical reactions go in spite of the fact that they may be up-hill processes, that is, energy-consuming processes. The required energy cannot simply be turned loose in the organism, it must be in a form that the organism can use.

In most cases this usable form of energy turns out to be the "high-energy molecule," adenosine triphosphate, otherwise known as ATP. ATP is the immediate source of energy for almost all energy-requiring biological processes, whether protein synthesis, nerve impulse transmission, or muscle contraction. We will consider just one example in enough detail to understand how the energy gets out of the ATP and into the process that needs the energy.

The energy contained in a molecule of ATP can be set free by almost any reaction that breaks one of the phosphoric anhydride bonds. Hydrolysis as in equation 15-2, for example, liberates about 7500 cal/mole. Just setting the energy free in the form of heat is all

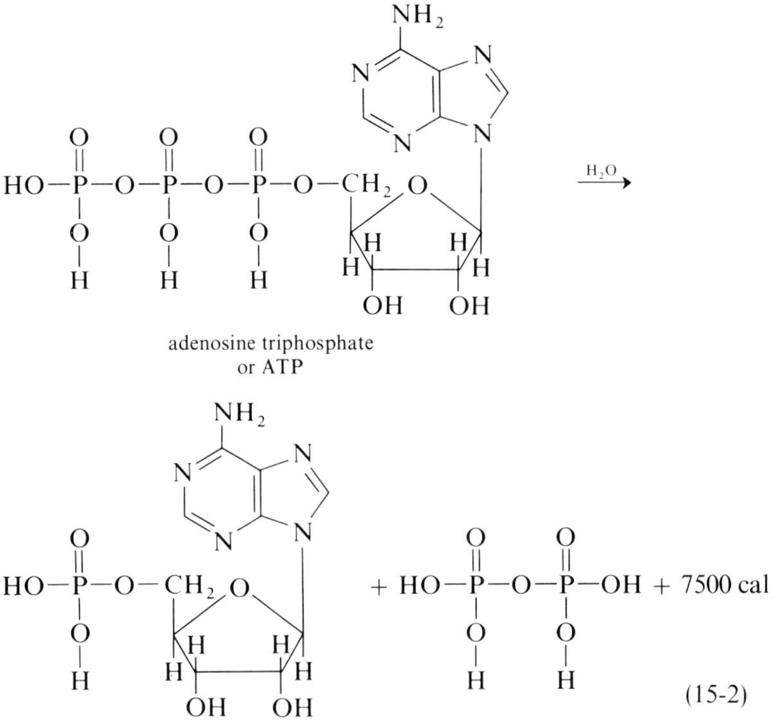

adenosine triphosphate or ATP

$\xrightarrow{H_2O}$

adenosine monophosphate, AMP, or adenylic acid + HO−P(=O)(O−H)−O−P(=O)(O−H)−OH + 7500 cal (15-2)

right in a crude machine like a steam engine, but in a living organism energy in that form would just be wasted. To use the energy, the energy-producing reaction (starting with ATP) has to be *coupled* with the reaction that needs the energy. There is only one way to do this: let the ATP make an intermediate product that is used as a reagent by the energy-consuming reaction.

Let us take as an example the energy-consuming reaction of ester synthesis, which ordinarily requires about 4000 cal/mole. The

$$4000 \text{ cal} + \text{RCOOH} + \text{R'OH} \rightleftharpoons \text{RCOR'} + H_2O \quad (15\text{-}3)$$

4000 cal on the left side of equation 15-3 means that, at equilibrium, we will have mostly acid and alcohol and very little ester. The cell, therefore, uses equation 15-4 instead; this is a down-hill reaction *liberating* 3500 cal and giving a good yield of ester at equilibrium.

Nucleic Acids

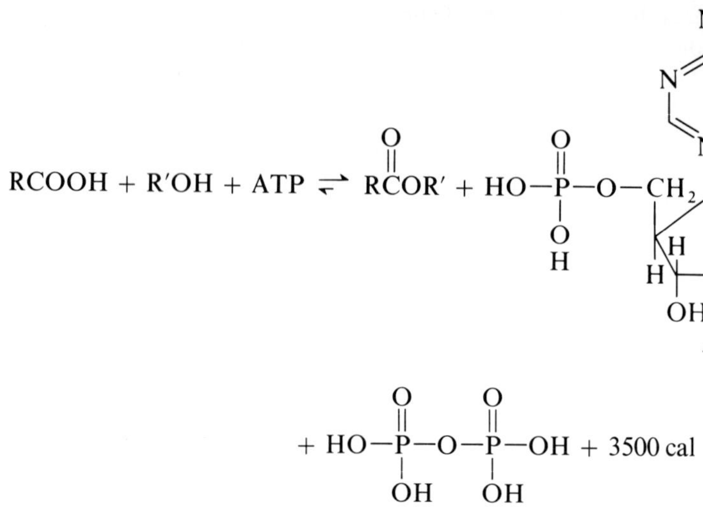

$$\text{RCOOH} + \text{R'OH} + \text{ATP} \rightleftarrows \text{RC}(=\!O)\text{OR'} + \text{AMP}$$

$$+ \text{ HO-P(=O)(OH)-O-P(=O)(OH)-OH } + 3500 \text{ cal} \quad (15\text{-}4)$$

Notice that reaction 15-4 amounts to a *hydrolysis* of ATP into AMP and the pyrophosphoric acid. The water required is the molecule of H_2O that is ordinarily formed in the synthesis of the ester from acid and alcohol.

How did the esterification reaction and the phosphoric anhydride hydrolysis reaction get coupled? The answer is that equation 15-4 is really just the sum of two reactions (equations 15-5 and 15-6).

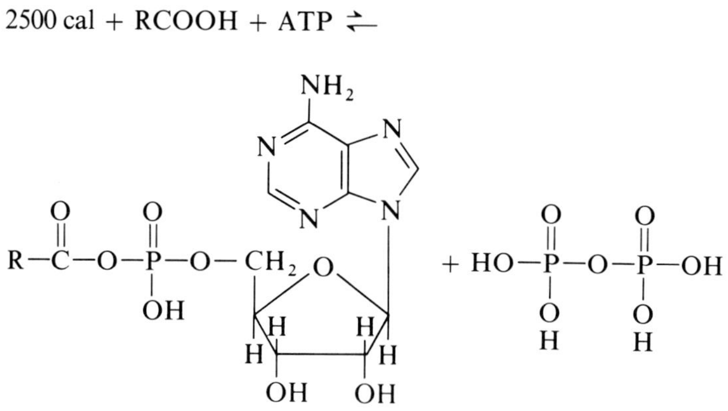

$$2500 \text{ cal} + \text{RCOOH} + \text{ATP} \rightleftarrows$$

$$\text{R-C(=O)-O-P(=O)(OH)-O-CH}_2\text{-(adenosine)} + \text{HO-P(=O)(OH)-O-P(=O)(OH)-OH} \quad (15\text{-}5)$$

an acyl adenylate

$$\text{the acyl adenylate} + \text{R'OH} \rightleftarrows \text{RC}(=\!O)\text{OR'} + \text{AMP} + 6000 \text{ cal} \quad (15\text{-}6)$$

Reaction 15-5 is actually slightly uphill, but reaction 15-6 makes it go anyway by gobbling up the intermediate acyl adenylate.

To summarize, **energy is obtained** from the hydrolysis of ATP for use in an up-hill reaction **by coupling the two reactions through a common intermediate**.

ATP itself is made by reactions in which the oxidation of a fat or carbohydrate is coupled (via intermediates) with the conversion of AMP to ATP.

Deoxyribonucleic Acid (DNA)

The structure of part of a deoxyribonucleic acid chain was shown in Figure 15-1. As we have already noted, DNA is a polydeoxyribonucleotide. To characterize a deoxyribonucleic acid more completely we need to know the relative amounts of each nitrogen base, the sequence in which they occur along the chain, and the secondary structure or conformation of the chain. That is, the task from this point is very much like the one that faced us in connection with polypeptides, and, indeed, these two different types of biological polymer turn out to be intimately related.

NUCLEOTIDE COMPOSITION

The proportion of each base or nucleotide is a constant for any given species, regardless of which part of the organism the DNA is taken from. Thus, in Table 15-2, we see that the percentages of the four bases are the same within experimental error in DNA from human liver cells and in DNA from human thymus cells. The DNA from most organisms contains only the four bases shown in Table 15-2, but a few higher plants and animals have methyl cytosine in place of cytosine and some viruses have a glycoside of 5-hydroxymethyl cytosine in place of the usual cytosine.

TABLE 15-2

Species	Mole % of each base				Ratios	
	Guanine	Adenine	Cytosine	Thymine	$\dfrac{\text{Adenine}}{\text{Thymine}}$	$\dfrac{\text{Guanine}}{\text{Cytosine}}$
Sarcina lutea	37.1	13.4	37.1	12.4	1.08	1.00
Escherichia coli K12	24.9	26.0	25.2	23.9	1.00	1.09
Human liver	19.5	30.3	19.9	30.3	1.00	0.98
Human thymus	19.9	30.9	19.8	29.4	1.05	1.01

The most striking thing about the DNA nucleotide composition is the tendency of the amount of purines to equal the amount of pyrimidines and, more specifically, for the amount of adenine to equal the amount of thymine and the amount of guanine to equal the amount of cytosine. You can check on this by looking at the numbers in Table 15-2. The ratios are consistently so close to unity that the differences are almost within experimental error. With these numbers, nature is trying to tell us something: deoxyadenylic acid is somehow *paired* with thymidylic acid, and deoxyguanidylic acid is somehow paired with deoxycytidylic acid.

One possibility, of course, might have been that the repeating units in a DNA chain are not mononucleotides at all but adenine-thymine dinucleotides and guanine-cytosine dinucleotides. Experimental evidence showed that this was not the case. In the mixture of dinucleotides obtained by partial hydrolysis of DNA, these combinations occur no more frequently than the others. Yet there is something about DNA or about the biosynthesis of DNA that says that for every adenine there is a thymine and for every guanine there is a cytosine. An answer was provided by the double helix hypothesis of Crick and Watson, and this hypothesis was confirmed by the data from an x-ray analysis of DNA carried out by Wilkins and coworkers.

THE DOUBLE HELIX

The α-helix, stabilized by hydrogen bonding, is a stable secondary structure of polypeptide chains and was well known at the time. If a single helix is good, perhaps a double helix might be even better.

The two polynucleotide strands making up the double helix are held together by hydrogen bonds between pairs of bases. The pairing seen in the nucleotide composition figures for adenine and thymine is due to the fact that adenine and thymine are uniquely constituted so as to make strong interchain hydrogen bonds with each other (Figure 15-5). An NH of a thymine group is hydrogen bonded to a

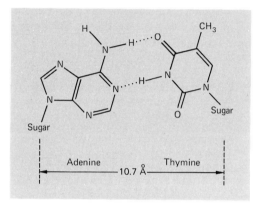

FIGURE 15-5. Adenine-thymine hydrogen bonded base pair. [After M. H. F. Wilkins and S. Arnott, *J. Mol. Biol.*, **11**:391 (1965)]

nitrogen of an adenine group. The NH_2 of the adenine group happens to be in just the right position to hydrogen bond to the oxygen atom of the thymine group. A further happy circumstance is that, thus connected, the thymine group of one DNA polynucleotide strand and the adenine group of the other, fit very well into the over-all double helical structure.

Similarly, cytosine and guanine fit together with not two but three hydrogen bonds (Figure 15-6). And again the connection between cytosine on one strand and guanine on the other proves to be compatible with, and strengthens, the over-all double helical structure of the two DNA strands.

There are several points that should be noted about the structure of the double helix (Figure 15-7).

1. The two chains are not identical, but *complementary*. Where one chain has adenine, the other has thymine; where one has guanine, the other has cytosine, and vice versa.

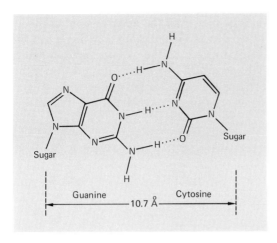

FIGURE 15-6. A cytosine-guanine base pair.

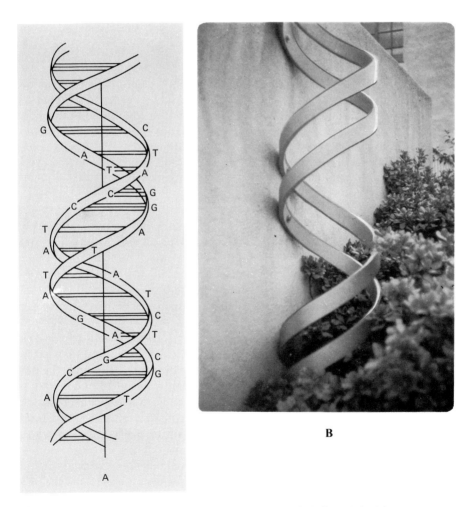

FIGURE 15-7. (A) Schematic representation of the double helix. (B) Architectural use of the double helix motif.

2. It would be nice if all the adenine were in one chain and all the thymine in the other chain, but this is not the case.
3. It would be nice if all the guanine were in one chain and all the cytosine in the other chain, but this is not the case either.
4. Both chains are right-handed helixes.
5. The chains run in opposite directions (Figure 15-8).

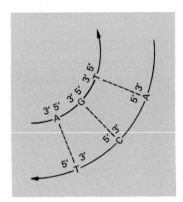

FIGURE 15-8. The two chains of the double helix are antiparallel: one runs in the 3′,5′ direction and the other in the 5′,3′ direction.

The magnitude of the task of completely determining the structure of any given DNA molecule is illustrated by what is known about the chromosomes of the bacterium *Escherichia coli*. Each chromosome of *E. coli* is a single molecule of DNA closed in a cyclic structure so that there are no special end nucleotides. *The molecule has about 200,000 turns of the helix*. Cut open and stretched out this cyclic double helix would be about half a millimeter long!

REPLICATION

The propagation of a species requires the reproduction of the genes, the heredity-determining elements of the chromosome, and requires that they be duplicated unchanged, or with only minor mutations. The chromosome is a nucleic acid molecule, and the genes are short sequences of nucleotides that make up the over-all sequence of the nucleic acid. Replicating genes is therefore a matter of replicating nucleic acid molecules.

The most likely mechanism for the replication of a nucleic acid is illustrated in Figure 15-9. First, the nucleic acid untwists to separate the two strands of the double helix. Although the two strands are complementary rather than identical, each contains all the information needed to construct the other. Each of the old strands acts as a template for the construction of a new strand complementary to it. The combination of each old strand with a new complementary strand gives a new double helix.

The reason why the new complementary strand exactly fits its template is quite simple: **the old template strand and the new complementary strand twist to form the double helix simultaneously with the growth of the new strand.** That is, as a new nucleotide is added to the growing chain it is required to fit into the structure

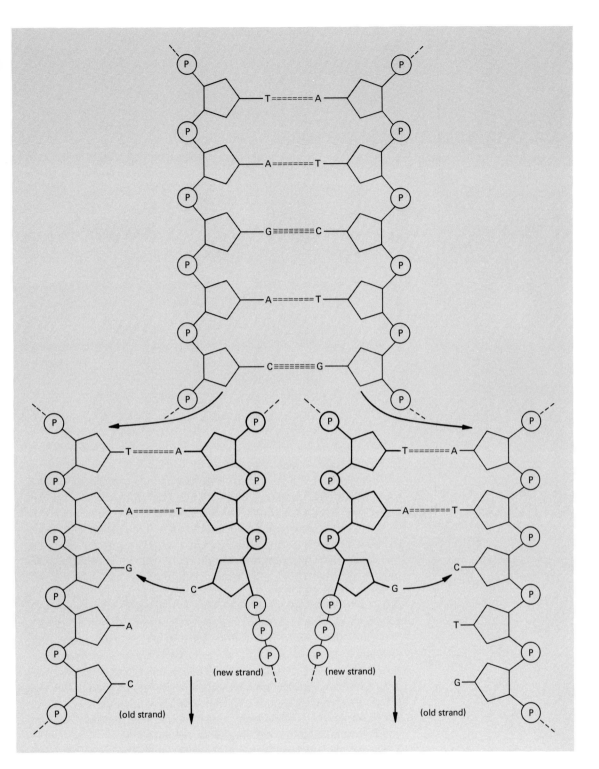

FIGURE 15-9. The two replicated double helixes each contain one of the original strands and one new complementary strand (bold face).

Nucleic Acids

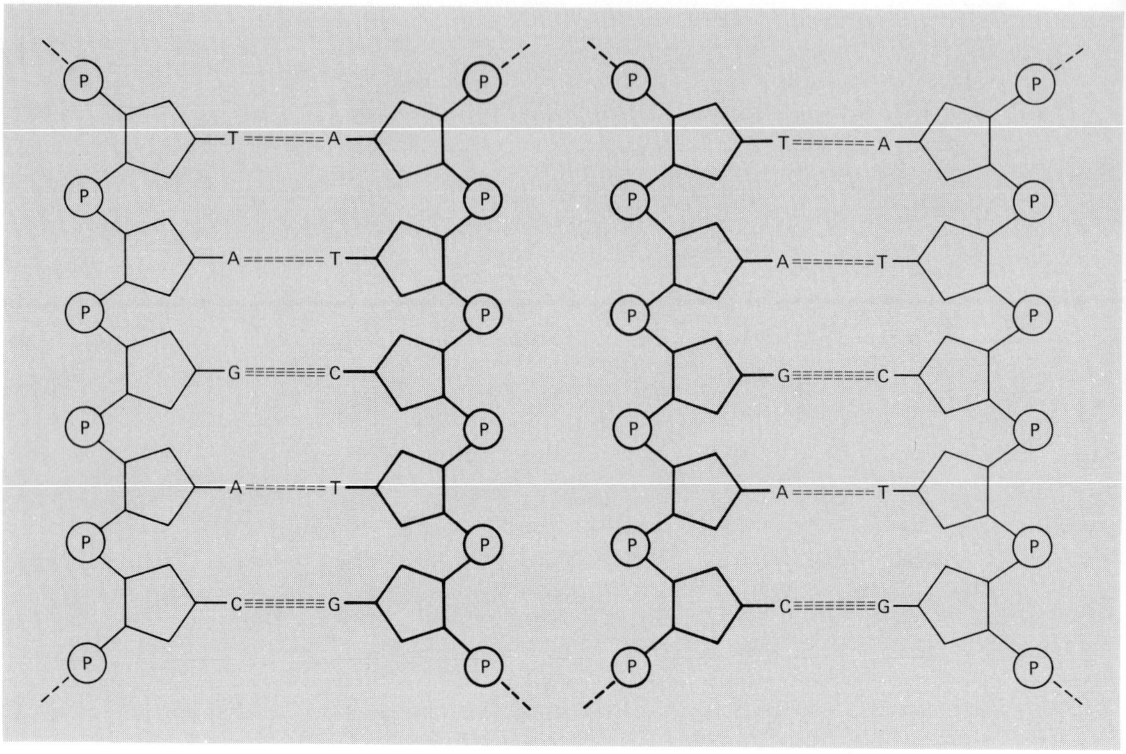

FIGURE 15-9. (*Continued*)

of the double helix as well. Only a nucleotide that is complementary to the next unit of the template chain will do this.

First of all, if the next unit of the template is a purine, the next unit added to the growing complementary chain has to be a pyrimidine. The backbones of the two strands of the double helix are separated by a constant distance of 10.85 Å. This distance is equal to the length of a purine plus a pyrimidine. Two purines would be too long and two pyrimidines would be too short. Thus if the next unit of the template is the purine guanine, the next unit of the growing chain has to be thymine or cytosine. Cytosine is preferred because it, like guanine, forms three hydrogen bonds whereas thymine can form only two.

Besides fitting into the double helix structure, each nucleotide unit has to fit into the "lock" of the DNA polymerase enzyme that catalyzes the growth of the chain. This double requirement is quite severe, but a few molecules have been found that are enough like the usual nitrogen base to fool both the template chain and the enzyme. An example is 2-aminopurine, which can be incorporated into the chain and introduces a copying error, or mutation.

The double helix mechanism for nucleic acid replication has been tested in two ways. First of all, if one strand of the new double helix comes from the parent organism and one strand is synthesized from the growth medium, then a growth medium labeled with isotopic nitrogen should give a first generation in which each nucleic acid molecule is exactly half labeled and half unlabeled. This has been tested with the bacterium *E. coli* and the result was as expected.

A second test of the hypothesis used a deoxyadenylic acid–thymidylic acid copolymer. When this synthetic material is treated with a mixture of all four of the usual nucleoside triphosphates in the presence of DNA polymerase, new DNA is formed but only adenosine and thymidine units are incorporated. The other nucleosides do not match the template and are rejected.

Nucleic Acids and Protein Synthesis

To each enzyme or protein synthesized by an organism there corresponds a particular part of the DNA molecule that directs the synthesis of that particular polypeptide. Such a region of the DNA molecule is called a gene. Each gene has a particular pattern or nucleotide sequence that is a code for the sequence of amino acid units in the particular polypeptide for which it is responsible.

In this section we will address ourselves to two questions:

1. What is the genetic code?
2. How is it translated into an actual polypeptide?

Proteins usually contain 20 different amino acids, and DNA contains only four kinds of nitrogen base; thus, it is clear that the "words" standing for individual amino acids have to contain more than one "letter" or nitrogen base. The "word," or coding unit is called a **codon** and consists of a sequence of three bases. With four different kinds of base, codons of this length are sufficient to indicate 64 different amino acids, and we have only 20 amino acids.[1]

The reading of the codons and their translation into the amino acid sequence of a polypeptide molecule is a complicated process requiring three different kinds of ribonucleic acid (RNA).

Messenger RNA or **mRNA** is a polyribonucleotide using the same bases as are found in DNA except that uracil appears wherever DNA would have thymine. In the synthesis of mRNA, the DNA double helix first untwists temporarily while one of its strands is **transcribed** as a complementary molecule of mRNA. The process is very much like the synthesis of a new complementary strand of DNA except that uracil is used rather than thymine wherever the complement of adenine is called for.

mRNA is sometimes called template RNA because it contains all the information of the original DNA strand, although the information has been transcribed in a different language. mRNA serves as a sort of assembly line for the construction of the polypeptide molecule. Each amino acid in the sequence is attached to the growing

```
START    ABC ABC ABC ABC  . . . .  ABC ABC ABC STOP
```

FIGURE 15-10. The mRNA assembly line. The ABC's represent the different codons while START and STOP represent special codons that signal the N-terminal and carboxyl-terminal ends of the polypeptide molecule.

---

[1] Doublet codons would give only 16 possibilities, which is not enough.

polypeptide chain by an operation that takes place at a corresponding location on the mRNA assembly line. The sequence of codons for amino acids in the mRNA chain is the same as the sequence of the amino acids themselves in the eventual polypeptide.

Ribosomes are molecules of ribonucleoprotein, the nucleic acid part being known as **ribosome RNA** or **rRNA**. The ribosome does not contain any genetic information or copies of the codons, and its nucleotide composition is about the same for different organisms. You can think of the ribosomes as machines that run down the mRNA tracks, adding the appropriate amino acid unit as they pass each codon (Figure 15-11).

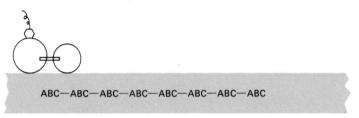

FIGURE 15-11. A ribosome, or ribonucleoprotein molecule, moving down a chain of mRNA codons. The smoke coming from the stack represents some unknown mechanism that moves the ribosome steadily down the track without skipping any of the codons.

Transfer RNA or **tRNA** is the third kind of RNA involved in polypeptide synthesis. Each molecule of tRNA has a site at which the amino acid is attached and a segment consisting of three nucleotides that constitute an **anticodon** or complementary codon for the amino acid. The anticodon permits the codon at each position along the mRNA track to recognize its particular kind of amino acid. Since there are 20 different amino acids, there are also 20 different kinds of tRNA in most cells; each tRNA has its own kind of amino acid group and the corresponding anticodon. Thus we have alanyl tRNA, glycyl tRNA, and so forth. In addition to the usual kinds of bases found in mRNA, tRNA has some rarer types as well; however, the most important nucleotides are the standard ones that make up its codon or label (Figure 15-12).

Polypeptide synthesis occurs when the ribosome moves down the mRNA track picking up the appropriate amino acid group from the tRNA hydrogen bonded at each codon. This is shown in Figure 15-13, where the ribosome is about to add an alanine unit to its growing polypeptide chain (~COOH). Refer again to Figure 15-13. We see that the next amino acid unit scheduled for incorporation in the protein is glycine. The glycine codon, GGG, which comes next on the track, has hydrogen bonded to the anticodon (CCC) of a glycyl tRNA molecule, thus assuring that a glycyl unit, *and nothing but a glycyl unit*, will be there.

After the tRNA has transferred its amino acid unit to the growing polypeptide, the tRNA departs from the mRNA chain. It will probably be used again for the same purpose. For example, an "empty" molecule of alanyl tRNA (that is, a molecule without its alanyl group) will react with a molecule of suitably activated alanine to pick up a new alanyl group. This reaction is catalyzed by an enzyme

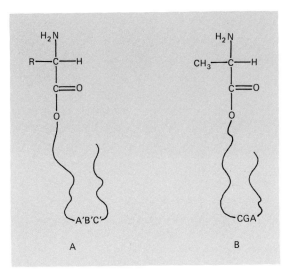

FIGURE 15-12. (A) Schematic representation of a molecule of tRNA. The anticodons are symbolized by the A'B'C'. (B) Schematic representation of alanyl tRNA; its anticodon is the trinucleotide sequence cytosine-guanine-adenine, which will hydrogen bond to the codon sequence guanine-cytosine-uracil on the mRNA chain.

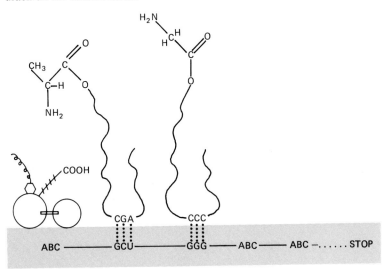

FIGURE 15-13. The ribosome is about to add an alanine unit to its growing polypeptide chain, represented by ⧣COOH. The next unit to be added will be a glycine unit.

specific enough so that there is no danger of the tRNA becoming attached to the wrong amino acid. The new alanyl tRNA will hydrogen bond to any vacant position of the mRNA molecule that happens to have the alanyl codon, GCU.

The ribosome continues down the chain in the same way, adding the appropriate amino acid unit at each codon, until eventually it comes up against a STOP codon such as UGA. This causes the ribosome and its polypeptide to leave the mRNA chain. The last codon before the UGA determines what the carboxy-terminal unit will be.

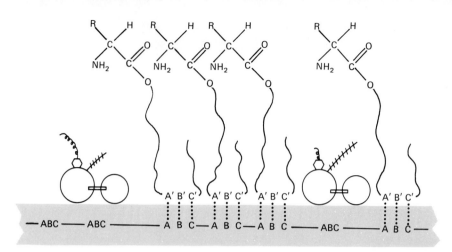

FIGURE 15-14. Notice that the first ribosome (the one on the right) has travelled farther and made a longer polypeptide chain.

Notice the resemblance between the synthesis of a polypeptide chain with one end fixed on a ribosome and the polymer bead method of Chapter 14 in which one end of the polypeptide is fixed on a crosslinked polymer. So far, nature's method is more highly automated than man's.

Up to this point we have concerned ourselves only with a single ribosome and a single molecule of growing polypeptide. Actually, it is believed that there are many of these ribosome production units moving down the mRNA assembly line at the same time (Figure 15-14). By the time a second ribosome reaches a position already traversed by the first, it is likely that the depleted tRNA molecules have been replaced.

The Code Dictionary and Mutations

In order to make it easier to discuss nucleotide sequences, we will use a commonly adopted set of abbreviations shown in Table 15-3.

TABLE 15-3. *Symbols and the Corresponding Nitrogen Bases*

A	Adenine	U	Uracil
G	Guanine	T	Thymine
C	Cytosine	p	Phosphate

The 5' end of the nucleotide is on the left and the 3' end is on the right. Thus p-A represents adenosine 5'-phosphate and A-p represents adenosine 3'-phosphate. The symbol in each case stands for the nitrogen base and the sugar. A small d prefix can be added in case the sugars are all deoxyribose rather than ribose. In representing amino acid codons it is customary to leave out the intervening phosphate groups, thus UUU and not pUpUpU.

A major step in breaking the amino acid code was to establish that all of the codons are triplets. A synthetic trinucleotide will act as "mRNA" to bind an amino acyl tRNA, but a dinucleotide will not. Another piece of evidence for triplet codons is the behavior of synthetic alternating polynucleotides. For example the polynucleotide

$$CUCUCUCUCU\ldots$$

when used as "mRNA," makes a polypeptide in which leucine and serine alternate.

    Leu   Ser   Leu   Ser   Leu $\cdots$
    CUC   UCU   CUC   UCU   CUC $\cdots$

Evidently CUC stands for Leu and UCU stands for Ser. If the codon consisted of two units or four units, the resulting polypeptide would have had only *one* kind of amino acid.

    X   X   X   X   X   X
   CU CU CU $\cdots$ or CUCU CUCU CUCU $\cdots$

The effect of other nitrogen base sequences in synthetic "mRNA" polynucleotides has led to the dictionary shown in condensed form in Table 15-4. To use the dictionary table to translate a symbol,

TABLE 15-4. *Partial Dictionary of Codons*

First position	Second position				Third position
	U	C	A	G	
U	Phe	Ser	Tyr	Cys	U
	Phe	Ser	Tyr	Cys	C
	Leu	Ser	(CT)	(CT)	A
	Leu	Ser	(CT)	Trp	G
C	Leu	Pro	His	Arg	U
	Leu	Pro	His	Arg	C
	Leu	Pro	Gln	Arg	A
	Leu	Pro	Gln	Arg	G
A	Ile	Thr	Asn	Ser	U
	Ile	Thr	Asn	Ser	C
	Ile	Thr	Lys	Arg	A
	Met (CI)	Thr	Lys	Arg	G
G	Val	Ala	Asp	Gly	U
	Val	Ala	Asp	Gly	C
	Val	Ala	Glu	Gly	A
	Val (CI)	Ala	Glu	Gly	G

(CI) = chain initiation.
(CT) = chain termination.

start with the first letter. For example, if the codon is UGU, the first letter U tells us to look in the upper quarter of the table. The fact that the second symbol is G narrows the possible meanings down to three. The third symbol, U, tells us that the answer is Cys, cysteine. Note that some amino acids are represented by more than one codon. Thus UCU, UCC, UCA, UCG, AGU, and AGC all represent serine. Some combinations are punctuation marks. For example GUG means "start a new chain with an *N*-terminal valine unit," and UGA means "stop, this is the carboxy-terminal end."

MUTATIONS

So far as is known, a single gene is a sequence in the DNA polynucleotide that is responsible for controlling the amino acid sequence of just a single kind of polypeptide. Often, the polypeptide is an enzyme and the effect of a single gene mutation is to make one enzyme ineffective and prevent one biochemical reaction from taking place.

Most biochemical syntheses are multistep processes and require a corresponding number of enzymes. They are therefore vulnerable to a mutation that affects *any* of the enzymes involved. The synthesis of arginine by the bread mold *Neurospora crassa* will serve as an example. The usual wild strain of *N. crassa* makes its own arginine from ammonia, but certain mutant strains require arginine in their growth medium. Still other mutant strains will grow on *either* arginine or on one of the intermediates of arginine biosynthesis (Figure 15-15). A mutant that lacks the enzymes required to make

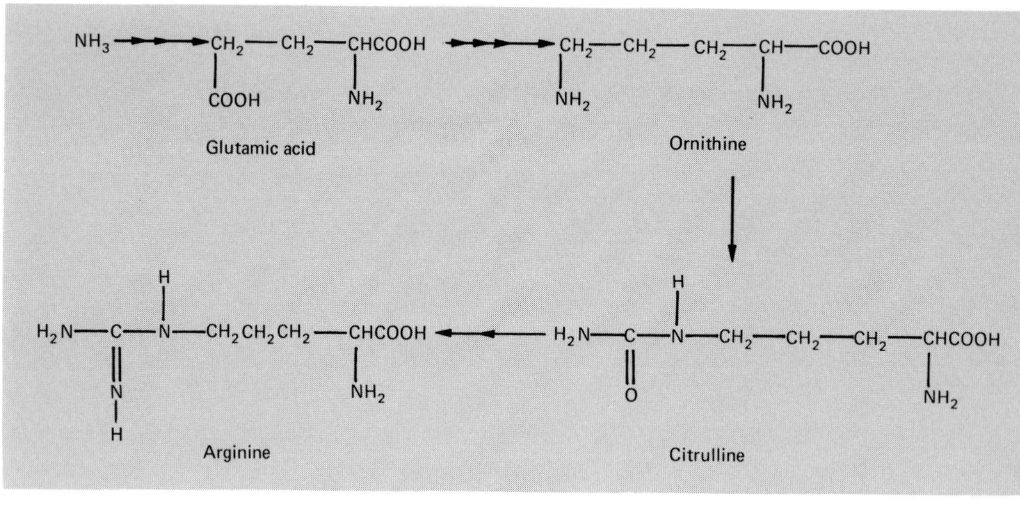

FIGURE 15-15. Biosynthesis of arginine by way of ornithine and citrulline as intermediates. Each step is vulnerable to a mutation that might spoil its enzyme.

ornithine, for example, will grow if supplied either with ornithine or any intermediate further along in the synthesis. The dietary requirement of a given mutant tells which enzyme, and therefore which gene, has been changed.

The long list of vitamins and essential amino acids required in the human diet reflects a long history of mutations that have eliminated the enzymes for one reaction after another. Fortunately, we live in an environment of other organisms that conduct these syntheses for us.

The hereditary disease hemophilia is due to a mutation that interferes with the reactions that cause blood to clot. Another hereditary disease, phenylketonuria is characterized by mental retardation and the appearance of phenylpyruvic acid in the urine. Phenylketonuria is caused by the absence of the enzyme phenylalanine hydroxylase that ordinarily diverts phenylalanine to other products. The mental retardation can be alleviated if the disease is recognized in time and the child put on a diet low in phenylalanine.

$$\text{C}_6\text{H}_5-\text{CH}_2-\underset{\underset{\text{NH}_2}{|}}{\overset{\overset{\text{H}}{|}}{\text{C}}}-\text{COOH} \rightarrow \text{C}_6\text{H}_5-\text{CH}_2-\overset{\overset{\text{O}}{\|}}{\text{C}}-\text{COOH}$$

phenylalanine phenylpyruvic acid

Sickle-cell anemia is caused by a mutation of the gene that controls the synthesis of one of the polypeptide chains of hemoglobin. The abnormal hemoglobin has just one amino acid unit that is different from the usual one at that position. Therefore the mutation involves a single codon. This small change in the polypeptide has two effects; one has some survival value, the other is dangerous. The valuable effect, especially for persons living in malarial parts of the world, is that the malarial parasite does not thrive very well in red cells containing sickle-cell hemoglobin. The tragic effect is due to a lowered solubility of the sickle-cell hemoglobin in its reduced form. If the hemoglobin is kept in its reduced form too long, it tends to crystallize out of solution and distort the blood cell into the shape of a sickle. These sickle cells not only fail to transport oxygen properly but also accumulate in narrow capillaries where they block circulation. Acute effects from sickle-cell hemoglobin are most likely under conditions of low oxygen pressure, and persons suffering from this disorder should stay away from mountain tops and unpressurized airplane cabins.

Some mutations are caused by the action of chemicals called **mutagens**. Some compounds mimic the usual nitrogen bases well enough so that they become incorporated in the DNA molecule. This would not matter perhaps, except that their mimicry is imperfect enough to cause mistakes in base pairing when the DNA is replicated. Examples of such mutagenic base analogs are 5-bromouracil, 5-chlorouracil, 8-azaguanine, and 6-thioguanine. The 5-halogenated uracils probably mimic thymine, which has a methyl group in that position.

Another cause of mutations is crosslinking between two nitrogen bases. This interferes with the normal double helix structure and can lead to errors in DNA replication or in transcription. A chemical crosslinking agent with mutagenic activity is nitrogen mustard, which alkylates two guanine bases and ties them together (Figure 15-16). X-rays have a similar effect by linking thymine units together (Figure 15-17).

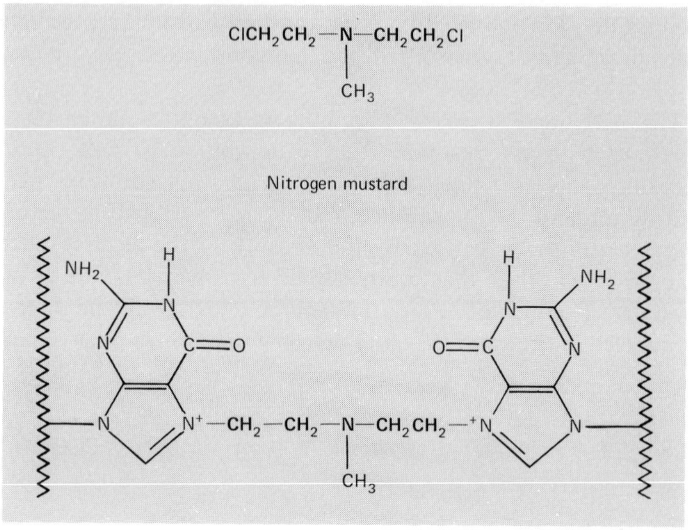

FIGURE 15-16. A crosslink between guanine bases. The wiggly line represents the portion of the polynucleotide chain to which the base is attached.

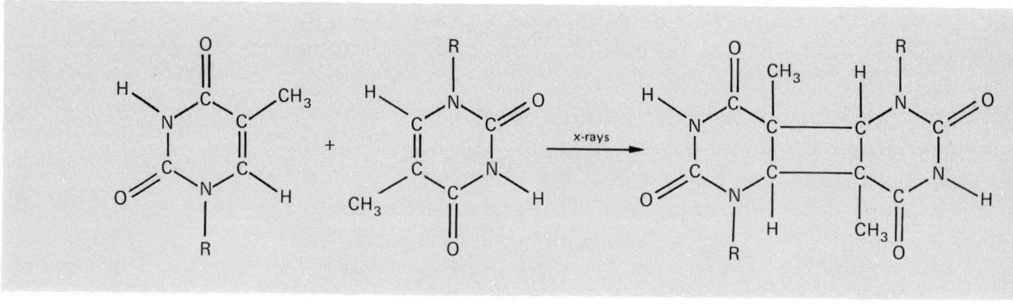

FIGURE 15-17. Mutagenic effect of x-rays caused by dimerizing thymine base units.

Viruses As Nucleoproteins

A virus is a molecule of nucleoprotein that can invade a host cell and reproduce itself therein, sometimes with harmful effects on the host. The mechanism of reproduction is like that by which the host cell reproduces its own genetic material and probably employs some of the apparatus used by the host. So far, viruses have not been observed to exhibit the lifelike activity of self-reproduction anywhere but in a host cell.

TOBACCO MOSAIC VIRUS (TMV)

The tobacco mosaic disease of tobacco plants is caused by a ribonucleoprotein that has been isolated in a pure crystalline form. The molecules of tobacco mosaic virus are long rods, big enough to be seen easily with an electron microscope (Figure 15-18). The ribonucleic acid part of the molecule runs down the center of the rod, and the protein is wrapped around the outside.

The nucleic acid and protein parts of tobacco mosaic virus can be separated and can also be recombined chemically to reconstitute

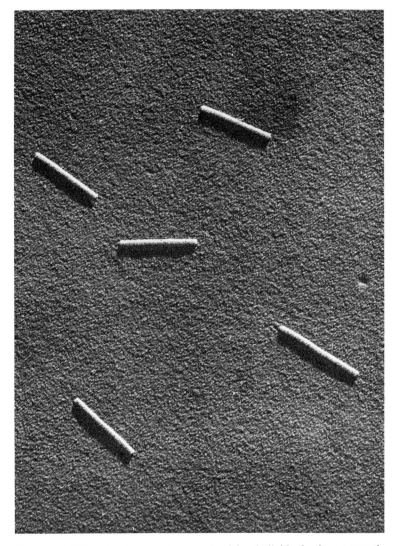

FIGURE 15-18. An electronmicrograph of five individual tobacco mosaic virus molecules, magnified 100,000 times. The virus molecules consist of rods 2800 Å long and 150 Å in diameter. The molecular weight is about 40,000,000. [Courtesy of Dr. R. C. Williams. From R. C. Williams, *Amer. J. Bot.*, **24**:59 (1937)]

the rodlike virus particles. The nucleic acid is infectious by itself and contains all of the genetic information needed to reproduce the entire nucleoprotein molecule.

SUMMARY

A. Nucleic acids are polymers whose backbone consists of alternating pentose and phosphoric acid units.
   1. The pentoses in a given nucleic acid are either all ribose (in RNA) or deoxyribose (in DNA).
B. Nitrogen bases are attached to each pentose unit by a glycoside linkage.
   1. The nitrogen bases are substituted purines and pyrimidines.
   2. The sequence of nitrogen bases along the chain contains the genetic code.

C. Hydrolysis of the nucleic acid polymer gives fragments called
1. Nucleotides, containing one phosphoric acid, one sugar molecule, and one nitrogen base.
2. Nucleosides, like nucleotides but without the phosphoric acid.
D. Adenosine triphosphate (ATP, an energy source) is an important biological compound related to the nucleotides.
E. DNA molecules associate to form double helixes in which two polymer strands are held together by hydrogen bonds.
1. Purines in one chain bond to pyrimidines in the other chain.
2. The two chains are complementary rather than identical.
3. Each chain is able to direct the synthesis of a duplicate of its complementary chain.
F. Synthesis of proteins, including enzymes, is controlled by the nucleic acids.
1. The amino acid code in DNA is translated into chains of messenger RNA (mRNA).
2. The polypeptide molecule is synthesized by ribosomes, using the information in the mRNA chain to control the amino acid sequence.
3. Transfer RNA (tRNA) transports a particular amino acid to the proper site on the mRNA chain and allows it to be recognized.
4. The recognition signal for a given amino acid is a sequence of three nucleotides in the nucleic acid chain, called a codon.
G. Mutations occur when something changes one of the amino acid codons or their sequence in the chain.
H. Viruses are infectious molecules of nucleic acid combined with protein. They reproduce within the cells of the host by using their own nucleic acid codons to direct the synthesis.

EXERCISES

1. Draw the structure of a segment of (a) a typical ribonucleic acid chain and (b) a typical deoxyribonucleic acid chain.

2. Give structural formulas for (a) two typical ribonucleotides and (b) two typical deoxyribonucleosides.

3. Classify each of the following as either a ribonucleotide, a deoxyribonucleotide, a ribonucleoside, or a deoxyribonucleoside, a purine, or a pyrimidine
   (a) adenosine 5'-phosphate
   (b) guanine
   (c) guanosine
   (d) deoxyuridine
   (e) 5'-adenylic acid
   (f) 5'-thymidylic acid
   (g) adenine
   (h) thymidine
   (i) deoxyuridine

4. Draw a structural formula for the trinucleotide GAG.

5. Give the structure of adenosine triphosphate.

6. Explain what is meant by a pair of *coupled* reactions and how they are used to make high-energy compounds.

7. (a) Arrange the four normally occurring bases of DNA into complementary pairs.
   (b) Identify each base as a purine or as a pyrimidine.

8. In Figure 15-7, why are some of the bases connected by two dotted lines and others by three?

9. Give a structural formula showing how the complementary bases on the two strands of DNA bond together.

10. Explain, with diagrams, how DNA replicates itself.

11. What is the effect of changing one codon in a DNA chain on the polypeptide synthesized by the cell?

12. In the hereditary disease sickle-cell anemia, the glutamic acid unit at a certain position in the hemoglobin molecule has been replaced by a valine unit. What change in just one nucleotide might have caused the synthesis of the abnormal hemoglobin? Show the codon (Table 15-4) before and after the one-nucleotide mutation.

13. Suppose that radiation or some chemical is able to *remove* nucleotides from a DNA chain. Compare the probable effects on the organism and its chemistry of removing (a) one nucleotide (b) three adjacent nucleotides. Explain your reasoning.

14. Give a brief description of protein synthesis, from nucleic acid to final product.

15. Prove that you can find your way through the ubiquitous alphabet soup by giving names or structural formulas for
    (a) DNA
    (b) RNA
    (c) mRNA
    (d) tRNA
    (e) rRNA
    (f) ATP
    (g) AMP
    (h) pApC
    (i) TMV
    (j) CUC

16. What kind of experiments were done to reveal (a) the length of a single codon, (b) the meaning of a given codon?

17. Explain why one variety of bread mold needs arginine in its diet whereas another can use either arginine or citrulline.

18. What makes it possible for an organism to survive when a mutation causes it to lose the ability to make one of its enzymes? Discuss the ecological implications of your answer.

19. Describe the process by which a virus causes a cell to produce new viruses.

Index

ABS, 260–61
Absorption spectrum, 39
Acetaldehyde, 138
Acetals, 145, 258–59, 280
Acetanilide, 200
Acetic acid, 189
  absorption of ir, 42
  absorption of $h\nu$, 39
  dissociation constant for, 19
Acetic anhydride, 201, 203, 266
Acetone, 138, 140
  bromination of, 125
  iodination of, 149
Acetyl chloride, 199
Acetylene, 90
Acetylides, 88–89
Acid anhydrides
  preparation, 200–201
  reactions of, 202–203
Acid-base
  indicator, 157, 202
  reactions, 17–27
Acid catalysis, 26, 75, 79, 80, 320–21
Acid chlorides, 198–200
Acidity, 19–22
  of carboxylic acids, 19, 21–22, 191–93
  of phenols, 130–31
  of terminal alkynes, 88–89
Aconitase, 323
Aconitic acid, 323
Acrylic acid
  esters, 259
  vinyl alcohol copolymer, 196
"Acrylic" polymers, 259
Acrylonitrile, 259
Adamantane, 60
Addition reactions
  of alkenes, 77–79
  of alkynes, 87–88
  of benzene, 99–100
  of carbonyl groups, 143–50
Adenine, 233, 336, 344
Adenosine, 340
  triphosphate (ATP), 340–42
Adenylic acid, 338t., 341
Adrenalin, 234–35
Aflatoxin, 211
Agar, 296
Aglycone, 278t.
Agricola, 202
Alanine, 306t.
  in silk, 303–304
Alcohol
  absolute, 118
  denatured, 118
  wood, 118
Alcohols, 116–26
  acid-base reactions, 20, 22–25, 120
  addition to C=O, 144–45
  dehydration of, 80–81
  oxidation of, 143, 152
  synthesis of, 123–25, 143, 146, 208–209
Aldehydes. *See also* Carbonyl group
  nomenclature of, 140–41
  oxidation of, 152

Aldehydes [cont.]
  preparation, 143
Aldohexoses, 277–84
Aldopentose, 278t., 284
Alice and alanine, xiv–xvii
Alizarin, 155
Alkaloids, 24, 231
Alkanes, 51–58, 62–64, 77, 87
Alkenes, 70–85
  hydration of, 124
Alkyd resins, 267
Alkyl halides
  hydrolysis, 123–24
Alkyl substituents
  names of, 54
Alkylation
  of alkenes, 79t.
  of aromatic compounds, 102–103
Alkylbenzenes, 103–105
Alkynes, 85–91
Allaric acid, 283
Allose, 282t., 283
Altrose, 282t.
Amber, 202
Amides, 199, 208, 211–13, 224, 307t.
Amines, 220–26
  acid-base reactions of, 17–19, 22–26
  reaction with acid chlorides, 200
  reaction with carbonyl compounds, 147, 224
Amino acids, 182–83, 304–309, 306–307t.
  codons for, 353
  sequence in proteins, 313–17
Ammonia, 220–21
  acid-base reactions of, 10, 19
  reactions with esters, 208
Ammonium ions, 10, 11
  quaternary, 223
  structure of, 221–22
AMP. *See* Adenylic acid
Amphetamine, 235
Amphoteric, defined, 18
Amygdalin, 148, 288
Amylopectin, 294
Amylose, 294
Anaesthetic, 63, 69, 118, 127
Analgesics, 133, 203–204
Anhydrides, 200–203
Aniline, 200, 221, 224, 225
Anisole, 129
Anomers, 278t., 279
Anosmia, 139
Anson's expedition, 126
Anthracene, 111–12
Antihistamines, 230
Ants
  sting of, 189
Apiose, 285
Arabic acid, 297
Arabinose, 282t., 285, 297
Arbutin, 287
Arginine, 307t.
  biosynthesis of, 354
Arrow poison, 287
Ascorbic acid, 126, 286
Asparagine, 307t.
Aspartic acid, 307t.
  isoelectric point of, 305
Aspirin, 203–204
Asymmetric carbon atoms, 170–73, 180–84, 255–57
Atactic polymer, 255
ATP. *See* Adenosine triphosphate
Atropine, 238

Axial hydrogens, 59
Azeotropic mixture (of ethanol and water), 118
Azo compounds, 226
Aztecs, 155, 237

Bananas
  odor of, 209
Barbital, 242
Barbiturates, 242–43
Barbituric acid, 242
Base pairing (in nucleic acids), 343–46
Base strength, 20, 22–23
  of ammonia, 18
  of nitrogen heterocycles, 228, 230–31
  of water, 18
Belladonna, 238
Benadryl, 230
Benzaldehyde, 148
Benz[a]pyrene, 113
Benzedrine, 235–36
Benzene, 95–103
  naming of substituted, 104
  oxidation of substituted, 197
Benzine, 99
Benzoic acid, 189–90
Benzoyl peroxide, 252
Benzyl (group), 108
Benzyl alcohol, 118
Benzyloxycarbonyl chloride, 326
Biot, Jean Baptiste, 182
Biphenyl, 107–108
Bisphenol A, 271
Bitumen, 268
Bollweevil attractant, 151, 210
Bombardier beetle, 134–35
Bonds
  in benzene, 95–98
  in cyclopropane, 68
  vibration of, 40–41
  double, 70–76
  triple, 85–86
Bromine, 77
  addition reactions of
    with alkenes, 77
    with isoprene, 262
  substitution reactions of, 129, 132t.
Bronsted acids, 18
Bubbles (plastic), 258
Bucarmé, Count, 232
Bufotenine, 237
Butterfly wing pigments, 234
Butyric acid, 52, 190t.

Cadaverine, 306
Caffeine, 233–34
Calcium carbide, 90
Calorie, 297
Cantharidine, 202
Caoutchouc, 262
Carbamic acid, 212
Carbanions, 10–12, 90, 149–50
Carbohydrate, 277
Carbohydrate terminology, 278t.
Carbolic acid, 131
Carbonium ions, 10–21, 27, 75, 79, 105–106, 121
Carbon tetrachloride, 62–63
Carbonyl group, 139–50
Carbowax, 261t., 267
Carboxylic acids, 17–25, 131, 189–95
Carcinogens, 113, 211
Carminic acid, 155

361

Carvone, 188
Catalysis, 80, 320–21
　by acids, 26, 75, 79
　by enzymes, 319–25
　by Lewis acids, 27
Catnip, v, 209
Cellobiose, 290–91
Cellophane, 293
Celluloid, 291–92
Cellulose, 291–92
　acetate, 292–93
　digestion of, 291
　nitrate, 291–92
　sponges from, 293
Charges
　on atoms in molecules, 9–12
Chemical energy, 107, 297, 340–42
Chicle, 261t., 264
Chirality, 174
Chitin, 296
Chloral hydrate, 144
Chlorine
　addition to alkenes, 77
　aliphatic substitution by, 34, 62
　aromatic substitution by, 102
Chloroform, 62–63
Chlorophyll, 228–29
Chloroprene, 265
Chlor-Trimeton, 230
Cholesterol, 133
Chromatography
　gas-liquid, 56
　on ion exchange resins, 309–10
　on paper, 195
Chromosomes, 336
Chrysanthemum, 142
Chymotrypsin, 324–25
Cineol, 128
Cinnamaldehyde, 150
Circumanthracene, 114
Cis-trans isomerism, 73–74, 85, 201–202
Clostridium acetobutylicum, 123
Coal tar, 99, 228, 231
Cocaine, 238–39
Coccus cacti, 155
Cochineal, 155
Codeine, 240
Codons, 349, 352–53
　table of, 353
Color, 39–40
　of aromatic hydrocarbons, 113
　of azo compounds, 227
　of quinones, 153–55
Combustion
　of alkanes, 63
　of carbohydrates and fats, 297–98
Composite materials, 268, 293
Condensation reactions, 150
Conformations
　of alkanes and cycloalkanes, 32–33, 58–59
　of polypeptides, 318–19
Coniferyl alcohol, 293
Coniine, 232
Conjugate acid-base pairs, 18–19
Copolymers, 260
Cortes, 154–55
Coumarin, 209
Covalence
　of charged atoms, 9–11
　orbitals and, 4, 7–8
　physical properties and, 3
Cracking
　of alkanes, 64t.
　of rubber, 262

Cracking [*cont.*]
　of vinyl polymers, 259
Cresols, 130–31
Crick, F., 344
Cross-links, 252–55
　mutations and, 355–56
　in nucleic acids, 344–45
　in polypeptides and nylon, 302–303, 311, 314–16
　in rubbers, 263–65
Crusoe, Robinson, 91
Cucumber, odor of, 141
Cyanin, 157, 287
Cyanogenetic glycosides, 148, 278t., 288
Cyanohydrins, 148
Cycloalkanes, 58–61
　small ring, 67–69
Cyclobutane
　reaction with hydrogen, 68
Cyclohexane, 58–59
Cyclopregnol, 69
Cyclopropane, 67–69
Cycolac, 261t.
Cysteine, 307t.
Cystine, 307t.
　cross-links from, 311, 314
Cytidine, 338t., 340
Cytochrome C, 316–17
Cytosine, 336, 345

Dacron, 261t., 269
D-amino acid oxidase, 322
DDT, 108, 232
Deadly nightshade, 238–39
Decalin, 110
Degree of polymerization, 251
Dehydration of alcohols, 80–81
Dehydrohalogenation, 81–82
Delrin, 261t., 266
Demons, conjuration of, 239
Deoxy-3'-adenylic acid, 340
Deoxyribonucleic acid (DNA), 336, 343–49
Detergents, 121–22, 290
Dextro, 163
Diabetes, 140, 280
Diamond, 60–61
Diastereomers, 176–77
Diazo-coupling, 132t., 227
Diazonium salts, 225, 227
Diborane, 124
Diene polymers, 262–65
Diethylenetriamine, 272
Dimethyl ether, 34, 45
Dimethyl phthalate, 258. *See also* Phthalic acid
Diphenylmethane, 107
Disaccharides, 288–90
Dissociation constants, 19–20
Divinylbenzene, 253–54
DNA (deoxyribonucleic acid), 336, 343–49
DNA-polymerase, 348
Doering, W. E., 244
Double bond. *See* Alkenes
　conjugated, 84
Double helix
　replication of, 346–49
　structure of, 344–46
Drip-dry fabrics, 269
Drugs, naming, 70
Dyes, 132, 155–57
　azo, 226–27
　and photography, 227
Dynamite, 123
　gelatin, 292

Ebonite, 263–64
Echinochrome A, 154–55
Elasticity
　of rubber, 263
Elastomers, 264–65
Electronegativity
　and base strength, 22–23
　and bond polarization, 20–22
Electrons
　shared and unshared, 8
Electrophilic reagents, 26, 75
Electrophilic substitution, 103
Electrophoresis, 305
Elimination reactions, 80–82
Emulsin, 287, 290
Enantiomers, 162
End groups, 252
　in polyoxymethylene, 266
　in polypeptides, 311
Energy
　ATP and, 340–41
　combustion and, 297–98
　explosives and, 107
　light and, 39–40
　photosynthesis and, 228–29, 297
Engine
　direct chemical, 195–96
　internal combustion, 63–64, 128
Enolization, 149
Enols, 87, 125
Enzymes, 319–25. *See also* specific enzymes
　hereditary diseases and, 354
　hydrolytic, 290
　mutations and, 354
　proteolytic, 314
　synthesis of, 309
Epimers, 278t., 283
Epinephrine, 234–35
Epoxides, 271
Epoxy resins, 271–72
Equatorial hydrogens, 59
Ergotism, 212
Erythrose, 282t.
Escherichia coli
　DNA of, 346
Esters, 204–11, 258
Estrone, 142–43
Ethane
　conformations of, 33
Ethanol, 36, 42–45, 118
Ether
　diethyl, 127
　origin of name, 52
Ethers, 126–29
Ethyl (gasoline), 63
Ethyl alcohol. *See* Ethanol
Ethylene, 70, 72
Ethylene glycol, 269
Ethylene oxide, 271
　polymer, 267
Eugenol, 133
Explosive limits, 63
Explosives, 105–107, 122, 291
　primary or primer, 89
Eye
　pigments of, 83–85
　vitreous humor, 297

Farnesol, 119
Fats and oils, 206–207
Fehling's solution, 280, 289
Fermentation, 123
Ferric chloride
　color reaction with enols and phenols, 126, 131
Finn, Michael, 144

Fischer, Emil, 164, 283
Fischer projection, 164
Fish oil
　amines in, 222
Flower pigments, 157
Fluorescein, 157–58
Fluorescence, 157–58
　of anthracene, 111–12
　of quinine, 180
Formaldehyde, 138, 139, 265
Formic acid, 189
Formvar, 261t.
Fortral, 261t.
Free radicals, 11–12
　addition of C=C, 76
　chlorination of alkanes, 62
　polymerization and, 252–53
Frequency
　of light, 38
Friedel-Crafts reaction, 200
Fructose, 285–86, 296
Fumaric acid, 201–202
Functional groups, 13–15
Fungi, 211
Furanoses, 278t., 285
Fusel oil, 123

Galactose, 282t., 284, 296, 297
Galactosidase, 288
Gasoline, 2, 63–64
Genes, 336
Geon, 261t.
Glucaric acid, 284f.
Gluconic acid, 280
Glucose, 277–84
α-Glucosidase, 321
Glucosides, 278t., 280
Glucuronic acid, 297
Glutamic acid, 307t.
　and Chinese restaurant
　　syndrome, 307
Glutamine, 307t.
Glyceraldehyde, 162–63, 282t.
Glycerol, 122
　in polyesters, 261t.
　trinitrate, 122
Glycine
　in silk, 303–304
Glycogen, 294–96
　combustion of, 298
Glycosidases, α and β, 290
Glycosides, 286–88
　α and β, 278t.
　cyanogenetic, 148, 278t.
　in nucleic acids, 336
Glyptal, 261t.
Goodyear, Charles, 263
Gout, 233
Grain alcohol, 118
Graphite, 114
Grignard reagents
　addition to C=O, 145
　alkanes from, 56
　carbonation of, 198–99
　reaction with acids, 22
　reaction with esters, 208–209
Guanine, 233, 336, 345
　8-aza-, 355
　6-thio-, 355
Gulose, 282t., 283–84
Gum
　arabic, 297
　benzoin, 190
　chewing, 264
Guncotton, 291
Gutta-percha, 261t., 263

Halogens
　addition of benzene, 99–100
　substitution reaction of, 101
Heliotrope, 151
Helix
　α, 318–19
　double, 344–49
Hemi-acetals, 144, 278–79
Hemicelluloses, 297
Hemin, 228
Hemlock, 232
Hemoglobin, 228
　end groups of, 312
　prosthetic group of, 228
　sickle-cell anemia, 355
Hemophilia, 355
Henbane, 239
Heroin, 240
Heterocycles, nitrogen, 227–34
Hexachlorophene, 129
3-Hexenal, cis, 141
Hippocrates, 203
Histamine, 230
Histidine, 307t.
Hofmann, A., 211
Hofmann reaction, 224
Hyaluronic acid, 297
Hybridization, 49
Hydrazine, 222
　as carboxy-terminal reagent, 312
Hydroboration, 124–25
Hydrogen bonds
　in alcohols, 119
　in DNA, 344
　in nylon, 302–303
　in water, 52, 120, 127
Hydrogen cyanide
　addition to carbonyl compounds, 148
Hydrogenation
　of alkenes, 77, 78
　of aromatic compounds, 95
　of carbonyl compounds, 143
Hydrolysis
　acid chlorides, 198
　amides, 198
　carbohydrates, 288–90, 294
　esters, 205–206
　polypeptides, 304
Hydronium ion, 11, 19
Hydroquinone, 134, 287
Hydroxy- (prefix), 118
o-Hydroxybenzyl alcohol, 118
Hydroxylamine, 224
Hyoscyamine, 238–39
Hypochlorous acid
　addition to alkenes, 79t.

Idose, 282t.
Imidazole, 230
Inclusion compounds, 294
Indigo, 155–56, 229
Indole, 229
Infra-red light, 40–42
　absorption by
　　esters, 204
　　carbonyl compounds, 140
　　hydroxyl compounds, 119
Ingrain dying, 227
Inhibitors, 321–22
Initiators, 252–53
Insects
　defenses of, 134–35, 148
　lures for, 9, 82–83, 151, 210
　poisons for, 211–12, 232
Insulin, 312
　structure of, 314–16

Inulin, 286, 296
Iodoform, 153
Ions
　dipolar, 304
Ion-exchange resins, 309
Ionic compounds
　physical properties, 2
Ips $confusus$, 82–83
Iso- (prefix), 190
Isoelectric point, 305
Isoleucine, 306t.
Isomerism, 33–36
　keto-enol, 87, 149, 240
Isopentane, 34–35
　infrared spectrum of, 41
Isoprene, 262
Isopropyl (group), 54t.
Isotactic polymers, 183, 255, 309

Jasmine, 118
Jimson weed, 239
Juglone, 154–55
Juvabione, 210–11

Keratin, 319
Ketene, 201
Ketohexoses, 278t., 285–86
Ketones, 138–52. See $also$
　Carbonyl group
　synthesis of, 143, 200
Knocking, 63–64

La Condamine, 262
Lactic acid, 165
Lactonic acid, 288
Lactose, 288–89
Laminac, 261t.
Laminating resin, 26t., 267–71
Latex, 262
Lauryl sulfate, sodium, 121
Leucine, 306t.
$Levo$, 163
Lewis acids, 26–27
　as catalysts, 101–102
Light, 36–45. See $also$ Fluorescence
　infrared, 40–42, 119, 140, 204,
　polarized, 165–68
　reactions caused by, 62, 84–85,
　　228–29, 355–56
　visible, 39, 84–85, 112–14
　X-rays, 37, 355–56
Lignin, 293
Linoleic acid, 206
Linseed oil, 274
Lucite, 261t.
Luminal, 243
Lysergic acid N,N-diethylamide
　(LSD), 211–12, 237
Lysine, 307t.
　degradation of, 306
　isoelectric point of, 305
　in peptides, 311
Lyxose, 282t.

Malaria, 232
Maleic acid, 201
Maltose, 290, 294
Mandrake, 239
Manioc, 148
Mannose, 282t., 283
Marijuana, 133
Markovnikov's rules, 78–79
Marlex, 261t.
Marsh gas, 50
Menthol, 119
Meprobamate, 212
Merrifield, R. B., 331

363

Mescaline, 24, 235–36
Meso compounds, 171
Meta (prefix), 104
Meta direction, 106
Methane, 48–51
  chlorinated, 31, 62–63
  occurrence of, 50
  tetrahedral orbitals in, 49
Methanol, 3, 4, 118, 123
  origin of name, 52
Methionine, 307t.
Methyl (group), 35
Methyl alcohol. *See* Methanol
Methyl chloride, 62
Methyl methacrylate, 259
Methylene (group), 35
Methylene chloride, 62
2-Methylheptadecane, 57
Methylmagnesium chloride, 56
Millipedes, 148
Miltown, 212
Mixtures, separation of. *See also* Chromatography
  by acid-base reactions, 24–26, 130, 212
  amino acids, 305, 309–10
Monomer unit, 251
Monosaccharides, 277
Moon, amino acids on, 308f.
Morphine, 240–41
Moses, 268
Moth repellent, 111
Munroe effect, 107
Murex, 157
Muscalure, 9
Muscone, 141–42
Mutagens, 348, 355–56
Mutarotation, 278t., 279
Mutations, 335, 354–56
  effect of, 354–55
  mechanisms of, 348
Mylar, 261t.

Naphthacene, 113
Naphthalene, 110–11
Naphthoic acids, 190
Neopentane, 34–36
Neoprene, 261t., 264–65
Nepetalactone, v, 209
Neurosterone, 69
New mown hay, 209
Nicotine, 231–32
Nicotinic acid, 231
  nicotinamide, 231
Nitration, aromatic, 102–103
Nitriles, 224
Nitrobenzene, 102–103
Nitrogen
  compounds, 219
  fixation, 219–20
Nitrogen mustard, 355–56
Nitroglycerin, 122
Nitrosamines, 225
Nitrous acid
  reaction with amines, 225–26
Nomenclature. *See also* specific compound classes
  of drugs, 70
Nonadienal, 141
Norepinephrine, 235
Normal alkanes, 51
Nuclear magnetic resonance, 42–45, 59, 83
Nucleic acids, 336–39, 343–56
Nucleophilic displacement reaction, 89–90, 124, 128, 213, 223

Nucleophilic reagents, 26, 75
Nucleoprotein
  hydrolysis of, 336
  viruses as, 356
Nucleosides, 336–38
  nomenclature, 338t., 339–40
Nucleotides, 336–38
  nomenclature, 338t., 339–40
Nylon, 302–303

Octane rating, 64
Octet rule, 5–6
Odor, 118, 139, 141, 142, 188
Odors of
  alcohols, 118
  carboxylic acids, 195
  cloves, 133
  musk, 142
  orchids, 128
  plants, 142, 203, 229
Oil of wintergreen, 203
Oleic acid, 206
Oligomer, 251
Opium, 240–41
Optical activity, 168–69
  amino acids, 308
  and enzymatic synthesis, 322–24
Oranges, odor of, 209
Orbitals
  atomic, 4
  bonding, 7
  hybrid, 49, 70, 86, 96
  light absorption and, 39
  molecular, 7
  pi (π), 72, 86, 96
  tetrahedral, 49
Orchids, fragrance of, 128
Orlon, 261t.
Ortho (prefix), 104
Ortho-para direction, 104–105
Osazones, 278t., 283
Oxidation
  of alcohols, 197
  of aldehydes, 197
  of aliphatic side chains, 197
  of alkenes, 79t.
  of benzene, 99
Oximes, 224

Paint, 274
Paper chromatography, 194–95
Para (prefix), 104
Paraformaldehyde, 265
Pasteur, Louis, 180-81
Pelican, 109
Pellagra, 231
Penicillium glaucum, 180–84
Pentane
  isomers, 34
Peptide linkage, 301
Perfumes, 142, 151f.
Peroxides, 252, 269
Perspex, 261t.
Peyote, 236
Phenanthrene, 112
Phenobarbital, 242–43
Phenol, 129, 202
Phenolphthalein, 202
Phenols, 129–34, 225
Phenyl (group), 107
Phenylalanine, 306t., 355
Phenylhydrazine, 147, 222, 283
Phenylhydrazones, 147
Phenylketonuria, 355
Phenylpyruvic acid, 355
Pheromone
  alkanes as, 57–58

Pheromone [*cont.*]
  carboxylic acids as, 195
  defined, 57
Phosgene, 63, 336
Phosphoric acid
  dissociation constant of, 20
  and nucleic acids, 336
Photons
  energy and frequency of, 38–39
Photosynthesis, 228–29, 297
Phthalic acid, 201
  anhydride, 201
  polyesters, 270
Pi electrons
  as nucleophiles, 75
Pi orbitals
  in alkenes, 72
  in benzene, 97
Picric acid, 132
Piperidine, 231
Piperine, 231
Piperonal, 151
pK, 192
Plane of symmetry, 170
Plasticizer, 258
Plastisol, 258
Plexiglas, 259, 261t.
Pneumococcus, type III, 297
Polarimeter, 168
Polarization
  of bonds, 21
  of light, 165–68
Polaroid, 167–68
Polyacetals, 265–67
Polyacrylonitrile, 260
Polyesters, 267–71
Polyethylene
  oxide, 267
  terephthalate, 269
Polyisoprene (*cis* and *trans*), 263
Polymer(s), 250
  common and trade-names of, 261t.
Polymerization, 250–74
  degree of, 251–52
Polymethyl methacrylate, 259
  how to make sheets, rods, or tubes, 259
Polynuclear hydrocarbons, 111–14
  colors of, 113
Polyox, 267
Polyoxymethylene, 266
Polypeptides, 309–19
  hydrolysis of, 304
  primary structure of, 309
  secondary structure of, 317–19
  separation of mixture, 305
  synthesis, 325–31, 349–51
Polypropylene, 256–57
Polysaccharides, 277, 291–97
Polystyrene, 251
  atactic, 255–56
  expanded, 251–52
  isotactic, 255–56
Polysulfide rubber, 267
Polytetrafluoroethylene, 260–61
Polythene, 261t.
Polyvinyl
  acetate, 258
  alcohol, 168, 258
  butyral, 259
  chloride (PVC), 257–58
Porphyrins, 228–29, 316
Primacord, 106
Primary, usage of, 116, 222
Primulin red, 227
Priority rules, 175

Proline, 307t.
  hydroxyproline, 307t.
Propylene, 70
Propylure, 210
Prosthetic groups, 316–17, 319
Propylure, 210
Proteins, 301–30. *See also* Polypeptides
  conjugated, 316–17, 319
  denaturation of, 319
  DNA-controlled synthesis of, 350–52
  separation of, 305
  structure of, 317–19
Protoporphyrin, 228
Psilocybe mexicana, 237
Psilocybine, 237
Psychomimetic drugs, 234–43
Pteridine, 234
Pterins, 234
Pteroylglutamic acid, 234
Ptomaines, 306
Purines, 233–34, 338–39
Putrescine, 306
PVC (polyvinyl chloride), 257–58
Pyranoses, 278t., 281
Pyrethrin, 210
Pyrex, 273
Pyribenzamine, 230
Pyridine, 230–31
Pyrimidines, 338–39
Pyrrole, 228

Quaternary ammonium salts, 223
Quinine, 179–80, 232
  synthesis of, 245
Quinoline, 233
Quinones, 134, 153–57
Quinuclidine, 233

R_f, 194
R and S classification, 173–75
Racemic acid, 180
Racemic mixtures, 169
  resolution of, 178–80
Racemization, 169
Radio waves
  absorption of, 42–45
Raspberries, odor, 142
Rayon, 292
Reduction. *See also* Hydrogenation
  of esters, 208
Reserve carbohydrate, 278t.
Resolution, 178–80
Resonance structures, 98, 105, 107, 193
Retinal, 84–85
Rhamnose, 297
Ribonucleic acids, 336
  messenger (mRNA), 349
  ribosome (rRNA), 350
  transfer (tRNA), 350
Ribose, 282t., 284–85, 336
  2-deoxy, 285, 336
RNA. *See* Ribonucleic acid
Rotation, specific, 165, 168
Rubber
  hard, 263–64
  natural, 261t., 262–64
  neoprene, 261t.

Saflex, 261t.
St. Anthony's fire, 212
Salicylates, 203–204
Salicylic acid, 191, 203
Salicilin, 203
Salp, 292f.

Sanger, F., 314
Saran, 261t.
Schardinger dextrins, 294
Schiff bases, 224
Scopolamine, 238–39
Scurvy, 126
Sea urchin, pigment of, 154–55
Secondary, usage of, 117, 222
Second Law, 263
Serine, 306t.
  in silk, 303–304
Serotonin, 237
Shellac, 267
Sickle-cell anemia, 355
Sigma bonds
  of alkenes, 71
  of benzene, 96
Silicones, 272–74
Silk fibroin, 303–304
Silly Putty, 273
Silver mirror test, 152
Skatole, 229
Smell, sense of, 139, 142
Smokeless powder, 291
Soap, 123, 206–207
Socrates, 232
Sodium, reactions of, 14–15, 88
Sodium bicarbonate
  reactions with acids, 22
Sodium chloride, 3f.
Sodium ethoxide, 120
Sodium lauryl sulfate, 121
Sodium methoxide, 15
Soothsaying, 239
Sorbose, 286
sp orbitals, 86
sp^2 orbitals, 70–71, 96
sp^3 orbitals, 49
Spanish fly, 202
Specific rotation, 165
Speed, 235–36
Spin, of protons, 42
Sponges
  artificial, 293
  D-arabinose in, 285
Starch, 293–94
Steroids, 69, 112, 133, 142–43, 287
STP, 235–36
Streptomycin, 284
Strophanthidin, 287
Structural formulas
  meaning of, 30–33
Styrax, 251
Styrene, 251
  polymerization of, 252–53
Styrofoam, 251–52, 261t.
Styron, 261t.
Substitution. *See also* Nucleophilic displacement
  aliphatic, 34, 62, 64t.
  aromatic, 100–103
  electrophilic, 100–103
Succinic acid, 201, 202
Sucrase, 289, 290
Sucrose, 123, 289
  detergents, 290
  octaacetate, 290
Sugar
  fermentation of, 123
Sulfanilamide, 321–22
Sulfonation, 102–103
Sulfuric acid
  addition to alkenes, 79t.
Superposition, 169
Sympathomimetic drugs, 235
Syndiotactic polymer, 257
Synthesis, art of, 243, 245

Tacticity, 183, 255
Talose, 282t.
Tartaric acid, 178
Teflon, 261
Terephthalic acid, 269
Tertiary, usage of
  alcohols, 117
  amines, 222
  proteins, 319
Terylene, 261t., 269
Testosterone, 142–43
Tetrachloroethylene, 72
Tetradecane, 52
Tetraethyllead, 63
Tetrafluoroethylene, 260–61
Tetrahedral bond angles, 31
Δ^1-Tetrahydrocannabinol, 133–34
Tetralin, 110
Thebaine, 241
Thermoplastics, 251
Thermosetting plastics, 255, 264
Threonine, 306t.
Threose, 282t.
Thujone, 69
Thymine, 336, 344
Thyroxine, 306t.
TNT (2,4,6-trinitrotoluene), 105–107
Toads, 237, 287
Toadstools, 237
Tobacco, 113, 231
  mosaic virus, 356–57
  nicotine, 231
Toluene, 102–103
Tranquilizers, 212
Trehalose, 289
Triboluminescence
  of quinine, 180
Trichloroacetic acid, 21
Tridecane, 57
Triglycerides, 206–207, 274, 298
1,3,5-Trioxane, 265–66
Triple bonds. *See also* Alkynes
  geometry, 85–86
  orbitals, 86
Tropane alkaloids, 238–39
Trypsin
  hydrolytic specificity of, 314
Tryptophan, 306t.
Tunicin, 291–92
Tyrian purple, 155, 157
Tyrosine, 306t.
  in silk, 303–304

Undecane, 57
Uracil, 336, 349
  5-bromo or chloro (-uracil), 355
Urea, vii, 242
Uric acid, 233
Urine
  phenylpyruvic acid in, 355
  pigment in, 234
  sugar in, 280

Valence
  of charged atoms, 9–11
  covalent, 7–9
  ionic, 5–6
Valeric acid, 190
Valine, 306t.
Vanillin, 151, 293
Vat dyeing, 156
Veronal, 242
Vinegar, 189
Vinyl
  polymerization, 252–53
  polymers, 257–61

Vinyl acetate, 258
Vinyl alcohol, 258
Vinyl ethyl ether, 129
Vinylidene chloride, 261t.
Vinyltrichlorosilane, 273
Viruses, 356–57
Vision, 39–40
Vitamin
  A, 84–85
  C, 126, 286
  K, 154–55
Vitamins
  and enzymes, 320
  and mutations, 354
Vitreous humor, 297
Vulcanization, 263–64

Water
  acid-base reaction of, 18–23
  addition to alkenes, 79
    to alkynes, 87–88
    to C=O compounds, 144
  physical properties of, 52
  reaction with sodium, 15
Watson, J. D., 344
Wave length
  units of, 38
Wavenumber, 38
Waves, permanent, 316
Waxes, 207
Wilkins, M. H. F., 344
Williamson synthesis, 128, 132t.
Woad, 155–56

Wood alcohol, 118
Woodward, R. B., 244–45

Xanthate, 292
Xanthopterin, 234
X-rays
  mutations from, 355–56
Xylan, 297
Xylene(s), 103–104
Xylose, 282t., 285

Ylang-ylang, 118

Ziegler-Natta catalysts, 256, 263
Zipper reaction, 266